Formulas from Analytic Geometry

Slope of line: $m = \dfrac{y_2 - y_1}{x_2 - x_1}$

Equation of line: $y - y_1 = m(x - x_1)$

Distance formula: $d = \sqrt{(x_2 - x_1)^2 + (y_2 - y_1)^2}$

Circle: $(x - x_0)^2 + (y - y_0)^2 = r^2$

Ellipse: $\dfrac{x^2}{a^2} + \dfrac{y^2}{b^2} = 1$

Formulas and Definitions from Differential Calculus

The statement $\lim_{x \to a} f(x) = L$ means that for any $\varepsilon > 0$ there is a $\delta > 0$ such that $|f(x) - L| < \varepsilon$ whenever $|x - a| < \delta$.

A function f is *continuous* at x if $\lim_{h \to 0} f(x + h) = f(x)$.

If $\lim_{h \to 0} \dfrac{f(x + h) - f(x)}{h}$ exists, it is denoted by $f'(x)$ or $\dfrac{d}{dx} f(x)$ and termed the *derivative* of f at x.

$(f + g)' = f' + g'$

$(fg)' = fg' + f'g$

$(f/g)' = (gf' - fg')/g^2$

$(f \circ g)' = (f' \circ g) \circ g'$

$\dfrac{d}{dx} x^\alpha = \alpha x^{\alpha - 1}$

$\dfrac{d}{dx} e^x = e^x$

$\dfrac{d}{dx} \sec x = \sec x \cdot \tan x$

$\sinh x = \frac{1}{2}(e^x - e^{-x})$

$\cosh x = \frac{1}{2}(e^x + e^{-x})$

$\dfrac{d}{dx} \ln x = x^{-1}$

$\dfrac{d}{dx} \sin x = \cos x$

$\dfrac{d}{dx} \cos x = -\sin x$

$\dfrac{d}{dx} \tan x = \sec^2 x$

$\dfrac{d}{dx} \text{Arcsin } x = (1 - x^2)^{-1/2}$

$\dfrac{d}{dx} \text{Arctan } x = (1 + x^2)^{-1}$

$\dfrac{d}{dx} \sinh x = \cosh x$

$\dfrac{d}{dx} \cosh x = \sinh x$

Numerical Mathematics and Computing

Numerical Mathematics and Computing

Ward Cheney David Kincaid
The University of Texas at Austin

Brooks/Cole Publishing Company
Monterey, California

CONTEMPORARY UNDERGRADUATE MATHEMATICS SERIES
Consulting Editor: Robert J. Wisner, New Mexico State University

Brooks/Cole Publishing Company
A Division of Wadsworth, Inc.

Printed in the United States of America

10 9 8 7 6 5 4 3 2

Library of Congress Cataloging in Publication Data

Cheney, Elliott Ward, (date)
Numerical mathematics and computing.

(Contemporary undergraduate mathematics series)
Bibliography: p.
Includes index.
1. Numerical analysis — Data processing. I. Kincaid, David Ronald, joint author. II. Title.
QA297.C426 519.4 79-17230
ISBN 0-8185-0357-2

Acquisition Editor: *Craig Barth*
Production Editor: *Marilu Uland*
Manuscript Editor: *Marion Tompkins*
Cover Design: *John Edeen*
Interior Design: *Jamie Sue Brooks*
Illustrations: *Graphic Design*
Composition: *Textype, Palo Alto, California*

To our wives, Denise and Carole

Preface

We have written this book with a specific objective in mind—namely, to introduce students of differing interests and backgrounds to the potentialities of the modern computer for solving problems in science and technology. We have been guided in choice of material, style, and level of presentation solely by *pedagogical* concerns. What these were can be summarized as follows.

1. The students we are addressing generally have not studied university mathematics beyond elementary calculus. This fact has placed a constraint upon the level of exposition and has induced us to review certain topics such as the Taylor series and the definition of the integral. Reinforcing an understanding of calculus (as well as algebra, trigonometry, and analytic geometry) while conveying some principles of scientific computing seemed to us to be a worthwhile secondary objective for the book.

2. Although the readers we have in mind have studied computer programming, we assume that they will welcome an opportunity to increase their proficiency. They are expected to develop further skills by carrying out a variety of programming exercises. A number of subroutines are presented to serve as models for programming, and many exercises are offered, ranging from the simple and routine to the complex and challenging. An appendix containing rudiments of Fortran is included for review, and a section of Chapter 1 is devoted to programming advice.

3. We believe that an introductory course in numerical mathematics should present a wide diversity of topics so that the student can see at once the immense range of applications for the subject. A detailed scholarly analysis of procedures is better left to subsequent study, after a survey of the subject.

4. We also believe that certain advanced topics, such as partial differential equations, should be included in a beginning course in order to give a more accurate impression of the real world of scientific computing.

5. Practical classroom experience dictated the order in which topics are presented. For example, it seemed natural to introduce students to numerical integration early because that subject is closely related to their prior experience. Boundary-value problems are placed later for the opposite reason.

6. The manner in which we approach various topics also has been determined by student needs. Thus we have exercised restraint in using matrix theory and linear algebra since these subjects are not assumed to be part of the reader's background. Differential equations are treated in a way appropriate for those who have not yet encountered the subject in a formal presentation.

We believe that in this ''Era of the Computer,'' university students should be introduced as early as possible to modern computing methods so that the tools thereby acquired will be available in their other endeavors. The study of programming alone is not sufficient to equip students for serious scientific computing because the principles of good numerical praxis cannot be expected to germinate spontaneously in their minds. Acquiring a mastery of these concepts requires attention to good exemplars, practice, and constructive criticism by an instructor.

This text can be used in a variety of ways, depending on the emphasis that the teacher prefers. Problems have been supplied in abundance to enhance the book's versatility; these are divided into two categories: "Problems" and "Computer Problems." In the first category we have approximately 600 exercises in analysis that require only pencil, paper, and possibly a pocket calculator. In the second category there are approximately 250 problems that involve writing a program and testing it on a high-speed computer. In some cases the reader can follow a programming model in the text, but in other cases he or she must proceed from a mathematical description of a procedure. In most of the computer exercises there is something to be learned—a *moral,* if you like—beyond simply writing code. Some computing problems are designed to give experience in using preprogrammed or "canned" library codes.

The text uses Fortran as its programming language. All codes presented are in conformity with standards set by the American National Standards Institute. Nevertheless, some users may need to modify certain features of the routines before they will run successfully with compilers of limited scope. Additional guidance on this matter is contained in the Appendix.

Our own recommendations for courses based on this book are as follows.

1. A one semester survey rapidly covering Chapters 1 through 15, slighting Sections 3.3, 6.3, 7.3, 7.4, 8.3, 10.2, 11.3, 12.2, 13.3, and 15.2.

2. A one semester course meticulously covering Chapters 1 through 8 and a selection of material from Chapters 9 through 15.

3. A two-semester course covering Chapters 1 through 15 carefully.

It is our pleasure to thank a number of persons for their kind assistance, suggestions, and advice. Barbara Allen and Dorothy Baker typed class notes and manuscripts with patience, forbearance, and attention to detail. Ian Barrodale of the University of Victoria, Joe Buhler of Pennsylvania State University, Charles Chui of Texas A&M University, James L. Cornette of Iowa State University, Philip Crooke of Vanderbilt University, David J. Rodabaugh of the University of Missouri at Columbia, Sherwood D. Silliman of Cleveland State University, Robert Webber of Longwood College, and Robert Wisner of the University of Michigan read the manuscript and offered helpful critiques. Carole Kincaid and William MacGregor gave us advice on various sections such as the one on programming suggestions. Randolph Bank and Andrew Sherman used a preliminary version of the book in their classes. The staff of Brooks/Cole has been most understanding and patient in bringing this book to fruition. In particular, we thank Marilu Uland and Craig Barth for their efforts on behalf of this project.

We would appreciate any comments, questions, criticisms, or corrections that readers may take the trouble of communicating to us.

Ward Cheney
David Kincaid

Contents

9 Monte Carlo Methods and Simulation 201

10 The Minimization of Multivariate Functions 223

11 Linear Programming 247

12 Smoothing of Data and the Method of Least Squares 267

13 Systems of Ordinary Differential Equations 285

14 Boundary Value Problems for Ordinary Differential Equations 301

1

Introduction

The Taylor series for $\ln(1 + x)$ gives us

$$\ln 2 = 1 - \frac{1}{2} + \frac{1}{3} - \frac{1}{4} + \frac{1}{5} - \frac{1}{6} + \frac{1}{7} - \frac{1}{8} + \cdots$$

Adding together the eight terms shown, we obtain 0.63452, which is a poor approximation to $\ln 2 \approx 0.69315$. However, the Taylor series for $\ln[(1 + x)/(1 - x)]$ gives us

$$\ln 2 = 2\left(3^{-1} + \frac{3^{-3}}{3} + \frac{3^{-5}}{5} + \frac{3^{-7}}{7} + \cdots\right)$$

Now adding only the four terms shown, we obtain 0.69313. This chapter reviews the use of Taylor series and offers suggestions for writing good computer codes.

The objective of this text is to acquaint the reader with some of the many types of scientific problems that the modern computer can solve. We intentionally limit ourselves to the typical problems that arise in science, engineering, and technology. Thus we do not touch upon problems of accounting, modeling in the social sciences, information retrieval, artificial intelligence, and so on.

Usually our treatment of these problems will not begin at the source, for that would take us far afield into areas of physics, engineering, chemistry, etc. Instead we consider problems after they have been cast into certain standard mathematical forms. The reader is asked, therefore, to accept on faith the assertion that the chosen topics are indeed important ones for scientific computing.

In order to survey a large number of topics, some must be covered briefly and therefore compressed somewhat to make them fit comfortably into this format. Obviously our treatment of some topics must be superficial. But it is hoped that the reader will acquire a good bird's eye view of the subject as a whole and therefore be better prepared for a further, deeper study of numerical analysis.

For each principal topic, we have listed good current sources for further information. In any realistic computing situation, considerable thought should be given to the choice of method to be employed. Although most procedures presented here are useful and important, they may or may not be the optimum choice for a particular problem. In choosing among available methods for solving a problem, the analyst or programmer should consult recent references.

Becoming familiar with basic numerical procedures without realizing their limitations would be foolhardy. Numerical computations are almost invariably contaminated by errors, and it is important to understand their source, propagation, and magnitude. While we cannot help but be impressed by the speed and accuracy of the modern computer, we should temper our admiration with generous measures of skepticism. As a wise observer commented upon the computer age: "Never in the history of mankind has it been possible to produce so many wrong answers so quickly." Thus one of our goals is to help the reader arrive at this state of skepticism, armed with methods of estimating and controlling errors.

1.1 Programming Suggestions

The reader is expected to be familiar with the rudiments of programming. The programming language adopted here is Fortran, the one most commonly used in scientific applications. Many other languages are suitable for the computing problems presented, but no attempt will be made to include them in the discussion. Since the programming of numerical schemes is essential to understanding them, we offer here a few words of advice on programming.

Strive to write programs carefully and correctly. Before beginning the Fortran coding, write out in complete detail the mathematical algorithm to be used. When writing the code, use a style that is easy to read and understand without its having to be run on a computer. Check code thoroughly for errors and omissions before beginning to keypunch or edit on a terminal. Every extra minute spent in checking the code is worth at least 10 minutes at the keypunch or computer terminal, in submitting the run, waiting for output, discovering an error, correcting the error, resubmitting the run, and so on *ad nauseam*.

If the code can be written so that it can handle a more general situation, then in many cases it will be worth the extra effort to do so. A program written for only a particular set of numbers must be completely rewritten for another set. For example, only a few additional Fortran statements are required to write a program with an arbitrary step size compared to a program in which the step size is fixed numerically, as will be shown later.

After writing a subprogram, check it by tracing through the code with pencil and paper on a typical and yet simple example. Checking boundary cases, such as the values of the first and second iteration in a loop and the processing of the first and last element in a data structure, will often disclose embarrassing errors. These same sample cases can be used as the first set of test cases on the computer.

Print out enough intermediate results to assist in debugging and understanding the program's operation. Fancy output formats are not necessary, but some simple labeling of the output is recommended. Never count data cards; instead terminate the input with a flag or end-of-file check. Always echo-print the input data.

It is often helpful to assign meaningful names to the variables. For example, `KOUNT`, `XMAX`, `NEW` may have greater mnemonic value than simply `K`, `X`, and `N`. On the other hand, the chances of making a typing or spelling error are slightly increased. There is perennial confusion between the characters O (letter "oh") and `0` (number zero) and between I (letter "eye") and 1 (number one). One way to diminish the chance of error is to use these symbols sparingly. A slash through a zero is quite popular.

Some comments within a routine are helpful in recalling what the code does at some later time. Extensive comments are not necessary, but a preface to each subprogram explaining usage and parameters is recommended. Inserting blank comment lines can greatly improve the readability of code. In this book comments are not included in the code because explanatory material is present in abundance near each routine.

Use data structures that are natural to the problem at hand. If the problem adapts more easily to a three-dimensional array than to several one-dimensional arrays, then a three-dimensional array should be used.

Build a large program in steps by writing and testing a series of subroutines or function subprograms. Try to keep subprograms reasonably small, less than a page whenever possible to make reading and debugging easier.

Do not sacrifice the clarity of a program in an effort to make the code run faster. Clarity of code is preferable to "optimized code" when the two criteria conflict. Optimization of code is not a concern of this book.

In preference to one you might write yourself, a *preprogrammed* routine from a program library should be used when applicable. Such routines can be expected to be state-of-the-art software, well tested, and, of course, completely debugged. The programmer should use elementary intrinsic subprograms, such as SIN and SQRT, when possible.

DATA statements should be used in preference to executable statements in setting constants. For example, use the statement DATA PI/3.14159265358979/ instead of PI = 3.14159265358979. Moreover, one should use 1.0E9, which is assigned at compile time, rather than 10.0**9, which involves a computation.

When values of a function at arbitrary points are needed in a program, several alternative ways of coding are available. For example, if values of the function $f(x) = e^x - \sin x + x^2$ are needed, a simple approach is to use the Fortran statement

```
Y = EXP(X) - SIN(X) + X*X
```

at appropriate places within the program. Equivalently, a statement function at the beginning of the program

```
F(X) = EXP(X) - SIN(X) + X*X
```

can be used with Y = F(2.5) or whatever value of X is desired. Moreover, a function subprogram

```
FUNCTION F(X)
F = EXP(X) - SIN(X) + X*X
RETURN
END
```

would accomplish the same objective. If the function subprogram F is passed as a parameter in another subprogram, then the statement EXTERNAL F is needed before the first executable statement in each routine that contains such a call. Which implementation is best? It depends on the situation. The Fortran statement is simple and safe. The statement function and function subprogram can be utilized to avoid duplicating code. In using program library routines, the user may be required to furnish a subprogram in order to communicate to the library routine function values for a particular problem. A separate subprogram is the best way to avoid difficulties that inadvertently occur when someone must insert code into another's routine.

Although the mathematical description of an algorithm may indicate that a sequence of values is computed and seems to imply the need for an array, it is often possible to avoid arrays when only the final value of a sequence is required. For example, the theoretical description of Newton's iteration reads

$$x_{n+1} = x_n - \frac{f(x_n)}{f'(x_n)}$$

but the Fortran code can be written simply as

```
X = X - F(X)/FP(X)
```

Such a statement automatically effects the replacement of the value of the "old" x with the numerical value of $x - f(x)/f'(x)$. Here the function subprograms F and FP compute f and f', respectively.

Some care should be exercised in writing Fortran statements that involve exponents. The general function x^y is computed on many machines as $\exp(y \ln x)$. Sometimes this is unnecessarily complicated and may contribute to roundoff errors. For example, X**5 is preferable to X**5.0 for this reason. Similarly, SQRT(X) is preferable to X**0.5.

Never put unnecessary statements within DO loops. Move expressions and variables that do not depend on a DO-loop index variable outside the loop. Indenting DO loops can add to the readability of code. CONTINUE statements for the terminator of DO loops allow programs to be easily altered and may contribute to good programming style. For economy, such CONTINUE statements are not used unless necessary in this book.

Do not write code for looping when a DO loop can be utilized, since unforeseen problems (e.g., programming errors and roundoff errors) could cause programmed loops to become infinite. For example, instead of writing

```
      X = 5.3
    2 CONTINUE
      .
      .
      .
      X = X - 0.001
      IF(X .LT. 4.3) GO TO 2
```

use a DO loop as follows.

```
      X = 5.3
      DO 2 K = 1,1000
      .
      .
      .
      X = X - 0.001
    2 CONTINUE
```

In an iterative process always limit the number of permissible steps. For example, in Newton's method we might write

```
      .
      .
      .
    2 PRINT 5,X
      X = X - F(X)/FP(X)
      IF(ABS(F(X)) .GE. 1.0E-8) GO TO 2
      .
      .
      .
```

If the function involves some erratic behavior, there is danger here in not limiting the number of repetitions of this loop. It is better to use a DO loop.

```
      .
      .
      .
      DO 2 N = 1,25
      PRINT 5,X
      X = X - F(X)/FP(X)
      IF(ABS(F(X)) .LT. 1.0E-8) STOP
    2 CONTINUE
      .
      .
      .
```

To evaluate the polynomial

$$p(x) = a_1 + a_2x + a_3x^2 + \cdots + a_nx^{n-1} + a_{n+1}x^n$$

we group the terms in *nested multiplication:*

$$p(x) = a_1 + x(a_2 + x(a_3 + \cdots x(a_n + x(a_{n+1})) \cdots))$$

The code which evaluates $p(x)$ starts with the innermost parentheses and looks like this:

```
      P = A(N+1)
      DO 2 I = 1,N
    2 P = X*P + A(N-I+1)
```

The final value of P is the value of the polynomial at X.

There is rarely any need for a calculation such as J = (-1)**N, since there are better ways of obtaining the same results. For example, in a DO loop we can write J = 1 before the loop and J = -J inside the loop.

Arrays in Fortran, whether one, two, or three dimensional, are stored in consecutive words of storage. Since the compiler maps two and three subscript expression values into a single subscript value that is used as a position index to determine the location of elements in storage, the use of two and three dimensional arrays can be considered a notational convenience for the user. For example, if IX, JX, KX are the dimensions of an array X in a dimension statement, and if I, J, K are subscript expressions that appear when the array is used, then the location of the element in storage is given by the following table.

Dimension of Array	*Element of Array*	*Index of Element of Array*
A(IA)	A(I)	I
B(IB, JB)	B(I, J)	I+IB*(J−1)
C(IC, JC, KC)	C(I, J, K)	I+IC*(J−1) + IC*JC*(K−1)

Any advantage in using only one-dimensional arrays and performing complicated subscript calculations is slight, and this matter is best left to the compiler.

When using DO loops, write the code so that fetches are made from *adjacent* words in memory. To illustrate, the following loops are correctly written so that the arrays are processed down the columns.

```
      DO 2 J = 1,500
      DO 2 I = 1,500
    2 X(I,J) = Y(I,J)*Z(I,J)
```

For some programs, this detail may be only a secondary concern. However, some computers have immediate access only to a portion or ''page'' of memory at a time; in this case, it is advantageous to process arrays down the columns.

When writing a subprogram, one often wants to dimension arrays for the largest problem that the routine will need to handle and use only a portion of the arrays for smaller problems. A two-dimensional array can be passed as an argument to a subroutine or function, along with its first dimension, to achieve this objective. When arrays are passed as parameters of a subprogram, only the extent—that is, column lengths for two-dimensional arrays—need be communicated to the subprogram, since the memory space has already been allocated. For example,

```
      DIMENSION X(75),A(251,75)
      IA = 251
      .
      .
      .
      CALL EXX(30,X,A,IA)
      .
      .
      .
      CALL EXX(40,X,A,IA)
      .
      .
      .
      END

      SUBROUTINE EXX(N,X,A,IA)
      DIMENSION X(N),A(IA,N)
      .
      .
      .
      RETURN
      END
```

This example shows the use of subroutine EXX with a problem of size 30 and one of size 40 without having to change any DIMENSION statements. The number 1 can be used in place of N (twice) in the DIMENSION statement within the subroutine; however, the compiler may produce diagnostic messages in this case.

In adding a long list of floating-point numbers in the computer, there will generally be less roundoff error if the numbers are added in order of increasing magnitude. (In Chapter 2 roundoff errors are discussed in detail.)

In some situations, notably in solving differential equations, a variable t assumes a succession of values equally spaced, a distance h apart, along the real line. Two ways of programming this are

```
      T = START
      DO 2 N = 1,500
      T = T + H
      .
      .
      .
    2 CONTINUE
```

and

```
      DO 2 N = 1,500
      T = H*FLOAT(N) + START
      .
      .
      .
    2 CONTINUE
```

In the first loop 500 additions occur, each with possible roundoff error. In the second this situation is avoided but at the cost of 500 multiplications. Which is better depends on the particular situation. (See Computer Problem 10.)

The sequence of steps in a program should not depend on whether two floating-point numbers are equal. Instead reasonable tolerances should be permitted to allow for floating-point arithmetic roundoff errors. For example, a suitable branching statement for 10 decimal digits of accuracy might be

```
      DATA EPSI/0.5E-10/
      .
      .
      .
      IF(ABS(X-Y) .LT. EPSI*AMAX1(ABS(X),ABS(Y))) GO TO 5
```

This statement will effect transfer to statement 5 when the relative error between x and y is within $\varepsilon = \frac{1}{2}(10^{-10})$.

These programming suggestions and techniques should be considered in their context. They are not intended to be complete, and many other good programming suggestions have been omitted in order to keep this discussion brief. Our purpose here is to encourage the reader to be attentive to considerations involving efficiency, economy, and roundoff errors. Moreover, we have illustrated several widely used coding techniques that are particularly useful in developing mathematical software.

We conclude with a short programming experiment involving a numerical computation. Here we consider, from the computational point of view, a familiar operation in calculus—namely, taking the derivative of a function. Recall that the derivative of a function f at a point x is defined by the equation

$$f'(x) = \lim_{h \to 0} \frac{f(x + h) - f(x)}{h}$$

A computer has the capacity of imitating the limit operation by using a sequence of numbers h such as

$$h = \frac{1}{4}, \frac{1}{4^2}, \frac{1}{4^3}, \ldots, \frac{1}{4^n}$$

for they certainly approach zero rapidly. Of course, many other simple sequences may suggest themselves, such as $1/n$, $1/n^2$, and $1/10^n$. The sequence $1/4^n$ consists of machine numbers in a binary computer and, for this experiment, will be sufficiently close to zero when n is 25.

Now we select a function to experiment with, say, $f(x) = \sin(x)$. Of course, $f'(x) = \cos(x)$, but we intend to compute $f'(x)$ by the limit definition. Here is a program to do so, with $x = 0.5$ radian.

```
      DATA N/25/,H/1.0/
      A = SIN(0.5)
      Z = COS(0.5)
      DO 2 I = 1,N
      H = H/4.0
      B = SIN(0.5 + H)
      C = B - A
      D = C/H
      E = ABS(Z - D)
    2 PRINT 3,H,C,D,E
    3 FORMAT(2X,4(2X,E20.13))
      END
```

We have not shown the output from this program nor explained the purpose of the experiment. We invite the reader to discover this by running the program or one like it. (See Computer Problems 1 and 2.)

Problems 1.1

1. Do you expect your computer to calculate $3 \cdot (1/3)$ with infinite precision? What about $2 \cdot (1/2)$ or $10 \cdot (1/10)$? Explain.
2. For small x, show that $(1 + x)^2$ can sometimes be more accurately computed by $(x + 2)x + 1$. Explain. What other expressions can be used to compute it?

3. Express in mathematical notation without parentheses the final value of Z in this program segment.

```
      Z = B(N) + 1.0
      DO 2 K = 1,N-2
    2 Z = Z*B(N-K) + 1.0
```

4. Criticize these three segments of code and write improved versions.

```
      DO 2 N = 1,100
      X = Z*Z + 5.7
    2 A(N) = X/FLOAT(N)

      DO 2 N = 1,100
      DO 2 M = 1,100
    2 A(N,M) = 1.0/FLOAT(N+M-1)

      DO 2 N = 1,100
      DO 2 M = 1,100
    2 A(M,N) = 1.0/FLOAT(N+M-1)
```

5. Fill in the code in the subprogram

```
      FUNCTION Z(N,A,B)
      DIMENSION A(N),B(N)
      .
      .
      .
      RETURN
      END
```

which evaluates the expression

$$z = \sum_{i=1}^{n} b_i^{-1} \prod_{j=1}^{i} a_j$$

Hint: Recall that

$$\prod_{k=1}^{m} x_k = x_1 \cdot x_2 \cdot x_3 \cdots x_m \quad \text{and} \quad \sum_{k=1}^{m} x_k = x_1 + x_2 + x_3 + \cdots + x_m$$

6. How many multiplications occur in executing the following code?

```
      DO 2 J = 1,N
      DO 2 I = 1,J
    2 X = X + A(I,J)*B(I,J)
```

7. Using the exponential function EXP(X), what is the best way to code the statement $y = 5e^{3x} + 7e^{2x} + 9e^{x} + 11$?

8. A doubly subscripted array A(I,J) can be added in any order. Write the Fortran code for each part. Which is best?

(a) $\sum_{i=1}^{n} \sum_{j=1}^{n} a_{ij}$ (b) $\sum_{j=1}^{n} \sum_{i=1}^{n} a_{ij}$

(c) $\sum_{i=1}^{n}\left[\sum_{j=1}^{i} a_{ij} + \sum_{j=1}^{i-1} a_{ji}\right]$ (d) $\sum_{k=0}^{n-1} \sum_{|i-j|=k} a_{ij}$

9. Count the number of operations involved in evaluating a polynomial using nested multiplication. Do not count subscript calculations.
10. Show how these polynomials can be efficiently evaluated.
    (a) $p(x) = x^{32}$
    (b) $p(x) = 3(x-1)^5 + 7(x-1)^9$
    (c) $p(x) = 6(x+2)^3 + 9(x+2)^7 + 3(x+2)^{15} - (x+2)^{31}$

Computer Problems 1.1

1. Run the program described in the text (p. 8) and *interpret the results.*
2. Select a function f and a point x and carry out a computer experiment like the one given. Interpret the results. Do not select too simple a function. For example, you might consider $1/x$, $\log x$, e^x, $\tan x$, $\text{Arctan } x$, $\cosh x$, or $x^3 - 23x$.
3. The number e in calculus is defined as a limit $e = \lim_{n\to\infty} (1 + 1/n)^n$. Estimate e by taking the values of this expression for $n = 8, 8^2, 8^3, \ldots, 8^{10}$. Compare with e obtained from the Fortran code `E=EXP(1.0)`. *Interpret the results.*
4. It is not difficult to see that the numbers $p_n = \int_0^1 x^n e^x \, dx$ satisfy the inequalities $p_1 > p_2 > p_3 > \cdots > 0$. Establish this fact. Next, use integration by parts to show that $p_{n+1} = e - (n+1)p_n$ and that $p_1 = 1$. In the computer, generate the first 20 values of p_n and explain why the inequalities above are violated. Do not use subscripted variables. (See Dorn and McCracken [1972], pp. 120–129.)
5. (Continuation) Let $p_{20} = 1/8$ and use the formula in Problem 4 to compute $p_{19}, p_{18}, \ldots, p_1$. Do the numbers generated obey the inequality $1 = p_1 > p_2 > p_3 > \cdots > 0$? Explain the difference in the two procedures. Repeat with $p_{20} = 20$ or $p_{20} = 100$. Explain what happens.
6. Write an efficient subroutine that accepts as input a list of real numbers $a_1, a_2, \ldots, a_n$ and then computes the following.

   Arithmetic mean $$m = \frac{(a_1 + a_2 + \cdots + a_n)}{n}$$

   Variance $$v = \frac{(a_1 - m)^2 + (a_2 - m)^2 + \cdots + (a_n - m)^2}{n}$$

   Standard deviation $$\sigma = \sqrt{v}$$

   Test the subroutine on a set of data of your choice.
7. (Continuation) Show that, in Problem 6, $v = (a_1^2 + a_2^2 + \cdots + a_n^2)n^{-1} - m^2$. Of the two given formulas for v, which is more accurate in the computer? Verify on the computer with a data set. *Hint:* Use a large set of real numbers which vary in magnitude from very small to very large.
8. Let a_1 be given. Write a program to compute for $n = 2, 3, \ldots, 1000$ the numbers $b_n = na_{n-1}$ and $a_n = b_n/n$. Print the numbers $a_{100}, a_{200}, \ldots, a_{1000}$. Do *not* use subscripted variables. What should a_n be? Account for the deviation of fact from theory. Determine four values for a_1 so that the computation does deviate from theory on your computer. *Hint:* Consider extremely small and large numbers and print to full machine precision.
9. In the computer, it can happen that $a + x = a$ when $x \neq 0$. Explain why. Describe the set of n for which $1 + 2^{-n} = 1$ in your computer. Write and run an appropriate program to illustrate the phenomenon.
10. Write a program to test the programming suggestion concerning the roundoff error in the computation of $t = t + h$ versus $t = nh + \text{start}$. For example, use $h = 1/10$ and compute $t = t + h$

in double precision for the correct single-precision value of t; print the absolute values of the differences between this calculation and the values of the two procedures. What is the result of the test when h is a machine number, such as $h = 1/128$, on a binary computer?

11. Consider the following code.

```
      DO 2 N = 1, 20
      X = 2.0 + 1.0/(8.0**N)
      Y = ATAN(X) - ATAN(2.0)
      Z = Y*(8.0**N)
    2 PRINT 3, X, Y, Z
    3 FORMAT(5X, 3E20.13)
```

What is the purpose of this program segment? Is it achieved? Explain.

12. What is the difference between

```
    X = FLOAT(M/N)
    Y = FLOAT(M)/FLOAT(N)
```

Illustrate with specific examples to show that sometimes Y = X and sometimes Y ≠ X.

1.2 Review of Taylor Series

Since the Taylor series is needed frequently in numerical mathematics we shall review this topic here. The most useful form of this series is

$$f(x+h) = f(x) + hf'(x) + \frac{h^2}{2!}f''(x) + \cdots + \frac{h^n}{n!}f^{(n)}(x) + \cdots = \sum_{k=0}^{\infty} \frac{h^k}{k!}f^{(k)}(x)$$

Here $f^{(k)}(x)$ denotes the kth derivative of $f(x)$. Only functions from a certain class ("analytic functions") satisfy this equation. This is not a serious drawback, for many of the common functions are indeed analytic, at least locally. In any event, the series cannot be used practically unless it is truncated (i.e., terminated) with some finite number of terms. In this case, an important result from calculus known as *Taylor's Theorem* tells us the nature of the error involved. Here is the precise statement.

TAYLOR'S THEOREM. If the function f possesses continuous derivatives of orders $1, 2, \ldots, n+1$ on an interval containing x and $x+h$, then

$$f(x+h) = \sum_{k=0}^{n} \frac{h^k}{k!}f^{(k)}(x) + E$$

where $E = h^{n+1}f^{n+1}(\xi)/(n+1)!$ and ξ is between x and $x+h$.

In this theorem h can be positive or negative. The error term E depends on h in two ways: first, h^{n+1} is explicitly present and second, the point ξ generally depends on h. As h converges to zero, E converges to zero with the same rapidity that h^{n+1} converges to zero. For large n, this is quite rapid. To express this quantitative fact, we write

$$E = \mathcal{O}(h^{n+1})$$

which is shorthand for the inequality

$$|E| \leqq C|h|^{n+1}$$

C being a constant. In the present circumstances, this constant could be any number for which $|f^{(n+1)}(t)|/(n+1)! \leqq C$, t being arbitrary in the initially-given interval.

It is important to realize that this is an entire sequence of theorems, one for each value of n. For example, we can write out the cases $n = 0, 1, 2$, as follows.

$$f(x+h) = f(x) + hf'(\xi_1) = f(x) + \mathcal{O}(h)$$

$$f(x+h) = f(x) + hf'(x) + \frac{h^2}{2}f''(\xi_2) = f(x) + hf'(x) + \mathcal{O}(h^2)$$

$$f(x+h) = f(x) + hf'(x) + \frac{h^2}{2}f''(x) + \frac{h^3}{6}f'''(\xi_3)$$

$$= f(x) + hf'(x) + \frac{h^2}{2}f''(x) + \mathcal{O}(h^3)$$

For a fixed value of n, the subscript on ξ is usually omitted.

As an example, we develop the Taylor series for $\ln(1+h)$.

$$\begin{aligned}
f(x) &= \ln x & f(1) &= 0 \\
f'(x) &= x^{-1} & f'(1) &= 1 \\
f''(x) &= -x^{-2} & f''(1) &= -1 \\
f'''(x) &= 2x^{-3} & f'''(1) &= 2 \\
f^{iv}(x) &= -6x^{-4} & f^{iv}(1) &= -6 \\
&\vdots & &\vdots \\
f^{(n)}(x) &= (-1)^{n-1}(n-1)!x^{-n} & f^{(n)}(1) &= (-1)^{n-1}(n-1)!
\end{aligned}$$

Therefore, we arrive at the formal series

$$\ln(1+h) = 0 + h - \frac{h^2}{2} + \frac{h^3}{3} - \frac{h^4}{4} + \cdots + (-1)^{n-1}\frac{h^n}{n} + \cdots$$

$$= \sum_{k=1}^{\infty}(-1)^{k-1}\frac{h^k}{k}$$

As another illustration of the use of Taylor's Theorem, consider the function $f(x) = \sqrt{x}$. The first few derivatives of f are $f'(x) = (1/2)x^{-1/2}$, $f''(x) = -(1/4)x^{-3/2}$, $f'''(x) = (3/8)x^{-5/2}$. Thus if a simple and accurate formula for f were needed in the vicinity of $x = 1$, we could let $n = 2$ in Taylor's Theorem and write

$$\sqrt{1+h} = 1 + \frac{1}{2}h - \frac{1}{8}h^2 + \frac{1}{16}h^3\xi^{-5/2}$$

where ξ is an unknown number satisfying $1 < \xi < 1 + h$. For example, if $h = 10^{-5}$, we get the approximate equality

$$\sqrt{1.00001} \approx 1 + 0.5 \times 10^{-5} - 0.125 \times 10^{-10} \approx 1.00000\ 49999\ 875$$

with an error at most $(1/16) \times 10^{-15}$, since $\xi^{-5/2} \leqq 1$. Similarly, we obtain

$$\sqrt{1-h} \approx 1 - \frac{1}{2}h - \frac{1}{8}h^2$$

by substituting $-h$ for h. Hence

$$\sqrt{0.99999} \approx 0.99999\ 49999\ 875$$

In this example it is important to notice that the function possesses derivatives of all orders at any point $x > 0$.

The special case $n = 0$ in Taylor's Theorem is known as the *Mean Value Theorem*. It is usually stated, however, in a somewhat more precise form.

MEAN VALUE THEOREM. If f is continuous on the closed interval $[a, b]$ and possesses a derivative at each point of the open interval (a, b), then

$$f(b) = f(a) + (b - a)f'(\xi)$$

for some ξ in (a, b).

From Taylor's *Theorem* it is easy to derive a condition on f that is strong enough to ensure the validity of Taylor's *series* for f. Suppose that on an interval $[a, b]$ we have

$$|f^{(n)}(x)| \leqq K \qquad (n = 1, 2, \ldots)$$

Then when x and $x + h$ belong to $[a, b]$,

$$f(x + h) = \sum_{k=0}^{\infty} \frac{h^k}{k!} f^{(k)}(x)$$

A result from calculus that can often be used to estimate the error in truncating alternating series is as follows.

Let $a_1 > a_2 > a_3 > \cdots > 0$, and $\lim_{n\to\infty} a_n = 0$. If $s = \sum_{k=1}^{\infty} (-1)^k a_k$ and $s_n = \sum_{k=1}^{n} (-1)^k a_k$, then $|s - s_n| < a_{n+1}$.

Speaking informally, we say that the error is less than the first neglected term.

As an application of this theorem, consider the series

$$\sin x = \sum_{k=0}^{\infty} (-1)^k \frac{x^{2k+1}}{(2k+1)!}$$

For x's such that $0 < x < 1/2$, how many terms are needed to compute $\sin x$ accurately to 12 decimals (rounded)? Since $|x| < 2^{-1}$, we require

$$\frac{2^{-(2n+1)}}{(2n+1)!} < \frac{1}{2} \times 10^{-12}$$

for this is a bound on the error when we use only terms to x^{2n-1}. Taking logarithms, we have

$$\log (2n + 1)! > -2n \log 2 + 12$$

By trial and error, we find that $n = 6$ is the first value to satisfy this inequality. A table of values for log (n!) can be found in most standard books of mathematical tables.

A word of caution is needed about this technique of calculating the number of terms to be used in a series by just making the $n + 1$st term less than some tolerance. This procedure is valid only for alternating series, although it is used to get rough estimates in other cases. For example, it can be used to reject a non-alternating series as converging too slowly. When this technique cannot be used, a bound on the remainder term of the Taylor series has to be established. Of course, this may be somewhat difficult if the $n + 1$st derivative is particularly complicated.

Problems 1.2

1. Verify that the Taylor series expansion about a

$$f(x) = f(a) + (x - a)f'(a) + \frac{1}{2}(x - a)^2 f''(a) + \cdots$$

is equivalent to that given in the text. Determine the Taylor series for cosh x about zero. Evaluate cosh 0.7 by summing four terms. Compare with the actual value.

2. (Continuation) The "Maclaurin series" is the Taylor series about zero

$$f(x) = f(0) + xf'(0) + \frac{x^2}{2!}f''(0) + \frac{x^3}{3!}f'''(0) + \cdots$$

Determine the first two nonzero terms of the series expansion for (a) $e^{\cos x}$, (b) sin (cos x), and (c) $(\cos x)^2 (\sin x)$.

3. The Taylor series for the function e^x is

$$e^x = 1 + x + \frac{x^2}{2!} + \frac{x^3}{3!} + \frac{x^4}{4!} + \cdots$$

Verify this. Determine how many terms are neeeded to compute $e = e^1$ correct to 15 decimal places (rounded).

4. (Continuation) If $x < 0$ in Problem 3, what are the signs of the respective terms in the series? Cancellation of significant digits can be a serious problem in using the series. Will the formula $e^{-x} = 1/e^x$ be helpful in circumventing the difficulty? Explain. (See Sec. 2.3 for further discussion of this type of difficulty.)

5. Verify the Taylor series

$$\ln (e + x) = 1 + \frac{x}{e} - \frac{x^2}{2e^2} + \frac{x^3}{3e^3} - \frac{x^4}{4e^4} + \cdots = 1 + \sum_{n=1}^{\infty} \frac{(-1)^{n-1}}{n}\left(\frac{x}{e}\right)^n$$

6. Determine a Taylor series to represent cos $[\pi/3 + h]$. Evaluate cos (60.001°) to eight decimal places. *Hint:* Use radians.

7. Find the smallest nonnegative integer m such that the Taylor series about m for $(x - 1)^{1/2}$ exists. Establish this series.

8. The Taylor series for $(1 + x)^n$ is also known as the *Binomial Theorem*. It states that

$$(1 + x)^n = 1 + nx + \frac{n(n - 1)}{2!}x^2 + \frac{n(n - 1)(n - 2)}{3!}x^3 + \cdots$$

Derive the series and give its particular form for $n = 2$, $n = 3$, and $n = 1/2$. Then use the last to compute $\sqrt{1.0001}$ correct to 12 decimal places.

9. (Continuation) Use the series of Problem 8 to obtain the series

$$(1 - x)^{-1} = \sum_{n=0}^{\infty} x^n$$

How could this series be used on a computing machine to produce x/y if only addition and multiplication were built in?

10. (Continuation) Use Problem 9 to obtain a series for $(1 + x^2)^{-1}$.

11. L'Hospital's Rule states that under suitable conditions

$$\lim_{x \to a} \frac{f(x)}{g(x)} = \frac{f'(a)}{g'(a)}$$

It is true, for instance, when f and g have continuous derivatives in an open interval containing a and $f(a) = g(a) = 0 \neq g'(a)$. Establish it by using the Mean Value Theorem.

12. (Continuation) Use Problem 11 to show that

$$\lim_{x \to 0} \frac{\sin x}{x} = 1.$$

Evaluate also

$$\lim_{x \to 0} \frac{\text{Arctan } x}{x}, \qquad \lim_{x \to \pi} \frac{\cos x + 1}{\sin x}$$

Supplementary Problems (Chapter 1)

1. The function $f(x) = \sin x$ is theoretically given by an infinite series

$$\sin x = x - \frac{x^3}{3!} + \frac{x^5}{5!} - \frac{x^7}{7!} + \frac{x^9}{9!} - \cdots$$

Verify that if we take only the terms up to and including $x^{2n-1}/(2n-1)!$ and if $|x| < \sqrt{6}$, then the error involved does not exceed $|x|^{2n+1}/(2n+1)!$. How many terms are needed to compute sin (23) with error at most 10^{-8}? What problems do you foresee in using the series to compute sin (23)? Show how to use periodicity to compute sin (23). Show that each term in the series can be obtained from the preceding one by a simple arithmetic operation.

2. Expand the *error function*

$$\text{erf}(x) = \frac{2}{\sqrt{\pi}} \int_0^x e^{-t^2}\, dt$$

in a series by using the exponential series and integrating. Obtain the Taylor series of erf (x) about zero directly from the Maclaurin series. (See Problem 2, p. 14.) Are the two series the same? Evaluate erf (1) by adding four terms of the series and compare to the value erf $(1) \approx 0.8427$, which is correct to four decimal places. *Hint:* Recall from the ***Fundamental Theorem of Calculus*** that

$$\frac{d}{dx} \int_0^x f(t)\, dt = f(x)$$

3. Establish the validity of the Taylor series

$$\text{Arctan}(x) = \sum_{k=1}^{\infty} (-1)^{k+1} \frac{x^{2k-1}}{2k-1}$$

Is it practical to use this series directly to compute Arctan (1) if ten decimals of accuracy are required? How many terms of the series would be needed? Will loss of significance occur? *Hint:* Start with the series for $1/(1 + x^2)$ and integrate term by term. Note that this procedure is only formal; the convergence of the resulting series can be proved by appealing to certain theorems of advanced calculus.

4. It is known that

$$\pi = 4 - 8 \sum_{k=1}^{\infty} (16k^2 - 1)^{-1}$$

Discuss the numerical aspects of computing π by means of this formula. How many terms would be needed to yield ten decimal places of accuracy?

5. Determine a Taylor series to represent $\sin [\pi/4 + h]$. Evaluate sin (45.0005°) to nine decimal places.
6. Write the Taylor series for the function $f(x) = x^3 - 2x^2 + 4x - 1$, using $x = 2$ as the point of expansion. That is, write a formula for $f(2 + h)$.
7. Determine the first four nonzero terms in the series expansion about zero for $(\sin x) + (\cos x)$ and $(\sin x)(\cos x)$. Use these results to find approximate values for $(\sin 0.001) + (\cos 0.001)$ and $(\sin 0.0006)(\cos 0.0006)$. Compare the accuracy of these approximations to those obtained from tables or via a calculator.
8. It is known that

$$\ln (1 + x) = \sum_{k=1}^{\infty} (-1)^{k-1} \frac{x^k}{k} \qquad \text{for } -1 < x \leqq 1$$

The series converges very slowly unless x is close to zero. What is the series for ln 2? How many terms are required to compute ln 2 with an error of less than 10^{-12}? Is this method practical?

9. (Continuation) Show how the simple equation $\ln 2 = \ln [e(2/e)]$ can be used to speed up the calculation of ln 2 in Problem 8.
10. (Continuation) What is the series for $\ln (1 - x)$? What is the series for $\ln [(1 + x)/(1 - x)]$?
11. (Continuation) Determine what value of x to use in the series for $\ln [(1 + x)/(1 - x)]$ if we wish to compute ln 2. Estimate the number of terms needed for ten digits (rounded) of accuracy. Is this method practical?
12. How many terms are needed in the series

$$\cos x = \sum_{k=0}^{\infty} (-1)^k \frac{x^{2k}}{(2k)!}$$

to compute $\cos x$ for $|x| < 1/2$ accurate to 12 decimals (rounded)?

13. A function f is defined by the series

$$f(x) = \sum_{k=1}^{\infty} (-1)^k \frac{x^k}{k^4}$$

Determine the minimum number of terms needed to compute $f(1)$ with error less than 10^{-8}.

14. Verify that the partial sums $s_k = \sum_{i=0}^{k} x^i/i!$ in the Taylor series for e^x can be written recursively as $s_k = s_{k-1} + t_k$, where $s_0 = 1$, $t_1 = x$, and $t_k = (x/k)t_{k-1}$.
15. What is the fifth term in the Taylor series of $(1 - 2h)^{1/2}$?
16. Show that if $|h| < 1$ and $E = \mathcal{O}(h^n)$, then $E = \mathcal{O}(h^m)$ for any nonnegative integer $m \leqq n$.
17. Why do the following functions not possess Taylor series expansions at $x = 0$?
    (a) $f(x) = \sqrt{x}$ (b) $f(x) = |x|$
    (c) $f(x) = \text{Arcsin}(x - 1)$ (d) $f(x) = \cot x$
    (e) $f(x) = \log x$ (f) $f(x) = x^\pi$

18. Show how the polynomial $p(x) = 6(x + 3) + 9(x + 3)^5 - 5(x + 3)^8 - (x + 3)^{11}$ can be efficiently evaluated in Fortran.
19. What is the second term in the Taylor series of $\sqrt[4]{4x - 1}$ about 4.25?
20. How would you compute a table of $\log n!$ for $1 \leqq n \leqq 1000$?

Supplementary Computer Problems (Chapter 1)

1. Verify that $x^y = e^{y \ln x}$. Try to find values of x and y for which these two expressions differ in your computer. *Interpret the results.*
2. (Continuation) Do Problem 1 for $\cos(x - y) = \cos x \cdot \cos y + \sin x \cdot \sin y$.
3. Design and carry out an experiment to determine whether your computer evaluates x^y correctly. *Hints:* Compare some examples, such as $32^{2.5}$ and $81^{1.25}$, to their correct values. A more elaborate test can be made by comparing single-precision results to double-precision results in various cases.
4. The roots of the quadratic equation $ax^2 + bx + c = 0$ are $(-b \pm \sqrt{b^2 - 4ac})/(2a)$. Using this formula, solve equation $x^2 + 10^8 x + c = 0$, by hand and by computer, for $c = 1$ and 10^8. *Interpret the results.*
5. The Fibonacci sequence 0, 1, 1, 2, 3, 5, 8, 13, 21, . . . is defined by the linear recurrence relation

$$\begin{cases} \lambda_0 = 0, & \lambda_1 = 1 \\ \lambda_n = \lambda_{n-1} + \lambda_{n-2} & (n \geqq 2) \end{cases}$$

A formula for the nth Fibonacci number is

$$\lambda_n = \frac{1}{\sqrt{5}}\left[\left(\frac{1 + \sqrt{5}}{2}\right)^n - \left(\frac{1 - \sqrt{5}}{2}\right)^n\right]$$

Compute λ_n $(0 \leqq n \leqq 50)$, using both the recurrence relation and the formula. Write a program employing integer, single-precision, and double-precision arithmetic. For each n, print the results using integer, single-precision, and double-precision formats, respectively.
6. (Continuation) Repeat the experiment of Problem 5 using the sequence given by the recurrence relation

$$\begin{cases} \alpha_0 = 1, & \alpha_1 = \left(\dfrac{1 + \sqrt{5}}{2}\right) \\ \alpha_n = \alpha_{n-1} + \alpha_{n-2} & (n \leqq 2) \end{cases}$$

A closed-form formula is

$$\alpha_n = \left(\frac{1 + \sqrt{5}}{2}\right)^n$$

7. (Continuation) Change $+\sqrt{5}$ to $-\sqrt{5}$ and repeat the computation of α_n. Explain the results.
8. The Russian mathematician Chebyshev spelled his name Чебищев. Many transliterations are possible. Cheb can be rendered Ceb, Tscheb, or Tcheb. The y can be rendered as i. Shev can be rendered as schef, cev, cheff, or scheff. Taking all combinations of these variants, program the computer to print all possible spellings. *Note:* This problem involves nonnumerical computation.

9. The number of combinations of n distinct items taken m at a time is given by the *binomial coefficient*

$$\binom{n}{m} = \frac{n!}{m!(n-m)!}$$

for integers $m, n, 0 \leqq m \leqq n$. Recall that $\binom{n}{0} = \binom{n}{n} = 1$.

(a) Write FUNCTION IBIN(N, M), which uses the definition, to compute $\binom{n}{m}$.

(b) Verify the formula

$$\binom{n}{m} = \prod_{k=1}^{\min(m, n-m)} \left(\frac{n-k+1}{k}\right)$$

for computing the binomial coefficients. Write FUNCTION JBIN(N, M) based on this formula.

(c) Verify the formulas

$$\begin{cases} a_{1j} = a_{j1} = 1 & (j \geqq 1) \\ a_{ij} = a_{i-1,j} + a_{i,j-1} & (i, j \geqq 2) \end{cases}$$

for computing an array of binomial coefficients (''Pascal's Triangle'')

$$\binom{i}{j} = a_{i-j+1,j+1}$$

Write FUNCTION KBIN(N, M, A) that does an array lookup after computing A.

10. Write a program in *double* precision to implement the following algorithm for computing π. Set $a = 0$, $b = 1$, $c = 1/\sqrt{2}$, $d = 1/4$, $e = 1$. Then repeatedly update five times (in the order given) by the formulas

$$a = b, \quad b = \frac{b+c}{2}, \quad c = \sqrt{ca},$$

$$d = d - e(b-a)^2, \quad e = 2e$$

After each cycle print $f = b^2/d$, $g = (b+c)^2/(4d)$, $|f - \pi|$, and $|g - \pi|$. Which converges faster, f or g? How accurate are the final values? Also compare with the following computation

```
DOUBLE PI, DATAN
PI = 4.0D0*DATAN(1.0D0)
```

Hint: The value of π correct to 36 digits is 3.14159 26535 89793 23846 26433 83279 50288.

Cultural note: A new formula for π was discovered in the early 1970s. This algorithm is based on that formula, which is a direct consequence of Gauss' method for calculating elliptic integrals and of Legendre's elliptic integral relation, both known for over 150 years! The error analysis shows that rapid convergence results in the computation of π with the number of significant digits doubling after each step. (The interested reader should see Salamin [1976] and Brent [1976].)

11. A fast algorithm for computing Arctan (x) to precision n, for x in the interval (0, 1], is as follows. Set $a = 2^{-n/2}$, $b = x/(1 + \sqrt{1+x^2})$, $c = 1$, $d = 1$. Then repeatedly update these variables by these formulas (in order from left to right)

$$c = \frac{2c}{1+a}, \quad d = \frac{2ab}{1+b^2}, \quad d = \frac{d}{1+\sqrt{1-d^2}},$$

$$d = \frac{b+d}{1-bd}, \quad b = \frac{d}{1+\sqrt{1+d^2}}, \quad a = \frac{2\sqrt{a}}{1+a}$$

After each sweep print $f = c \ln [(1 + b)/(1 - b)]$. Stop when $1 - a \leqq 2^{-n}$. Write a double-precision routine to implement this algorithm and test it for various values of x. Compare the results to those obtained via `DATAN`.

Cultural note: These fast multiple-precision algorithms depend on the theory of elliptic integrals, using the arithmetic-geometric mean iteration and ascending Landen transformations. For other fast algorithms for trigonometric functions, see Brent [1976].

12. Write `FUNCTION ALOG36(X)` to compute the logarithm with base 36 of a floating-point argument x. *Hint:* Use the natural logarithm function `ALOG`.
13. Write `FUNCTION SINF(X)` to compute the sine of a floating-point argument x in radians as follows. First, using the periodicity of the sine function, reduce the range so that $-\pi/2 \leqq x \leqq \pi/2$. Then if $|x| < 10^{-8}$, set $\sin(x) \approx x$; if $|x| > \pi/6$, set $u = x/3$, compute $\sin(u)$ by the formula below, and then set $\sin(x) \approx [3 - 4\sin^2(u)]\sin(u)$; if $|x| \leqq \pi/6$, set $u = x$ and compute $\sin(u)$ as follows.

$$\sin(u) \approx u\left[\frac{1 - \left(\frac{325523}{2283996}\right)u^2 + \left(\frac{34911}{7613320}\right)u^4 + \left(\frac{479249}{11511339840}\right)u^6}{1 + \left(\frac{18381}{761332}\right)u^2 + \left(\frac{1261}{4567992}\right)u^4 + \left(\frac{2623}{1644477120}\right)u^6}\right]$$

Try to determine if the sine function `SIN(X)` on your computer system uses this algorithm.

Cultural note: This is the Padé rational approximation for sine.

14. Write `FUNCTION ALOGF(X)` to compute the natural logarithm of a floating-point argument x by the algorithm outlined here based on telescoped rational and Gaussian continued fractions for $\ln(x)$ and test for several values of x. First check whether $x = 1$ and return zero if so. Reduce the range of x by determining n and r such that $x = r2^n$ with $1/2 \leqq r < 1$. Next, set $u = (r - \sqrt{2}/2)/(r + \sqrt{2}/2)$ and compute $\ln[(1 + u)/(1 - u)]$ by the approximation

$$\ln\left(\frac{1+u}{1-u}\right) \approx u\left[\frac{20790 - 21545.27u^2 + 4223.9187u^4}{10395 - 14237.635u^2 + 4778.8377u^4 - 230.41913u^6}\right]$$

which is valid for $|u| < 3 - 2\sqrt{2}$. Finally, set $\ln(x) \approx (n - 1/2)\ln 2 + \ln[(1 + u)/(1 - u)]$.

15. Write and test `FUNCTION ATANF(X)` to compute the Arctangent of x in radians by the following algorithm. If $0 \leqq x \leqq 1.7 \times 10^{-9}$, set $\text{Arctan}(x) \approx x$. If $1.7 \times 10^{-9} < x \leqq 2 \times 10^{-2}$, use the series approximation

$$\text{Arctan}(x) \approx x - \frac{x^3}{3} + \frac{x^5}{5} - \frac{x^7}{7}$$

Otherwise set $y = x$, $a = 0$, $b = 1$ if $0 \leqq x \leqq 1$ and set $y = 1/x$, $a = \pi/2$, $b = -1$ if $1 < x$. Then set $c = \pi/16$, $d = \tan(c)$ if $0 \leqq y \leqq \sqrt{2} - 1$ and $c = 3\pi/16$, $d = \tan(c)$ if $\sqrt{2} - 1 < y \leqq 1$. Now compute $u = (y - d)/(1 + dy)$ and the approximation

$$\text{Arctan}(u) \approx u\left[\frac{135130 + 175196.246u^2 + 52490.4832u^4 + 2218.1u^6}{135130 + 217007.46u^2 + 97799.3033u^4 + 10721.3745u^6}\right]$$

Finally, set $\text{Arctan}(x) \approx a + b[c + \text{Arctan}(u)]$.

Cultural note: This algorithm uses telescoped rational and Gaussian continued fractions.

16. The Bessel functions J_n are defined by

$$J_n(x) = \frac{1}{\pi}\int_0^{\pi} \cos(x \sin\theta - n\theta)\, d\theta$$

Establish that $|J_n(x)| \leqq 1$. It is known that $J_{n+1}(x) = 2nx^{-1}J_n(x) - J_{n-1}(x)$. Use this equation to compute $J_0(1), J_1(1), \ldots, J_{20}(1)$, starting from known values $J_0(1) \approx 0.76519\ 76865$ and $J_1(1) \approx 0.44005\ 05857$. Account for the fact that the inequality $|J_n(x)| \leqq 1$ is violated.

2

Number Representation and Errors

Computers usually do not use base 10 arithmetic for their internal computations. Hence, it is necessary to know how to represent numbers in different bases. Also, knowledge about floating-point number systems is helpful in debugging programs. Loss of significance may occur when subtracting two machine numbers, thereby resulting in catastrophic cancellation. Methods of averting this problem are discussed here, and the conversion of numbers between bases is outlined.

2.1 Representation of Numbers in Different Bases

The familiar decimal notation for numbers employs the digits 0, 1, 2, 3, 4, 5, 6, 7, 8, 9. When we write down a whole number such as 37294, the individual digits represent coefficients of powers of 10 as follows.

$$\begin{aligned} 37294 &= 4 + 90 + 200 + 7000 + 30000 \\ &= 4 \times 10^0 + 9 \times 10^1 + 2 \times 10^2 + 7 \times 10^3 + 3 \times 10^4 \end{aligned}$$

Thus, in general, a string of digits represents a number according to the formula

$$a_n a_{n-1} \cdots a_2 a_1 a_0 = a_0 \times 10^0 + a_1 \times 10^1 + \cdots + a_{n-1} \times 10^{n-1} + a_n \times 10^n$$

This takes care of the positive whole numbers. A number between 0 and 1 is represented by a string of digits to the right of a decimal point. For example,

$$\begin{aligned} 0.7215 &= \frac{7}{10} + \frac{2}{100} + \frac{1}{1000} + \frac{5}{10000} \\ &= 7 \times 10^{-1} + 2 \times 10^{-2} + 1 \times 10^{-3} + 5 \times 10^{-4} \end{aligned}$$

In general, we have the formula

$$0.b_1 b_2 b_3 \cdots = b_1 \times 10^{-1} + b_2 \times 10^{-2} + b_3 \times 10^{-3} + \cdots$$

Note that there can be an infinite string of digits to the right of the decimal point, and indeed there *must* be an infinite string to represent some numbers. For example,

$$\tfrac{1}{3} = 0.3333\ 3333\ 3333\ 3333\ 3333\ 3333\ 3333\ 3333\ \ldots$$

$$\sqrt{2} = 1.4142\ 1356\ 2373\ 0950\ 4880\ 1688\ 7242\ 0969\ \ldots$$

$$e = 2.7182\ 8182\ 8459\ 0452\ 3536\ 0287\ 4713\ 5266\ \ldots$$

$$\pi = 3.1415\ 9265\ 3589\ 7932\ 3846\ 2643\ 3832\ 7950\ \ldots$$

$$\ln 2 = 0.6931\ 4718\ 0559\ 9453\ 0941\ 7232\ 1214\ 5817\ \ldots.$$

For a real number of the form

$$a_n a_{n-1} \cdots a_1 a_0 . b_1 b_2 b_3 \cdots = \sum_{k=0}^{n} a_k 10^k + \sum_{k=1}^{\infty} b_k 10^{-k}$$

the "integer part" is the first summation in the expansion and the "fractional part" is the second.

The foregoing discussion pertains to the usual representation of numbers with base 10. Other bases are also used, especially in computers. The *binary* system uses 2 as the base; the *octal* system uses 8.

In the octal representation of a number, the digits used are 0, 1, 2, 3, 4, 5, 6, 7. A number represented by a string of digits in octal is signified, if ambiguity may arise, by enclosing it in parentheses and adding the subscript "8." Thus

$$\begin{aligned}(21467)_8 &= 7 + 6 \cdot 8 + 4 \cdot 8^2 + 1 \cdot 8^3 + 2 \cdot 8^4 \\ &= 7 + 8(6 + 8(4 + 8(1 + 8\,(2)))) \\ &= 9015\end{aligned}$$

A number between 0 and 1, if expressed in octal, is represented with combinations of 8^{-1}, 8^{-2}, and so on. For example,

$$\begin{aligned}(0.36207)_8 &= 3 \times 8^{-1} + 6 \times 8^{-2} + 2 \times 8^{-3} + 0 \times 8^{-4} + 7 \times 8^{-5} \\ &= 8^{-5}(3 \times 8^4 + 6 \times 8^3 + 2 \times 8^2 + 7) \\ &= 8^{-5}(7 + 8^2(2 + 8(6 + 8(3)))) \\ &= \frac{15495}{32768} \\ &= 0.4728\ 6987\ \ldots\end{aligned}$$

We shall see presently how to convert easily to decimal form without having to find a common denominator.

If we use another base, say β, then numbers represented in the β system look like this:

$$(a_n a_{n-1} \cdots a_1 a_0 . b_1 b_2 b_3 \cdots)_\beta = \sum_{k=0}^{n} a_k \beta^k + \sum_{k=1}^{\infty} b_k \beta^{-k}$$

The digits are 0, 1, . . . , $\beta - 1$ in this representation. If $\beta > 10$, it is necessary to introduce symbols for 10, 11, . . . , $\beta - 1$. In the system based on 16 (as some computers are), we use a, b, c, d, e, f for 10, 11, 12, 13, 14, 15, respectively. Thus, for example, $(2bed)_{16} = d + e \times 16 + b \times 16^2 + 2 \times 16^3 = 11245$.

Since any number used in calculations within a computer system must conform to the format of numbers in that system, it must have a ***finite expansion***. Numbers having a nonterminating expansion cannot be accommodated precisely. Moreover, a number possessing a terminating expansion in one base may have a nonterminating expansion in another. A good example of this is the simple fraction,

$$\frac{1}{10} = (0.1)_{10} = (0.0631\ 4631\ 4631\ 4631\ \ldots)_8$$

$$= (0.0001\ 1001\ 1001\ 1001\ \ldots)_2$$

The important point here is that in a computer many numbers are not representable exactly.

We now consider the problem of converting a number from one base to another. It is advisable to consider separately the integer and fractional parts of a number. Consider, then, an integer N given to us in the number system with base α:

$$N = (a_n a_{a-1} \ldots a_1 a_0)_\alpha = \sum_{k=0}^{n} a_k \alpha^k$$

Suppose that we wish to convert this to the number system with base β and that the calculations are to be performed in arithmetic with base β. Write N in its nested form:

$$N = a_0 + \alpha(a_1 + \alpha(a_2 + \alpha(\cdots)) \cdots)$$

and then replace each of the numbers appearing here by its representation in base β. Next, carry out the arithmetic in β arithmetic. The replacement of the a_k's and α by equivalent base β numbers requires a table showing how each of the numbers 0, 1, 2, . . . , α appears in the β system. Moreover, a base β multiplication table may be required.

To illustrate this procedure, consider the conversion of the decimal number 3781 to binary form. The following table shows binary equivalents.

Decimal	0	1	2	3	4	5	6	7	8	9	10
Binary	0	1	10	11	100	101	110	111	1000	1001	1010

Thus in our example we have, using longhand multiplication in base 2,

$$3781 = 1 + 10(8 + 10(7 + 10(3)))$$

$$= (1)_2 + (1010)_2[(1000)_2 + (1010)_2[(111)_2 + (1010)_2(11)_2]]$$

$$= (111011000101)_2$$

This arithmetic calculation in binary is easy for a computer that operates in binary but tedious for humans. Another procedure should be used for hand calculation.

Write down an equation containing the coefficients $c_0, c_1, \ldots$ that we seek:

$$N = (c_m c_{m-1} \ldots c_1 c_0)_\beta = c_0 + \beta(c_1 + \beta(c_2 + \beta(\cdots)) \cdots)$$

Next, observe that if N is divided by β, the *remainder* in this division is c_0 and the *quotient* is

$$c_1 + \beta(c_2 + \beta(c_3 + \beta(\cdots)) \cdots)$$

If *this* number is divided by β, the remainder is c_1 and so on. Thus we divide repeatedly by β, saving remainders $c_0, c_1, \ldots$ and quotients.

This procedure can be illustrated with the previous example.

Quotients	*Remainders*
2)3781	
2)1890	$1 = c_0$
2)945	$0 = c_1$
2)472	$1 = c_2$
2)236	$0 = c_3$
2)118	$0 = c_4$
2)59	$0 = c_5$
2)29	$1 = c_6$
2)14	$1 = c_7$
2)7	$0 = c_8$
2)3	$1 = c_9$
2)1	$1 = c_{10}$
0	$1 = c_{11}$

Again, to illustrate, consider the conversion of the binary number $N = (1110\text{-}11000101)_2$ to decimal form, arithmetic being performed in decimal.

$$\begin{aligned} N &= 1 + 0 \times 2 + 1 \times 2^2 + 0 \times 2^3 + 0 \times 2^4 + 0 \times 2^5 + 1 \times 2^6 + \\ &\quad 1 \times 2^7 + 0 \times 2^8 + 1 \times 2^9 + 1 \times 2^{10} + 1 \times 2^{11} \\ &= 1 + 2(0 + 2(1 + 2(0 + 2(0 + 2(0 + 2(1 + 2(1 + 2(0 + 2(1 + \\ &\quad 2(1 + 2 \times 1)))))))))) \\ &= 3781 \end{aligned}$$

Another conversion problem exists in going from an integer in base α to one in base β, using calculations in base α. As before, the unknown coefficients in the equation

$$N = c_0 + c_1\beta + c_2\beta^2 + \cdots$$

are determined by a process of successive division, and this arithmetic is carried out in the α system. At the end the numbers c_k are in base α, and a table of $\alpha - \beta$ equivalents is used.

For example, we can convert a binary integer into decimal form by repeated division by $10 = (1010)_2$, carrying out the operations in binary. The table of binary-decimal equivalents is used at the end. However, binary division is easy only for computers and so we will develop another procedure presently.

Consider next the conversion problem for "fractions"—that is, numbers in the range $0 < x < 1$. Suppose that the coefficients in the representation

$$x = \sum_{k=1}^{\infty} c_k\beta^{-k} = (0.c_1c_2c_3 \ldots)_\beta$$

are to be determined. Observe that

$$\beta x = (c_1.c_2c_3c_4 \ldots)_\beta$$

for it is only necessary to shift the "radix" point when multiplying by base β. Thus the unknown coefficient c_1 can be described as the *integer part* of βx. It is denoted by $\mathcal{I}(\beta x)$. The fractional part, $(0.c_2c_3c_4 \ldots)_\beta$, is denoted by $\mathcal{F}(\beta x)$. The process is repeated in the following pattern.

$$
\begin{aligned}
d_0 &= x \\
d_1 &= \mathcal{F}(\beta d_0) \qquad c_1 = \mathcal{I}(\beta d_0) \\
d_2 &= \mathcal{F}(\beta d_1) \qquad c_2 = \mathcal{I}(\beta d_1)
\end{aligned}
$$

etc.

As an example, consider the conversion of the ordinary decimal 0.372 into binary form. We have

$$
\begin{array}{lll}
 & d_0 = 0.372 & \\
2d_0 = 0.744 & d_1 = 0.744 & c_1 = 0 \\
2d_1 = 1.488 & d_2 = 0.488 & c_2 = 1 \\
2d_2 = 0.976 & d_3 = 0.976 & c_3 = 0 \\
2d_3 = 1.952 & d_4 = 0.952 & c_4 = 1 \\
2d_4 = 1.904 & d_5 = 0.904 & c_5 = 1
\end{array}
$$

etc.

Thus,

$$(0.372)_{10} = (0.010\ 111 \ldots)_2$$

Here the arithmetic was carried out in the decimal system. If the same conversion is to be performed with binary arithmetic, we could proceed as follows.

$$
\begin{aligned}
(0.372)_{10} &= 3 \times 10^{-1} + 7 \times 10^{-2} + 2 \times 10^{-3} \\
&= \frac{1}{10}\left(3 + \frac{1}{10}\left(7 + \frac{1}{10}(2)\right)\right) \\
&= \frac{1}{1010}\left(11 + \frac{1}{1010}\left(111 + \frac{1}{1010}(10)\right)\right)
\end{aligned}
$$

Dividing in binary arithmetic is not straightforward, so we consider an easier way of doing this conversion.

In converting numbers between decimal and binary form by hand (in contrast to what the computer does), it is convenient to use octal representation as an intermediate step. In the octal system the base is 8 and, of course, digits 8 and 9 are not used. Conversion between octal and decimal proceeds according to the principles already stated. Conversion between octal and binary is especially simple: groups of three binary digits can be translated directly to octal according to the table below.

Binary	000	001	010	011	100	101	110	111
Octal	0	1	2	3	4	5	6	7

This grouping starts at the "decimal" point (binary point) and proceeds in both directions. Thus

$$(101101001.1100101)_2 = (101\ 101\ 001.110\ 010\ 100)_2$$
$$= (551.624)_8$$

In order to justify this convenient sleight of hand, consider, for instance, a fraction expressed in binary form:

$$x = (0.b_1b_2b_3b_4b_5b_6 \ldots)_2$$
$$= b_1 2^{-1} + b_2 2^{-2} + b_3 2^{-3} + b_4 2^{-4} + b_5 2^{-5} + b_6 2^{-6} + \ldots$$
$$= (4b_1 + 2b_2 + b_3)8^{-1} + (4b_4 + 2b_5 + b_6)8^{-2} + \cdots$$

In the last line of this equation, the parentheses enclose numbers from the set $\{0, 1, 2, 3, 4, 5, 6, 7\}$, since the b_i are either 0 or 1. Hence, this must be the octal representation of x.

Continuing with more examples, let us convert $(0.276)_8$, $(0.11001)_2$, and $(492)_{10}$ into different number systems. We show one way for each number and invite the reader to work out the details for the others and also to verify the answers by converting them back to the original base.

1. $(0.276)_8 = 2 \times 8^{-1} + 7 \times 8^{-2} + 6 \times 8^{-3}$

   $$= \frac{2}{8} + \frac{7}{64} + \frac{6}{512} = \frac{190}{512}$$
   $$= 0.3710\ 9375$$

   Hence $(0.276)_8 = (0.3710\ 9375)_{10}$.

2. $(0.110\ 010)_2 = (0.62)_8 = 6 \times 8^{-1} + 2 \times 8^{-2}$

   $$= \frac{6}{8} + \frac{2}{64} = \frac{50}{64}$$
   $$= 0.78125$$

   Hence $(0.11001)_2 = (0.78125)_{10}$.

3. $492 \div 8 = 61$ with remainder 4; $61 \div 8 = 7$ with remainder 5; $7 \div 8 = 0$ with remainder 7. Therefore $(492)_{10} = (754)_8 = (111\ 101\ 100)_2$.

It might seem that four or five different procedures exist for converting between number systems. Actually, there are only *two* basic techniques. The first procedure for converting the number $(N)_\alpha$ to base β can be outlined as follows. (a) Express $(N)_\alpha$ in nested form, using powers of α, (b) replace each digit by the corresponding base β number, and (c) carry out the indicated arithmetic in base β. This outline holds whether N is an integer or a fraction. The second procedure is either the divide-by-β and "remainder-quotient-split" process for N an integer or the multiply-by-β and "integer-fraction-split" process for N a fraction. The first procedure is preferred when $\alpha < \beta$, the second is preferable for $\alpha > \beta$. Of course, the $2-8-10$ base conversion procedure should be used whenever possible, for it is the easiest way to convert numbers.

Problems 2.1

1. Convert from binary to decimal form: $(110111001.101011101)_2$.
2. Convert $(14326)_7$ to decimal form. Convert the result to binary.
3. Convert the following decimal numbers to octal: (a) 23.58, (b) 75.232, and (c) 57.321.
4. Convert the binary number 1001100101.01101 to octal and to decimal.
5. Convert the following numbers.
   (a) $(100101101)_2 = (\ )_{10}$ (b) $(0.782)_{10} = (\ )_2$
   (c) $(47)_{10} = (\ )_2$ (d) $(0.47)_{10} = (\ )_2$
   (e) $(51)_{10} = (\ )_2$ (f) $(0.694)_{10} = (\ )_2$
   (g) $(11011.1110101101101)_2 = (\ )_8$ (h) $(361.4)_8 = (\ )_2$
6. Convert $(0.4)_{10}$ first to binary and then to octal. Check by converting first to octal and then to binary.
7. Describe the set of all decimal numbers that convert to binary numbers with infinitely many digits. Do the same problem when the words "decimal" and "binary" are interchanged.
8. Explain the algorithm for converting an integer in base 10 to one in base 2, assuming that the calculations will be performed in binary arithmetic. Illustrate by converting 479 to binary form.
9. Justify for integers the rule given for the conversion between octal and binary numbers.
10. Find the binary representation of $e = 2.71828\ldots$ Check by reconverting to decimal representation.

Computer Problems 2.1

1. Try the code

```
      I = .50*14
      J = 50*.14
      IF(J+1.EQ. I) PRINT 2
    2 FORMAT(5X,8HSURPRISE)
      END
```

   Explain in detail what happened. Print the intermediate values of the numbers involved to help explain.
2. Write and test a subroutine for converting integers into octal and binary.
3. (Continuation) Write and test a subroutine for converting decimal fractions into octal and binary.
4. (Continuation) Using the two subroutines of the preceding problems, write and test a program that reads decimal numbers from data cards and prints out the decimal, octal, and binary representation of these numbers.

2.2 Floating-Point Number System

The standard way to represent a real number in decimal form is with an integer part, a decimal point, and a fractional part—as for example, 37.21829 or 0.002271828 or 3,000,527.11059.

Another standard form, often called *normalized scientific notation,* is obtained by shifting the decimal point and supplying appropriate powers of 10. Thus the preceding numbers have alternative representations as

$$37.21829 = 0.3721829 \times 10^2$$

$$0.002271828 = 0.2271828 \times 10^{-2}$$

$$3000527.11059 = 0.300052711059 \times 10^7$$

In normalized scientific notation the number is represented by a fraction multiplied by 10^n, and the leading digit in the fraction is not zero (except when the number involved *is* zero). Thus, we write 79,325 as 0.79325×10^5, not 0.079325×10^6 or 7.9325×10^4.

Normalized scientific notation is, in the context of computer science, also called *normalized floating-point representation*. In the decimal system any real number x (other than zero) can be represented in normalized floating-point form as

$$x = \pm 0.d_1 d_2 d_3 \ldots \times 10^n$$

where $d_1 \neq 0$ and n is an integer. The numbers $d_1, d_2, \ldots$ are decimal digits $0, 1, \ldots, 9$.

Stated another way, the real number x, if different from 0, can be represented in normalized floating-point decimal form as

$$x = \pm r \times 10^n \qquad \left(\frac{1}{10} \leqq r < 1\right)$$

This representation consists of three parts: a sign that is either + or −, a number r in the interval $[1/10, 1)$, and an integer power of 10. The number r is called the *mantissa* and n the *exponent*.

The floating-point representation in the binary system is similar to the decimal system in several ways. If $x \neq 0$, it can be written as

$$x = \pm q \times 2^m \qquad \left(\frac{1}{2} \leqq q < 1\right)$$

The mantissa q would be expressed as a sequence of bits (zeros or ones) in the form $q = (0.b_1 b_2 b_3 \cdots)_2$ with $b_1 \neq 0$. Hence $b_1 = 1$ and then necessarily $q \geqq \frac{1}{2}$.

A floating-point number system within a computer is similar to what we have just described with one important difference: every computer has only a finite word length and a finite total capacity. So only numbers with a finite number of digits can be represented. A number is allotted only one word of storage in single-precision mode (two words in double precision). In either case, the degree of precision is strictly limited. Clearly irrational numbers cannot be represented, nor even those rational numbers that do not fit the finite format imposed by the computer. Furthermore, numbers may be either too large or too small to be representable. The effective number system for a computer is thus not a continuum but a rather peculiar discrete set. For example, there are only 31 numbers in the floating-point number system of a minicomputer with the representation

$$x = \pm(0.b_1 b_2 b_3)_2 \times 2^m$$

where the b_i are either 0 or 1 and m can have only the values $-1, 0, 1$. The numbers of such a restricted binary system are symmetrically distributed about zero but rather unevenly. The positive ones are depicted in Fig. 2.1. For a computer with such a floating-point number system, any number closer to zero than ± 0.0625 would "underflow" to zero, and any number outside the range ± 1.75 would "overflow" to machine infinity.

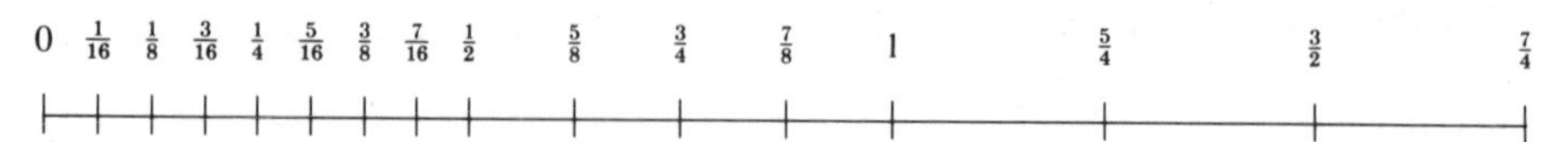

Figure 2.1

A computer operating in single-precision floating-point mode represents numbers as described earlier except for the limitations imposed by the finite word length. A binary computer might have a word length of 60 bits (binary digits), and as much of the floating-point number $\pm q \times 2^m$ must be contained in those 60 bits as possible. One way of allocating the 60 bits is as follows.

sign of x	1 bit
sign of m	1 bit
the integer $\lvert m \rvert$	10 bits
the number q	48 bits

By adopting such a scheme, real numbers with m as large as $2^{10} - 1 = 1023$ can be represented. Henceforth we call this fictional computer the MARC 60.

The real numbers representable by a computer are called *machine numbers*.

We now describe how the hypothetical MARC 60 computer represents a machine number. Assume that the number x has been converted to base 2. The leftmost 12 bits of the 60-bit word are used to represent the "biased" exponent; thus the mantissa is 48 bits long. If x is negative, we first determine the machine representation for the corresponding positive number. Normalize $(x)_2$ by shifting the binary point so that the first bit to the right of the point is one and all bits to the left are zero. To compensate for this shift in the position of the binary point, multiply by the appropriate power of 2. To obtain the biased exponent, convert the exponent to octal and add the bias $(2000)_8$ or $(1777)_8$ to it if it is nonnegative or negative, respectively. Here $(2000)_8 = 2^{10}$ and $(1777)_8 = 2^{10} - 1$. If the original number is positive, the internal representation is a set of 60 zeros and ones partitioned as in Fig. 2.2. If the original number is negative, however, we complement this internal representation.

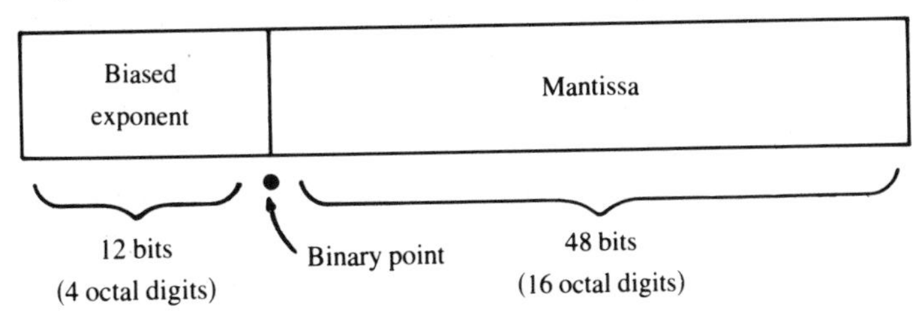

Figure 2.2

Consider, for example, the decimal number 52.234375. Since $(52.)_{10} = (64.)_8 = (110100.)_2$ and $(0.234375)_{10} = (0.17)_8 = (0.001111)_2$, we have $(52.234375)_{10} = (110100.001111)_2 = (0.110100001111)_2 \times 2^6$, which is the corresponding normalized number in base 2. Now $(6)_{10} = (6)_8$, and the biased exponent is $(2000)_8 + (6)_8 = (2006)_8$. The mantissa in base 8 is $(6417000000000000)_8$. Hence, the internal representation of 52.234375 is 20066417000000000000 in octal. If the original number was -52.234375, the internal representation is the sevens complement of the octal representation we obtained. Thus -52.234375 is 57711360777777777777 in octal.

We can deduce that the leftmost octal digit of a positive number is 0, 1, 2, or 3 and that of a negative number is 4, 5, 6, or 7. So the first bit is 0 for a positive number and 1 for a negative number. In binary the internal representation of 52.234375 is

010 000 000 110 110 100 001 111 000 000 000 000 000 000 000 000-
000 000 000 000

and -52.234375 is

101 111 111 001 001 011 110 000 111 111 111 111 111 111 111 111-
111 111 111 111

Determination of Machine Precision

In the MARC 60 the mantissa is 48 bits, and since $2^{-48} \approx 0.3553 \times 10^{-14}$, we infer that in a simple computation approximately 14 significant decimal digits of accuracy should be obtained in single precision (and approximately 28 significant decimal digits in double precision). For integers, only the 48-bit mantissa is used, and the decimal integers range from $-(2^{48} - 1)$ to $(2^{48} - 1) = 281\ 474\ 976\ 710\ 655$. Since six bits are needed to represent an alphanumerical character, the MARC 60 has ten characters per word.

In an actual computer the precise manner of allocating the bits in a word and many other details will differ from the preceding idealized description. *Fortunately, one need not know all these details in order to use the computer intelligently.* Nevertheless, in debugging, it may be helpful to understand the internal representation of numbers in the computer.

We turn now to the errors that can occur when we attempt to represent a given real number x in the computer. We will continue to use a model, the MARC 60 machine, with its 60-bit word length. Suppose first that $x = 2^{5,321,697}$ or $x = 2^{-32,591}$. These numbers would overflow and underflow, respectively, and the relative error in replacing x by the closest machine number will be very large. Such numbers are *outside the range* of the MARC 60.

Consider next a positive number x in normalized floating-point form:

$$x = q \times 2^m \quad \text{with} \quad \frac{1}{2} \leqq q < 1, \qquad |m| \leqq 1023$$

The process of replacing x by its nearest machine number is called *rounding,* and the error involved is called *roundoff error.* We want to know how large it can be. We suppose that q is expressed in normalized binary notation so that

$$x = (0.1a_2a_3a_4 \ldots a_{48}a_{49}a_{50} \ldots)_2 \times 2^m$$

One nearby machine number can be obtained by simply dropping the excess bits $a_{49}a_{50}$ $\ldots$, since only 48 bits have been allocated to q. This machine number is

$$x' = (0.1a_2a_3a_4 \ldots a_{48})_2 \times 2^m$$

It lies to the left of x on the real number axis. Another machine number x'' is just to the right of x on the real axis and is obtained by rounding up. It is found by adding one unit to a_{48} in the expression for x'. Thus

$$x'' = [(0.1a_2a_3a_4 \ldots a_{48})_2 + 2^{-48}]2^m$$

The closer of these machine numbers is the one chosen to represent x.

There are two situations, conveniently described by the simple diagrams in Fig. 2.3.

Figure 2.3

If x lies closer to x' than to x'', then

$$|x - x'| \leqq \frac{1}{2}|x'' - x'| = 2^{-49+m}$$

In this case, the relative error is bounded as follows.

$$\left|\frac{x - x'}{x}\right| \leqq \frac{2^{-49+m}}{(0.1a_2a_3 \ldots) \times 2^m} \leqq \frac{2^{-49}}{\frac{1}{2}} = 2^{-48}$$

On the other hand, if x lies closer to x'' than to x', then

$$|x - x''| \leqq \frac{1}{2}|x'' - x'|$$

and the same analysis shows the relative error to be no greater than 2^{-48}.

We note in passing that in certain machines proper rounding is not followed, but excess digits or bits are discarded in *all* cases. This process is called *chopping*. If the hypothetical MARC 60 were designed to chop numbers, the bound on the relative roundoff error would be twice as large as above or 2^{-47}.

We say that a number x is *chopped* to n digits or figures when all digits following the nth digit are discarded and none of the remaining n digits is changed. Conversely, x is *rounded* to n digits or figures when x is replaced by an n digit number that approximates x with minimum error. The question of rounding up or down an $(n + 1)$-digit number that ends with a 5 is best handled by always selecting the rounded n-digit number with an *even* nth digit.

Suppose that α and β are two numbers, of which one is regarded as an approximation to the other. The ***absolute error*** of β as an approximation to α is $|\alpha - \beta|$. The ***relative error*** of β as an approximation to α is $|\alpha - \beta|/|\alpha|$. Notice that in computing the absolute error, the roles of α and β are the same, whereas in computing the relative error it is essential to distinguish one of the two numbers as "correct." For practical reasons, the relative error is usually more meaningful than the absolute error. For example, if $\alpha_1 = 1.333$, $\beta_1 = 1.334$, and $\alpha_2 = 0.001$, $\beta_2 = 0.002$, then the absolute error of β_i as an approximation to α_i is the same in both cases—namely, 10^{-3}. However, the relative errors are $(3/4)10^{-3}$ and 1, respectively. The relative error clearly indicates that β_1 is a good approximation to α_1 but that β_2 is a poor approximation to α_2.

For the MARC 60, the relative error in approximating a real number (within the range of the machine) by its nearest machine number is not greater than $2^{-48} \approx 0.35 \times 10^{-14}$, and thus 14 significant decimal digits should be preserved in most operations. The number 2^{-48} is the "unit roundoff error" for this machine.

If, in the course of a computation, a number x is produced of the form $\pm q \times 2^m$ with m outside the computer's permissible range, then we say that an *overflow* or an *underflow* has occurred or that "x is outside the range of the computer." Generally an overflow results in a fatal error and the execution of the program stops. An underflow, however, is usually treated automatically by setting x to zero without any interruption of the program and, in many computers, without any warning message.

Next we turn to the errors that are produced in the course of elementary arithmetic operations. In order to illustrate the principles, suppose that we are working with a five-place decimal machine. Two typical machine numbers in normalized floating-point form might be

$$x = 0.37218 \times 10^4 \qquad y = 0.71422 \times 10^{-1}$$

Most computers perform arithmetic operations in a double-length work area; so let us assume that our minicomputer will have a ten-place accumulator. The exponent of the smaller number is adjusted so that both exponents are the same; then the numbers are added in the accumulator and the (normalized) results are placed in a computer word.

$$\begin{aligned} x &= 0.37218\ 00000 \times 10^4 \\ y &= \underline{0.00000\ 71422 \times 10^4} \\ x + y &= 0.37218\ 71422 \times 10^4 \end{aligned}$$

The nearest machine number is $z = 0.37219 \times 10^4$, and the relative error involved in the machine addition is

$$\frac{|x + y - z|}{|x + y|} = \frac{0.0000028578 \times 10^4}{0.3721871422 \times 10^4} \approx 0.77 \times 10^{-5}$$

This relative error would be regarded as acceptable on a machine of such low precision.

In order to facilitate the analysis of such errors, it is convenient to introduce the notation $\mathrm{fl}\,(x)$ to denote the floating-point machine number corresponding to the real number x. Of course, the function fl depends on the particular computer involved. The hypothetical five-decimal machine of the previous example would give

$$\mathrm{fl}\,(0.3721871422) = 0.37219$$

For the hypothetical MARC 60 computer, we established previously that if x is any real number within the range of the computer, then

$$\frac{|x - \mathrm{fl}\,(x)|}{|x|} \leqq 2^{-48}$$

This inequality can also be expressed in the more useful form

$$\mathrm{fl}\,(x) = x(1 + \delta), \qquad |\delta| \leqq 2^{-48}$$

In order to see that these two inequalities are equivalent, simply let $\delta = (\mathrm{fl}\,(x) - x)/x$. Then by the first inequality $|\delta| \leqq 2^{-48}$; solving for $\mathrm{fl}\,(x)$ yields $\mathrm{fl}\,(x) = x\delta + x$.

Now let the symbol $\odot$ denote any one of the arithmetic operations $+, -, \times, \div$. Suppose the MARC 60 has been designed so that whenever two *machine* numbers, x and y, are to be combined arithmetically, the computer will produce $\mathrm{fl}\,(x \odot y)$ instead of $x \odot y$. We can imagine that $x \odot y$ is first *correctly* formed and then rounded to become a machine number. Under this assumption the relative error (in the MARC 60) will not exceed 2^{-48} by the previous analysis:

$$\mathrm{fl}\,(x \odot y) = (x \odot y)(1 + \delta), \qquad |\delta| \leqq 2^{-48}$$

This equation can be interpreted in the following way. The computer result $\mathrm{fl}\,(x \odot y)$ is the *exact* solution of a related problem involving slightly perturbed data. Thus

$$\text{fl}\,(x \pm y) = (x \pm y)(1 + \delta) = [x(1 + \delta)] \pm [y(1 + \delta)]$$

$$\text{fl}\,(xy) = xy(1 + \delta) = [x\sqrt{1 + \delta}][y\sqrt{1 + \delta}] = [x(1 + \delta)]y$$
$$= x[y(1 + \delta)]$$

$$\text{fl}\left(\frac{x}{y}\right) = \frac{x}{y}(1 + \delta) = \frac{[x\sqrt{1 + \delta}]}{[y/\sqrt{1 + \delta}]} = \frac{[x(1 + \delta)]}{y}$$

In each case, $x(1 + \delta)$, $y\sqrt{1 + \delta}$, and so on can be regarded as a small perturbation of x, y, and so forth.

This interpretation is an example of *inverse error analysis*. It attempts to determine what perturbation of the original data would cause the *computer* answers to be the exact answers for a perturbed problem. In contrast, a *direct error analysis* would attempt to determine how computed answers differ from exact answers based on the same data. In this aspect of numerical analysis, computers have stimulated a new way of looking at computational errors.

Now for an example. If x, y, and z are machine numbers, what upper bound can be given for the relative roundoff error in computing $x(y + z)$? To answer this using the principles developed here, we write

$$\begin{aligned} \text{fl}\,(x(y + z)) &= (x \cdot \text{fl}\,(y + z))(1 + \delta_1) \\ &= x((y + z)(1 + \delta_2))(1 + \delta_1) \\ &\approx x(y + z)(1 + \delta_1 + \delta_2) \\ &= x(y + z)(1 + \delta_3) \end{aligned}$$

Since $|\delta_1| \leqq 2^{-48}$ and $|\delta_2| \leqq 2^{-48}$, we neglect the $\delta_1\delta_2$ term and note that $|\delta_3| \leqq 2^{-47}$ Thus the relative error due to roundoff would be about 2^{-47} in the worst case.

Problems 2.2

5, 7, 8, 9

1. Generally, in adding a list of floating-point numbers, less roundoff error will occur if the numbers are added in order of increasing magnitude. Give some examples to illustrate this principle.
2. (Continuation) The principle of Problem 1 is not *universally* valid. Consider a decimal machine with two decimal digits allocated to the mantissa. Show that the four numbers 0.25, 0.0034, 0.00051, 0.061 can be added with less roundoff if *not* added in ascending order.
3. In the case of machine underflow, what is the relative error involved in replacing a number x by 0?
4. If a certain computer operates in base β and carries n digits in the mantissa of its floating-point numbers, show that the rounding of a real number x to the nearest machine number x' involves a relative error of at most $\frac{1}{2}\beta^{1-n}$. *Hint:* Imitate the argument in the text.
5. Consider a decimal machine with five decimal digits allocated to the mantissa. Give an example, avoiding overflow or underflow, of a real number x whose closest machine number x' involves the greatest possible relative error.
6. In a five-decimal machine that correctly rounds numbers to the nearest machine number, what real numbers x will have the property $\text{fl}\,(1.0 + x) = 1.0$?
7. Consider a computer operating in base β. Suppose that it "chops" or "truncates" numbers instead of correctly rounding them. If its floating-point numbers have a mantissa of n digits, how large is the relative error in storing a real number in machine format?

8. If x and y are real numbers within the range of the MARC 60 and if xy is also within the range, what relative error can there be in the machine computation of xy? ***Hint:*** The machine produces $\mathrm{fl}\,[\mathrm{fl}\,(x) \cdot \mathrm{fl}\,(y)]$.
9. Let x and y be real numbers that are not machine numbers but within the exponent range of the MARC 60. What is the largest possible relative error in the machine representation of $x + y^2$? Include errors made to get the numbers in the machine as well as errors in the arithmetic. Use the fact that roundoff errors in the MARC 60 are at most 2^{-48}.
10. Show that if x and y are positive real numbers having the same first n digits in their decimal representation, then y approximates x with relative error better than 10^{1-n}. Is the converse true?
11. Suppose that in floating-point arithmetic $\mathrm{fl}\,(x_i \odot x_j) = (x_i \odot x_j)(1 + \delta)$ for an operation $\odot$ and all machine numbers x_i and x_j. Show that a rough bound on the relative error in computing

$$x_1 \odot x_2 \odot \cdots \odot x_n$$

for machine numbers x_k is $(n - 1)\delta$.

12. Determine the octal representation of the following decimal numbers in the MARC 60.

(a) 1.0 (b) -1.0 (c) 0.0 (d) 0.234375
(e) 492.78125 (f) 64.37109375 (g) -285.75 (h) 10^{-2}

Computer Problems 2.2

1. Print several numbers, both integers and reals, in octal format and try to explain the machine representation used in your computer.
2. The harmonic series $1 + \frac{1}{2} + \frac{1}{3} + \frac{1}{4} + \cdots$ is known to diverge to $+\infty$. In fact, Euler's constant $\approx 0.57721 \approx \lim_{m\to\infty} [\sum_{k=1}^{m} 1/k - \ln (m)]$. If your computer were allowed to run for a week on the code

```
     X = 1.0
     S = 1.0
   2 X = X + 1.0
     S = S + 1.0/X
     GO TO 2
```

what is the largest value of S it would obtain? Write and test a program using a `DO` loop of 5000 steps to estimate Euler's constant. Print intermediate answers at every 100 steps.

3. Let A denote the set of positive integers whose decimal representation does not contain the digit 0. The sum of the reciprocals of the elements in A is known to be 23.10345. Can you verify this numerically?
4. Write a function subprogram `NDIG(N, X)` that returns the nth nonzero digit in the decimal expression for the real number x.

2.3 Loss of Significance

Another source of errors in complicated calculations is the loss of significant digits involved in subtraction. This effect is potentially quite serious and can be catastrophic. The more nearly equal the two numbers whose difference is being computed, the more pronounced is the effect. To illustrate the phenomenon, consider the computation of $x - y$, where $x = 0.3721448693$ and $y = 0.3720214371$. Suppose that the computation will be carried out by a minicomputer with five decimal digits. The numbers would first be rounded to $x' = 0.37214$ and $y' = 0.37202$. Then $x' - y' = 0.00012$, while the correct answer is $x - y = 0.0001234322$. The relative error involved is

$$\frac{|(x - y) - (x' - y')|}{|x - y|} = \frac{0.0000034322}{0.0001234322} \approx 3 \times 10^{-2}$$

This magnitude of relative error must be judged to be quite large when compared to the relative errors of x' and y'. (They cannot exceed 5×10^{-5} by the coarsest estimates, and in this example are, in fact, approximately 1.3×10^{-5}.)

It should be emphasized that this discussion does not pertain to the operation

$$x - y \rightarrow \text{fl}\,(x - y)$$

but rather to the operation

$$x - y \rightarrow \text{fl}\,[\text{fl}\,(x) - \text{fl}\,(y)]$$

Roundoff error in the former case was governed by the equation

$$\text{fl}\,(x - y) = (x - y)(1 + \delta)$$

where $|\delta| \leqq 2^{-48}$ on the hypothetical MARC 60 computer and where $|\delta| \leqq (1/2)10^{-4}$ on a five-digit decimal computer.

In the preceding numerical illustration we observe that the computed difference of 0.00012 has only two significant figures of accuracy, whereas in general, one expects the numbers and calculations in this minicomputer to have five significant figures of accuracy.

The remedy for this difficulty consists first in anticipating that it may occur, and then reprogramming. The simplest technique may be to carry out part of a computation in double-precision arithmetic (that means roughly twice as many significant digits). But often a slight change in the formulas is what is required. Several illustrations of this will be given here, and the reader will find additional ones among the problems.

Consider the previous numerical example but imagine that double-precision calculations are being used to obtain x, y, and $x - y$. Then suppose that single-precision arithmetic is used thereafter. In the computer all ten digits of x, y, and $x - y$ will be retained, but at the end $x - y$ will be rounded to its five-digit form: 0.12343×10^{-3}. This answer has five significant digits of accuracy, as we would like. Of course, the programmer or analyst must know in advance where the double-precision arithmetic will be necessary in the computation. Programming everything in double precision is very wasteful if it is not needed. Another drawback to this approach is that there may be such serious cancellation of significant digits that even double precision will not help.

The previous remark can be illustrated by a function such as

$$f(x) = \sqrt{x^2 + 1} - 1$$

whose values may be required for x near 0. Since $\sqrt{x^2 + 1} \approx 1$ when $x \approx 0$, we see that there is a potential loss of significance in the subtraction. However, the function can be rewritten in the form

$$f(x) = (\sqrt{x^2 + 1} - 1)\left(\frac{\sqrt{x^2 + 1} + 1}{\sqrt{x^2 + 1} + 1}\right) = \frac{x^2}{\sqrt{x^2 + 1} + 1}$$

and this avoids the trouble. For example, if we use five-decimal arithmetic and if $x = 10^{-3}$, then $f(x)$ will be computed incorrectly as 0 by the first formula but as 0.5×10^{-6} by the second formula. If we use the first formula together with double precision, the difficulty is ameliorated but not circumvented altogether. For example, in double precision we have the same problem when $x = 10^{-6}$

As another example, suppose that the values of

$$f(x) = x - \sin x$$

are required near $x = 0$. The careless programmer may write the Fortran statement function

```
F(X) = X-SIN(X)
```

not realizing that serious loss of accuracy will occur. One should recall from calculus that

$$\lim_{x \to 0} \frac{\sin x}{x} = 1$$

in order to see that $\sin x \approx x$ when $x \approx 0$. One cure for this problem is to use the Taylor series for $\sin x$.

$$\sin x = x - \frac{x^3}{3!} + \frac{x^5}{5!} - \frac{x^7}{7!} + \cdots$$

This series is known to represent $\sin x$ for all real values of x. For x near zero, it converges quite rapidly. Using this series, the function f can be written

$$f(x) = x - \left(x - \frac{x^3}{3!} + \frac{x^5}{5!} - \frac{x^7}{7!} + \cdots\right)$$

$$= \frac{x^3}{3!} - \frac{x^5}{5!} + \frac{x^7}{7!} - \cdots$$

We see in this equation where the source of the original difficulty arose—namely, for small values of x, the term x in the sine series is much larger than $x^3/3!$ and thus more important. But in forming $f(x)$, this dominant x term disappears, leaving only the lesser terms. The series starting with $x^3/6$ is very effective for calculating $f(x)$ when x is small. Thus $f(x)$ should be calculated from the formula $f(x) = x - \sin x$ if $|x| \geqq \frac{1}{2}$ and from the formula $f(x) = x^3/3! - x^5/5! + \ldots$ if $|x| < \frac{1}{2}$. The number of terms to be taken in the series depends on the precision of the machine. For example, in order to ensure that the term $x^{2n+1}/(2n+1)!$ is less than 2^{-48}, n should be at least 7 for $|x| < \frac{1}{2}$. (Verify this statement.)

Problems 2.3

1. Indicate how the following formulas may be useful in arranging computations so as to avoid loss of significant digits.

$$\sin x - \sin y = 2 \sin \frac{1}{2}(x - y) \cos \frac{1}{2}(x + y)$$

$$\log x - \log y = \log \frac{x}{y}$$

$$e^{x-y} = \frac{e^x}{e^y}$$

$$1 - \cos x = 2 \sin^2 \frac{x}{2}$$

$$\text{Arctan } x - \text{Arctan } y = \text{Arctan} \left(\frac{x - y}{1 + xy} \right)$$

2. Calculate $f(10^{-2})$ for the function $f(x) = e^x - x - 1$. The answer should have five significant figures and can easily be obtained by pencil and paper. Contrast it with the straightforward evaluation of $f(10^{-2})$ using $e^{0.01} = 1.0101$.
3. What is a good way to compute values of the function $f(x) = e^x - e$ if full machine precision is needed? Note the difficulty when $x = 1$.
4. What problem could the Fortran statement `Y = 1.0 - SIN(X)` cause? Circumvent it without resorting to Taylor series if possible.
5. The hyperbolic sine function is defined by the equation $\sinh x = \frac{1}{2}(e^x - e^{-x})$. What drawback could there be in using this formula to obtain values of the function? How can values of $\sinh x$ be computed to full machine precision when $|x| \leqq \frac{1}{2}$?
6. On your computer determine the range of x for which $(\sin x)/x \approx 1$ with full machine precision. *Hint:* Use Taylor series.
7. In solving the quadratic equation $x^2 - 10^5 x + 1 = 0$ with a machine that carries eight digits, use of the familiar quadratic formula, $x = (-b \pm \sqrt{b^2 - 4ac})/(2a)$, will cause a problem. Investigate the example, observe the difficulty, and propose a remedy. *Hint:* An example in the text is similar.

Computer Problems 2.3

1. Suppose that we wish to evaluate the function $f(x) = (x - \sin x)/x^3$ for values of x close to zero.
   (a) Write a program with statement function `F(X) = (X - SIN(X))/X**3`. Evaluate `F(X)` for each of the following: initially, let `X=1.0` and divide `X` by `10.0` fifteen times. Explain the results. *Note:* L'Hospital's rule indicates that $f(x)$ should tend to $\frac{1}{6}$.
   (b) Write a function subprogram that produces more accurate values of $f(x)$ for all values of x. Test this code.
2. Write a program to print a table of the function $f(x) = 5 - \sqrt{25 + x^2}$ for $x = 0$ to 1 with steps of 0.01. Be sure that your program yields "full machine precision" but do not program the problem in double precision. Explain the results.
3. The accurate computation of the absolute value $|z|$ of a complex number $z = x + iy$ is quite important in many numerical calculations. Design and carry out a computer experiment comparing the following three schemes.

$$|z| = [x^2 + y^2]^{1/2}$$

$$|z| = v\left[1 + \left(\frac{w}{v}\right)^2\right]^{1/2}$$

$$|z| = 2v\left[\frac{1}{4} + \left(\frac{w}{2v}\right)^2\right]^{1/2}$$

where $v = \max \{|x|, |y|\}$ and $w = \min \{|x|, |y|\}$. Use very small and large numbers for the experiment.

4. For what range of x is the approximation $(e^x - 1)/2x \approx 0.5$ correct to 15 decimals of accuracy? Using this information, write a function subroutine for $(e^x - 1)/2x$.
5. In the theory of Fourier series some numbers known as *Lebesgue constants* play a role. A formula for them is

$$\rho_n = \frac{1}{2n + 1} + \frac{2}{\pi} \sum_{k=1}^{n} \frac{1}{k} \tan \frac{\pi k}{2n + 1}$$

Write and run a program to compute $\rho_1, \ldots, \rho_{100}$ with eight decimals of accuracy. Then test the validity of the inequality

$$0 \leqq \frac{4}{\pi^2} \ln (2n + 1) + 1 - \rho_n \leqq 0.0106$$

Supplementary Problems (Chapter 2)

1, 6, 8, 12, 16
Ans: 3, 10, 11, 13, 17, 19

1. Show by an example that in computer arithmetic $a + (b + c)$ may differ from $(a + b) + c$.
2. Consider a decimal machine whose floating-point numbers have 13 decimal places. Suppose that numbers are correctly rounded up or down to the nearest machine number. Give the best bound for the roundoff error, assuming no underflow or overflow. Use relative error of course. What if the numbers are always chopped?
3. Consider a computer that uses five-decimal numbers. Let $\text{fl}(x)$ denote the floating-point machine number closest to x. Show that if $x = 0.5321487513$ and $y = 0.5321304421$, then the operation $\text{fl}(x) - \text{fl}(y)$ involves a large relative error. Compute the relative error for this numerical example.
4. Two numbers x and y that are not machine numbers are read into the MARC 60. The machine computes xy^2. What sort of relative error can be expected? Assume no underflow or overflow.
5. Let x, y, and z be three machine numbers in the MARC 60 computer. By analyzing the relative error in the worst case, determine how much roundoff error should be expected in forming $(xy)z$.
6. How can values of the function $f(x) = \sqrt{x + 2} - \sqrt{x}$ be computed accurately when x is large?
7. On most computers a highly accurate subroutine for $\cos x$ is provided, such as `COS(X)`. It is proposed to base a subroutine for $\sin x$ upon the formula $\sin x = \pm \sqrt{1 - \cos^2 x}$. From the standpoint of precision (not efficiency), what problems do you foresee and how can they be avoided if we insist on using the subroutine `COS(X)`?
8. Criticize and recode the Fortran statement

```
Z=SQRT(X**4 + 4.0)-2.0
```

assuming that Z will sometimes be needed for an X close to 0.

9. For what X values will the Fortran statement `X=ALOG(X)-1.0` produce inaccurate values of X? Recode this statement to eliminate this difficulty.
10. Find a way to calculate $f(x) = (\cos x - e^{-x})/(\sin x)$ correctly. Determine $f(0.008)$.
11. Determine the first two nonzero terms in the expansion about zero for $f(x) = (\tan x - \sin x)/(x - \sqrt{1 + x^2})$. Give an approximate value for $f(0.0125)$.
12. Find a method for computing $(\sinh x - \tanh x)/x$ that avoids loss of significance when x is small. Find appropriate identities to solve this problem rather than using Taylor series.
13. Find a way to calculate accurate values for

$$f(x) = \frac{\sqrt{1 + x^2} - 1}{x^2} - \frac{x^2 \sin x}{x - \tan x}$$

Determine $\lim_{x \to 0} f(x)$.

14. For some values of X, the Fortran statement `Y=1.0-COS(X)` involves a difficulty. What is it, what values of X are involved, and what remedy do you propose?
15. For some values of x, the function $f(x) = \sqrt{x^2 + 1} - x$ cannot be accurately computed by using this formula. Explain and find a way around the difficulty.
16. The inverse hyperbolic sine is given by the formula $f(x) = \ln (x + \sqrt{x^2 + 1})$. Show how to avoid loss of significance in computing $f(x)$ when x is negative. *Hint:* Find and exploit the relationship between $f(x)$ and $f(-x)$.
17. Calculate the relative error involved in the approximation $\pi \approx 22/7$.
18. Let x and y be machine numbers in the MARC 60. What relative roundoff error should be expected in the computation of $x + y$? If x is around 30 and y is around 250, what absolute error should be expected in the computation of $x + y$?
19. A real number x is represented approximately by 0.6032, and we are told that the relative error is 0.1%. What is x? Note that there are *two* answers.
20. What is the relative error involved in rounding 4.9997 to 5.000?
21. Write a function subprogram that computes accurate values of $f(x) = \sqrt[4]{x + 4} - \sqrt[4]{x}$ for positive x.
22. Determine the octal representation in the MARC 60 of each:
    (a) 0.125 (b) −0.125 (c) 0.5 (d) −0.5

Supplementary Computer Problems (Chapter 2)

1. Write `SUBROUTINE QUAD(A,B,C,X1,X2,FX1,FX2)` for computing the two roots x_1 and x_2 of the quadratic equation $f(x) = ax^2 + bx + c = 0$ for real constants a, b, c and evaluating $f(x_1)$ and $f(x_2)$. Use formulas that reduce roundoff errors and write efficient code. Test your subroutine on the following (a, b, c) values.

   $(0, 0, 1), (0, 1, 0), (1, 0, 0), (0, 0, 0), (1, 1, 0),$

   $(2, 10, 1), (1, -8.001, 16.004), (2 \cdot 10^{17}, 10^{18}, 10^{17}),$

   $(10^{-17}, -10^{17}, 10^{17}), (1, -4, 3.99999)$

2. Write `FUNCTION EX(N,X,EPSI)` that computes EX=e^x by summing N terms of the Taylor series until the addition of the $(N + 1)$st term T is such that $|T| <$ EPSI $= 10^{-6}$. Use the reciprocal of e^x for negative values of x. Test EX on the following data: 0, +1, −1, 0.5, −0.123, −25.5, 3.14159, −1776. Compute the relative error, the absolute error, and N for each case, using EXP(X). Sum no more than 25 terms.
3. (Continuation) The computation of e^x can be reduced to computing e^u for $|u| < (\ln 2)/2$ only. This algorithm removes powers of 2 and computes e^u in a range where the convergence of the series is very fast. It is given by

   $$e^x = 2^m e^u$$

   where m and u are computed by the steps
   (a) $z = x/\ln 2$
   (b) $m =$ integer part of $(z \pm \frac{1}{2})$
   (c) $w = z - m$
   (d) $u = w \ln 2$
   Here the minus sign is used if $x < 0$, since $z < 0$. Incorporate this "range reduction" technique into the code for EX above.

4. (Continuation) Write `FUNCTION EXF(X)` that uses this "range reduction" procedure and computes e^x from the even part of the Gaussian continued fraction—that is,

$$e^x = \frac{s+x}{s-x} \qquad \text{where} \qquad s = 2 + x^2 \left\{ \frac{2520 + 28x^2}{15120 + 420x^2 + x^4} \right\}$$

Test `EXF` on the data given for `EX` above.

5. Write `FUNCTION ASINF(X)` to compute the Arcsine of a real argument `X` based on the following algorithm, using telescoped polynomials for Arcsine. If $|x| < 10^{-8}$, set Arcsin $(x) \approx x$. Otherwise if $0 \leqq x \leqq \frac{1}{2}$, set $u = x$, $a = 0$, $b = 1$; if $\frac{1}{2} < x \leqq \sqrt{3}/2$, set $u = 2x^2 - 1$, $a = \pi/4$, $b = \frac{1}{2}$; if $\sqrt{3}/2 < x \leqq (\frac{1}{2})\sqrt{2 + \sqrt{3}}$, set $u = 8x^4 - 8x^2 + 1$, $a = 3\pi/8$, $b = \frac{1}{4}$; if $(\frac{1}{2})\sqrt{2 + \sqrt{3}} < x \leqq 1$, set $u = \sqrt{(\frac{1}{2})(1 - x)}$, $a = \pi/2$, $b = -2$. Now compute the approximation

$$\text{Arcsin}\,(u) \approx u\Big[1.0 + \frac{1}{6}u^2 + 0.075u^4 + 0.04464286u^6 + 0.03038182u^8$$
$$+ 0.022375u^{10} + 0.01731276u^{12} + 0.01433124u^{14} + 0.009342806u^{16}$$
$$+ 0.01835667u^{18} - 0.01186224u^{20} + 0.03162712u^{22}\Big]$$

Finally, set Arcsin $(x) \approx a + b$ Arcsin (u). Test this routine for various values of x.

6. Write `FUNCTION TANF(X)` to compute the tangent of x in radians, using the algorithm below. Test the resulting routine over a range of values of x. First, the argument x is reduced to $|x| \leqq \pi/2$ by adding or subtracting multiples of π. If $0 \leqq |x| \leqq 1.7 \times 10^{-9}$, set tan $(x) \approx x$. If $|x| > \pi/4$, set $u = \pi/2 - x$; otherwise set $u = x$. Now compute the approximation

$$\tan\,(u) \approx u\left[\frac{135135 - 17336.106u^2 + 379.23564u^4 - 1.0118625u^6}{135135 - 62381.106u^2 + 3154.9377u^4 + 28.17694u^6}\right]$$

Finally, if $|x| > \pi/4$, set tan $(x) \approx 1/\tan\,(u)$; and if $|x| \leqq \pi/4$, set tan $(x) \approx \tan\,(u)$. *Cultural note:* This algorithm is obtained from the telescoped rational and Gaussian continued fraction for the tangent function.

3

Locating Roots of Equations

An electric power cable is suspended (at points of equal height) from two towers 100 meters apart. The cable is allowed to dip 10 meters in the middle. How long is the cable?

It is known that the curve assumed by a suspended cable is a *catenary.* When the y axis passes through the lowest point, we can assume an equation of the form $y = \lambda \cosh (x/\lambda)$. Here λ is a parameter to be determined. The conditions of the problem are that $y(50) = y(0) + 10$. Hence

$$\lambda \cosh \frac{50}{\lambda} = \lambda + 10$$

From this equation λ can be determined by the methods of this chapter. The result is $\lambda = 126.632$. From the equation of the catenary the arc-length is easily computed by a standard method of calculus:

$$\begin{aligned} S &= 2 \int_0^{50} ds = 2 \int_0^{50} \sqrt{1 + \left(\frac{dy}{dx}\right)^2}\, dx \\ &= 2 \int_0^{50} \sqrt{1 + \sinh^2 \frac{x}{\lambda}\, dx} \\ &= 2 \int_0^{50} \cosh \frac{x}{\lambda}\, dx = 2\lambda \sinh \frac{50}{\lambda} \\ &= 102.619 \text{ meters} \end{aligned}$$

Let f be a real-valued function of a real variable. Any real number r for which $f(r) = 0$ is called a *root* of that equation or a *zero* of f. For example, the function

$$f(x) = 6x^2 - 7x + 2$$

has $\frac{1}{2}$ and $\frac{2}{3}$ as zeros, as can be verified by direct substitution or by writing f in its factored form

$$f(x) = (2x - 1)(3x - 2)$$

For another example, the function

$$g(x) = \cos 3x - \cos 7x$$

has not only the obvious zero $x = 0$ but every integer multiple of $\pi/5$ and of $\pi/2$ as well, which we discover by applying the trigonometric identity

$$\cos A - \cos B = 2 \sin \frac{1}{2}(A + B) \sin \frac{1}{2}(B - A)$$

Why is the problem of locating roots important? Frequently the solution to a scientific problem is a number about which we may have little information other than the fact that it satisfies some equation. Since every equation can be written so that a function stands on one side and zero on the other, the desired number must be a zero of the function. Thus, if we possess an arsenal of methods for locating zeros of functions, we shall be able to solve such problems.

We illustrate with a concrete problem whose solution is the root of some equation. We want to build a rectangular tank without a lid such that the volume is 21 cubic meters, the height 3 meters, and the surface area is minimal. (The cost depends directly on the area, for the area determines the quantity of material needed to build the tank.) If one of the dimensions of the bottom is x, then the other must be $7/x$ in order for the volume to be correct: $(x)(7/x)(3) = 21$. The area is then $f(x) = (2)(3)(7/x) + (2)(3)(x) + 7$, and the minimum occurs when $f'(x) = 0$. Thus a zero of the function $f'(x) = -42x^{-2} + 6$ is required.

The examples cited so far could be solved easily and quickly by pencil and paper. But how do we locate zeros of functions that are more complicated, such as these?

$$f(x) = 3.24x^8 - 2.42x^7 + 10.37x^6 + 11.01x^2 + 47.98$$

$$g(x) = 2^{x^2} - 10x + 1$$

$$h(x) = \cosh(\sqrt{x^2 + 1} - e^x) + \log|\sin x|$$

What is needed to find roots for such equations is a general numerical method that does not depend on special properties of our functions. Of course, continuity and differentiability are special properties, but they are common attributes of functions usually encountered. The sort of special property that we probably *cannot* easily take advantage of in general-purpose codes is typified by the trigonometric identity mentioned previously.

Myriads of methods are available for locating zeros of functions, and three of the most useful have been selected for study. These are the bisection method, Newton's method, and the secant method.

3.1 Bisection Method

Let f be a function that has values of opposite sign at the two ends of an interval. Suppose also that f is continuous on that interval. To fix the notation, let $a < b$ and $f(a) \cdot f(b) < 0$. It then follows that f has a root in the interval $[a, b]$. In other words, there must exist a number r satisfying the two conditions $a < r < b$ and $f(r) = 0$. How is this conclusion reached? One must recall the ***Intermediate-Value Property of Continuous Functions***. If x traverses an interval $[a, b]$, then the values of $f(x)$ completely fill out the interval between $f(a)$ and $f(b)$. No intermediate values can be "skipped." So our function f must take on the value zero somewhere in the interval $[a, b]$.

The bisection method exploits this property of continuous functions. At each step in this algorithm we have an interval $[a, b]$ and the values $u = f(a)$, $v = f(b)$. The numbers u and v satisfy $uv < 0$. Next, we construct the midpoint of the interval, $c = \frac{1}{2}(a + b)$ and compute $w = f(c)$. It can happen fortuitously that $f(c) = 0$. If so, the objective of the algorithm has been fulfilled. In the usual case, $w \neq 0$, and either $wu < 0$ or $wv < 0$. (Why?) If $wu < 0$, we can be sure that a root of f exists in the interval $[a, c]$. Consequently, we store the value of c in b and w in v. If $wu > 0$, then we cannot be sure that f has a root in $[a, c]$; but since $wv < 0$, f must have a root in $[c, b]$. So in this case we store the value of c in a and w in u. In either case, the situation at the end of this step is just like that at the beginning except that the final interval $[a, b]$ is half as large as the initial interval. This step can now be repeated until the interval is satisfactorily small, say $|b - a| < 10^{-9}$. At the end the best estimate of the root would be $(a + b)/2$.

At this juncture, the reader is invited to program the algorithm as outlined in Computer Problem 3 at the end of this section.

As a general rule, in programming subroutines for locating roots of arbitrary functions, unnecessary evaluations of the function should be avoided, for a given function may be costly to evaluate in terms of computer time. Thus any value of the function which may be needed later should be stored rather than being recomputed. A careless programming of the bisection method might violate this desideratum.

In writing a Fortran subroutine to operate on an arbitrary function F, the parameter list following the subroutine name should contain F. For example, a subroutine to carry out the bisection method might begin with the Fortran statement

```
SUBROUTINE BISECT(F, A, B, EPSI)
```

where F is the function name, A and B are the ends of the initial interval, and EPSI is a tolerance prescribing the desired accuracy.

In using such a subroutine, the calling program would then contain statements of the form

```
EXTERNAL G
. . .
CALL BISECT(G, 1.2, 2.3, 1.0E-9)
. . .
```

In addition to the calling program (or main program) and the subroutine BISECT, a function subprogram with name G is needed. It might have the form

```
FUNCTION G(X)
G = SIN(X) - EXP(X*X) - 2.0
RETURN
END
```

Any program that uses the function can reference it in a statement such as

```
Y = G(X+3.0) + COS(Z)
```

In this way, we can use the BISECT subroutine on several different functions during one run. Of course, we must have a function subprogram for each; their names, separated by commas, must appear in the EXTERNAL statement.

Suppose that the function $f(x)$ has one real root in the interval $[a, b]$. After one iteration of the bisection method the root is contained in an interval of width $(b - a)/2$. After two iterations the interval of interest is now $(b - a)/2^2$ units wide. If we stop after n iterations, then the last interval has a width of $(b - a)/2^n$. The error is now at most one-half the width of the final interval. Hence, if we wish the estimated root to differ from the true root by at most ε, we will require $(b - a)2^{-n-1} \leqq \varepsilon$. This inequality can be solved by taking logarithms; the result is

$$n \geqq \frac{\log (b - a) - \log 2\varepsilon}{\log 2}$$

Problems 3.1

1. Find all the roots of the function $f(x) = \cos x - \cos 3x$.
2. Give a graphical demonstration that the equation $\tan x = x$ has infinitely many roots. Determine one root precisely and another approximately by using a graph.
3. Demonstrate graphically that the equation $50\pi + \sin x = 100 \text{ Arctan } x$ has infinitely many roots.
4. Find the root or roots of $\ln ((1 + x)/(1 - x^2)) = 0$.
5. If f has an inverse, then the equation $f(x) = 0$ can be solved by writing $x = f^{-1}(0)$. Does this remark eliminate the problem of finding roots of equations? Illustrate with $\sin x = 1/\pi$.
6. Find where the graphs of $y = 3x$ and $y = e^x$ intersect by finding the root of $e^x - 3x = 0$ correct to four decimals.
7. How many binary digits of precision are gained in each step of the bisection method? How many steps are required for each decimal digit of precision?
8. Try to devise a stopping criterion for the bisection method to guarantee that the root is determined with *relative* error at most ε.
9. Denote the successive intervals that arise in the bisection method by $[a_0, b_0]$, $[a_1, b_1]$, $[a_2, b_2]$, etc.
   (a) Show that $a_0 \leqq a_1 \leqq a_2 \leqq \cdots$ and that $b_0 \geqq b_1 \geqq b_2 \geqq \cdots$.
   (b) Show that $b_n - a_n = 2^{-n}(b_0 - a_0)$.
   (c) Show that, for all n,
   $$a_n b_n + a_{n-1} b_{n-1} = a_{n-1} b_n + a_n b_{n-1}$$
10. (Continuation) Using the notation of Problem 9, let $c_n = (a_n + b_n)/2$. Show that
    $$\lim_{n\to\infty} c_n = \lim_{n\to\infty} a_n = \lim_{n\to\infty} b_n$$
11. (Continuation) Consider the bisection method with the initial interval $[a_0, b_0]$. After ten steps with this method, show that
    $$\left| \frac{1}{2}(a_{10} + b_{10}) - \frac{1}{2}(a_9 + b_9) \right| = 2^{-11}$$
    Also, show that 20 steps are required to guarantee an approximation of a root to six decimal places (rounded).
12. By graphical methods, locate approximations to all roots of the equation $\ln (x + 1) + \tan (2x) = 0$.

Computer Problems 3.1

1. Select a program from your computing center library to solve polynomial equations and use it to find the roots of this equation.

   $$x^8 - 36x^7 + 546x^6 - 4536x^5 + 22449x^4 - 67284x^3 + 118124x^2 - 109584x + 40320 = 0$$

   The correct roots are the integers 1, 2, . . . , 8. Next, solve the same equation when the coefficient of x^7 is changed to -37. Observe how a minor perturbation in the coefficients can cause massive changes in the roots. Thus the roots are *unstable* functions of the coefficients. (Be sure to program the problem to allow for complex roots.)
2. Write a Fortran program to find a zero of each of the following functions, using the bisection method.
   (a) $x^3 - 3x + 1$ on the interval $[0, 1]$
   (b) $x^3 - 2 \sin x$ on the interval $[\frac{1}{2}, 2]$
3. Write a subroutine called `BISECT` which accepts as input a function `F`, an interval $[A, B]$, and a tolerance `EPSI`, and then carries out the bisection method on `F`. The subroutine should print the endpoints of the successive intervals and the values of `F` at these endpoints. Include tests to ensure that the numbers `U = F(A)` and `V = F(B)` have opposite signs at each step. Do so by testing `SIGN(1.0,U)`$\neq$`SIGN(1.0,V)`. Test your subroutine on $f(x) = x^3 + 2x^2 + 10x - 20$, with $a = 1$ and $b = 2$. The zero is 1.368808108. In programming this polynomial function, use nested multiplication: $((x + 2)x + 10)x - 20$.
4. Write a program to find a zero of a function f in the following way. In each step an interval $[a, b]$ is given, and $f(a)f(b) < 0$. Then c is computed as the root of the linear function that agrees with f at a and b. We retain either $[a, c]$ or $[c, b]$, depending on whether $f(a)f(c) < 0$ or $f(c)f(b) < 0$. Test your program on several functions.

3.2 Newton's Method

The procedure known as Newton's method is also called the *Newton–Raphson iteration*. It exists in a more general form than the one seen here, and the more general form can be used to find roots of *systems* of equations. Indeed, it is one of the important procedures in numerical analysis, and its applicability extends to differential equations and integral equations. Here it is being applied to a single equation of the form $f(x) = 0$. As before, we seek one or more points at which the value of the function f is zero.

In Newton's method it is assumed at once that the function f is differentiable. This implies that the graph of f has a definite *slope* at each point and hence a unique tangent line. Now let us pursue the following simple idea. At a certain point $(x_0, f(x_0))$ on the graph of f there is a tangent, which is a rather good approximation to the curve in the vicinity of that point. Analytically, it means that the linear function

$$l(x) = f'(x_0)(x - x_0) + f(x_0)$$

is close to the given function f near x_0. At x_0 the two functions l and f agree. We take the zero of l as an approximation to the zero of f. The zero of l is easily found. It is

$$x_1 = x_0 - \frac{f(x_0)}{f'(x_0)}$$

Thus, starting with point x_0 (which we may interpret as an approximation to the root sought), we pass to a new point x_1 obtained from the formula above. Naturally the process can be repeated ("iterated") to produce a sequence of points:

$$x_2 = x_1 - \frac{f(x_1)}{f'(x_1)}$$

$$x_3 = x_2 - \frac{f(x_2)}{f'(x_2)}$$

etc.

With good luck the sequence of points will approach the zero of f.

The geometry of Newton's method is shown in Fig. 3.1. The line $y = l(x)$ is tangent to the curve $y = f(x)$. It intersects the x axis at a point x_1. The slope of $l(x)$ is $f'(x_0)$.

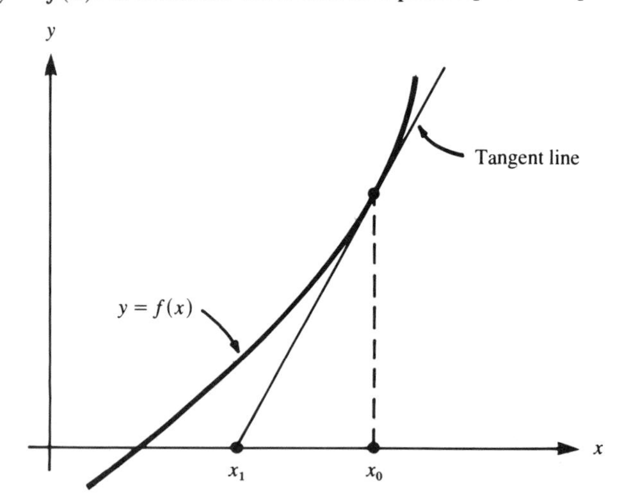

Figure 3.1 Newton's Method

There are other ways of interpreting Newton's method. Suppose that f is represented by a finite or infinite Taylor series near x_0:

$$f(x_0 + h) = f(x_0) + hf'(x_0) + \frac{h^2}{2} f''(x_0) + \cdots$$

If we choose h so that the first two terms of the series vanish, then $x_0 + h$ will be an approximate root of f, since $f(x_0 + h) = \frac{1}{2}h^2 f''(x_0) + \cdots \approx 0$ for small h. We are led to the equation

$$f(x_0) + hf'(x_0) = 0$$

from which we obtain $h = -f(x_0)/f'(x_0)$. Thus the correction h to be added to the initial estimate x_0 is exactly the one derived previously.

If we describe the *Newton iteration* in terms of a sequence $x_0, x_1, x_2, \ldots$, the sequence is defined inductively ("step by step") by the equation

$$x_{n+1} = x_n - \frac{f(x_n)}{f'(x_n)}$$

Now we illustrate Newton's method with a simple program to locate a root of the equation $x^3 + x = 2x^2 + 3$. We apply the method to the function $f(x) = x^3 - 2x^2 + x - 3$, starting with $x = 4$. Of course, $f'(x) = 3x^2 - 4x + 1$, and these two functions should be arranged in "nested" form for efficiency:

$$f(x) = ((x - 2)x + 1)x - 3$$

$$f'(x) = (3x - 4)x + 1$$

To see in greater detail the rapid convergence of Newton's method, we use double-precision arithmetic in the program.

```
      DOUBLE PRECISION  X, F, FP
      DATA  X/4.0D0/
      DO 2 N = 1,8
      F = ((X - 2.0D0)*X + 1.0D0)*X - 3.0D0
      FP = (3.0D0*X - 4.0D0)*X + 1.0D0
      X = X - F/FP
    2 PRINT 3,N,X,F
    3 FORMAT(5X,I5,F35.30,5X,E10.3)
      END
```

The program gives the following set of successive x values.

n	x_n	$f(x_{n-1})$
1	3.000000000000000000000000000000	3.300E+01
2	2.437500000000000000000000000000	9.000E+00
3	2.213032716315098014092654921114	2.037E+00
4	2.175554938721475650709180627018	2.564E-01
5	2.174560100666440121131017804145	6.463E-03
6	2.174559410293298356009472627192	4.479E-06
7	2.174559410292971506351022981107	2.156E-12
8	2.174559410292971506351022981107	4.994E-25

Figure 3.2 is a computer plot of three iterations of Newton's method for this sample problem.

Anyone who has experimented with Newton's method—for instance, by working one of the problems—will have observed the remarkable rapidity in the convergence of the sequence to the root. This phenomenon is also noticeable in the example just given. Indeed, the number of correct figures in the answer is nearly *doubled* at each successive step. Thus in the example we have first 0 and then 1, 2, 3, 6, 13, . . . accurate digits. Four or five steps of Newton's method often suffice to yield full machine precision in the determination of a root. There is a theoretical basis for this dramatic performance, as we shall now see.

Let the function f, whose zero we seek, possess two continuous derivatives f' and f'' and let r be a zero of f. Assume further that r is a *simple* zero—that is, $f'(r) \neq 0$. Then Newton's method, if started sufficiently close to r, converges *quadratically* to r. This means that the errors in successive steps obey an inequality of the form

$$|x_{n+1} - r| \leqq c|x_n - r|^2$$

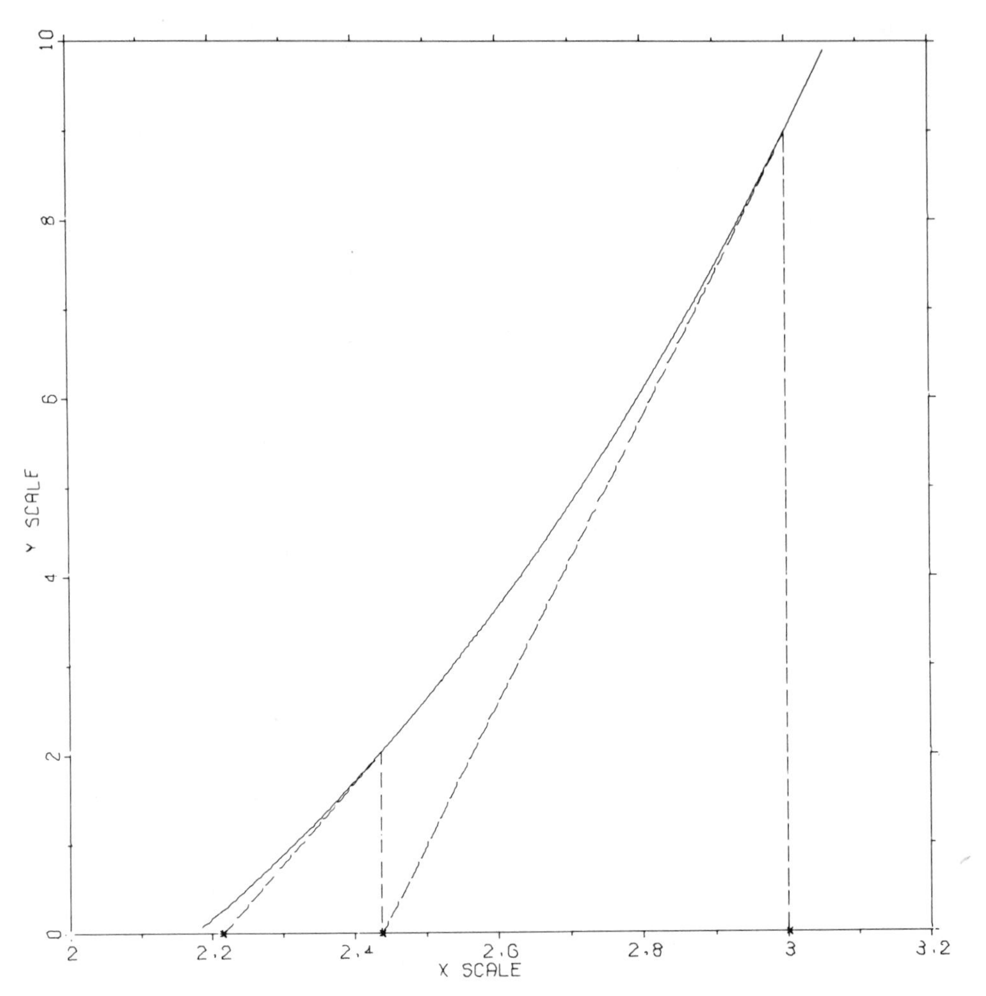

Figure 3.2 Three steps of Newton's method for $f(x) = x^3 - 2x^2 + x - 3$.

We will establish this presently. But first, an interpretation of the inequality may be helpful. Suppose that $c = 1$ for simplicity. Let x_n be an estimate of r which is correct to k decimal places. Then $|x_n - r| \leqq 10^{-k}$. So by the preceding inequality $|x_{n+1} - r| \leqq 10^{-2k}$. In other words, x_{n+1} will be correct to approximately $2k$ decimal places. This is the doubling of the number of significant digits alluded to previously.

In order to establish the quadratic convergence of Newton's method, let $e_n = x_n - r$. The formula that defines the sequence $\{x_n\}$ then gives

$$e_{n+1} = x_{n+1} - r = x_n - \frac{f(x_n)}{f'(x_n)} - r = e_n - \frac{f(x_n)}{f'(x_n)} = \frac{e_n f'(x_n) - f(x_n)}{f'(x_n)}$$

By Taylor's theorem (refer to Sec. 1.2), there exists a point ξ_n situated between x_n and r for which

$$0 = f(r) = f(x_n - e_n) = f(x_n) - e_n f'(x_n) + \frac{1}{2} e_n^2 f''(\xi_n)$$

(The subscript on ξ_n emphasizes the dependence on x_n.) This last equation can be rearranged to read

$$e_n f'(x_n) - f(x_n) = \frac{1}{2} e_n^2 f''(\xi_n)$$

and if this is used in the previous equation for e_{n+1}, the result is

$$e_{n+1} = \frac{1}{2}\left(\frac{e_n^2 f''(\xi_n)}{f'(x_n)}\right) \tag{1}$$

This is, at least qualitatively, the sort of equation we want. Continuing the analysis, we define a function

$$c(\delta) = \frac{\frac{1}{2}\max\limits_{|x-r|\leqq\delta} |f''(x)|}{\min\limits_{|x-r|\leqq\delta} |f'(x)|} \qquad (\delta > 0)$$

By virtue of this definition, we can assert that, for any two points x and ξ within distance δ of the root r, the inequality $\frac{1}{2}|f''(\xi)/f'(x)| \leqq c(\delta)$ is true. Now select δ so small that $\rho = \delta c(\delta) < 1$. This is possible because as δ converges to 0, $c(\delta)$ converges to $\frac{1}{2}|f''(r)/f'(r)|$, and so $\delta c(\delta)$ converges to 0. Remember that f' and f'' are continuous, and that $f'(r) \neq 0$. In the remainder of the argument, δ, $c(\delta)$, and ρ remain fixed.

Suppose now that our initial point x_0 is within distance δ of r. Then $|e_0| = |x_0 - r| \leqq \delta$ and $|\xi_0 - r| \leqq \delta$. Hence $\frac{1}{2}|f''(\xi_0)/f'(x_0)| \leqq c(\delta)$. By Eq. (1) it follows that

$$|x_1 - r| = |e_1| = e_0^2 \left|\frac{\frac{1}{2} f''(\xi_0)}{f'(x_0)}\right| \leqq e_0^2 c(\delta) \leqq e_0 \delta c(\delta) = \rho |e_0| \leqq |e_0| \leqq \delta$$

This shows that x_1 is also within distance δ of r. Therefore, *the preceding argument can be repeated*. The results will be

$$|e_2| \leqq \rho|e_1| \leqq \rho^2|e_0|$$
$$|e_3| \leqq \rho|e_2| \leqq \rho^3|e_0|$$

and so forth. Since

$$|e_n| \leqq \rho^n|e_0|$$

and

$$\rho < 1$$

we see that e_n converges to 0. In other words, x_n converges to r.

To summarize: there is a positive number δ such that, if the initial point x_0 satisfies $|x_0 - r| \leqq \delta$, then the subsequent points in Newton's iteration satisfy the same inequality, converge to r, and do so in accordance with the inequality

$$|e_{n+1}| \leqq c(\delta) e_n^2$$

Although Newton's method is truly a marvelous invention, its convergence depends upon hypotheses that are difficult to verify *a priori*. A graphic example will show what can happen. Consider a function f whose graph has the appearance shown in Fig. 3.3.

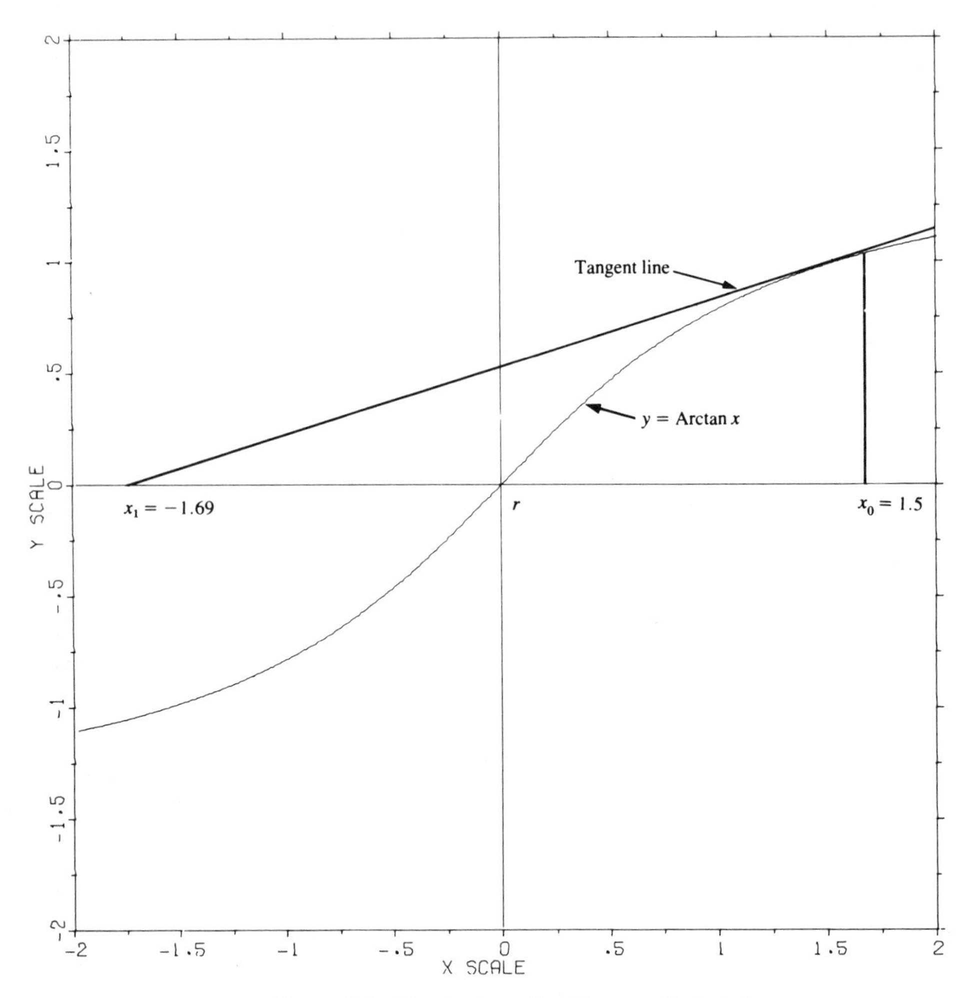

Figure 3.3 Newton's method for y = Arctan x.

for artanx, x_0 can never be close enough to r to make Newton's iteration converge.

The tangent to the graph at x_0 intersects the x axis at a point remote from the root r, and successive points in Newton's iteration *recede* from r instead of converging to r. The difficulty can be ascribed to a poor choice of the initial point x_0; it is *not* sufficiently close to r.

In using Newton's method, consideration must be given to the proper choice of a starting point. Usually one must have some insight as to the shape of the graph of the function. Sometimes a coarse graph is adequate, but in other cases a step-by-step evaluation of the function at many points may be necessary to find two points a, b such that $f(a)f(b) < 0$. Often the bisection method is used initially to obtain a suitable starting point, and Newton's method is used to improve the precision.

The analysis that established the quadratic convergence will disclose another troublesome hypothesis—namely, that $f'(r) \neq 0$. If $f'(r) = 0$, then r is a zero of f *and* f'. Such a zero is termed a ***multiple*** zero of f—in this case, at least a double zero. Ordinarily one would not know in advance that the zero sought was a multiple zero. If one knew this, however, Newton's method could be accelerated by modifying the equation to read

$$x_{n+1} = x_n - \frac{mf(x_n)}{f'(x_n)}$$

in which m is the *multiplicity* of the zero in question. The multiplicity is the least m such that $f^{(k)}(r) = 0$ for $0 \leqq k < m$, but $f^{(m)}(r) \neq 0$.

Problems 3.2

1. Write Newton's method in simplified form for determining the reciprocal of the square root of a positive number. Perform two iterations to approximate $1/\pm\sqrt{4}$, starting with $x_0 = 1$ and $x_0 = -1$.
2. Two of the four zeros of $x^4 + 2x^3 - 7x^2 + 3$ are positive. Find them by Newton's method, correct to two significant figures.
3. Derive a formula for Newton's method for the function $F(x) = f(x)/f'(x)$, where $f(x)$ is a three times continuously differentiable function with simple zeros. Show that the convergence of the resulting method to any zero r of $f(x)$ is at least quadratic. ***Hint:*** Apply the result in the text to F, making sure that F has the required properties.
4. The equation $x - Rx^{-1} = 0$ has $x = \pm R^{1/2}$ for its solution. Establish Newton's iterative scheme, in simplified form, for this situation. Carry out five steps for $R = 25$ and $x_0 = 1$.
5. The Taylor series for a function f looks like this:

$$f(x + h) = f(x) + hf'(x) + \frac{h^2}{2}f''(x) + \frac{h^3}{6}f'''(x) + \cdots$$

   Suppose that $f(x)$, $f'(x)$, and $f''(x)$ are easily computed. Derive an algorithm like Newton's method but using more terms in the Taylor series than in Newton's method. The algorithm should take as input an approximation to the root and produce as output a better approximation to the root. Show that the method is cubically convergent. ***Hint:*** Use $e_n = e_{n+1} - h$ and ignore e_{n+1}^2 terms as being negligible.
6. Using a calculator, observe the sluggishness with which Newton's method converges in the case of $f(x) = (x - 1)^m$ with $m = 8$ or 12. Reconcile this with the theory. Use $x_0 = 1.1$.
7. What linear function $y = ax + b$ approximates $f(x) = \sin x$ best in the vicinity of $x = \pi/4$? How does this problem relate to Newton's method?
8. In Problems 8 to 11 of the Supplementary Problems of Chapter 1, several methods are suggested for computing ln 2. Compare them to the use of Newton's method applied to the equation $e^x = 2$.
9. Refer to the discussion of Newton's method and establish that

$$\lim_{n\to\infty} (e_{n+1}e_n^{-2}) = \frac{1}{2}\left(\frac{f''(r)}{f'(r)}\right)$$

   How can this be used to test, in a practical case, whether the convergence is quadratic? Devise an example in which r, $f'(r)$, and $f''(r)$ are all known and test numerically the convergence of $e_{n+1}e_n^{-2}$.
10. Show that in the case of a zero of multiplicity m the modified Newton's method

$$x_{n+1} = x_n - \frac{mf(x_n)}{f'(x_n)}$$

   is quadratically convergent. ***Hint:*** Use Taylor series for $f(r + e_n)$ and $f'(r + e_n)$.
11. A method due to Steffensen for solving the equation $f(x) = 0$ employs the formula

$$x_{n+1} = x_n - \frac{f(x_n)}{g(x_n)}$$

in which $g(x) = [f(x + f(x)) - f(x)]/f(x)$. It is quadratically convergent, like Newton's method. How many function evaluations are necessary per step? Using Taylor series, show that $g(x) \approx f'(x)$ if $f(x)$ is small and thus relate Steffensen's iteration to Newton's. What advantage does it have? Establish the quadratic convergence.

Computer Problems 3.2

1. Write a simple program to apply Newton's method to the function $f(x) = x^3 + 2x^2 + 10x - 20$, starting with $x = 2$. Note that a subscripted variable is not necessary. Evaluate f and f', using nested multiplication. Stop the computation when two successive points in the process differ by 10^{-9} or some other convenient tolerance. Print all intermediate points with the corresponding values of f. Put an upper limit of 20 on the number of steps (to avoid an infinite computing loop in case of difficulty).
2. Find the root of the equation

$$2x[1 - x^2 + x] \ln x = x^2 - 1$$

in the interval $[0, 1]$ by Newton's method, using double precision. Make a table showing the number of correct digits in each step.
3. In 1685 Wallis published a book called *Algebra,* in which he described a method devised by Newton for solving equations. In slightly modified form, this method was also published by Raphson in 1690. This form is the one now commonly called Newton's method or the Newton–Raphson method. Wallis illustrated it on the equation $x^3 - 2x - 5 = 0$. Find a root of this equation in double precision, thus continuing the tradition that every numerical analysis student should solve this venerable equation.
4. In celestial mechanics, Kepler's equation is important. It reads $x = y - \varepsilon \sin y$, in which x is a planet's mean anomaly, y its eccentric anomaly, and ε the eccentricity of its orbit. Taking $\varepsilon = 0.9$, construct a table of y for 30 equally spaced values of x in the interval $0 \leqq x \leqq \pi$. Use Newton's method to obtain each value y. The y corresponding to an x can be used as the starting point for the iteration when x is changed slightly.
5. In Newton's method we progress in each step from a given point x to a new point $x - h$, where $h = f(x)/f'(x)$. A refinement which is easily programmed is this: If $|f(x - h)|$ is not smaller than $|f(x)|$, then reject this value of h and use $h/2$ instead. Test this refinement.
6. Write a brief program to compute a root of the equation $x^3 = x^2 + x + 1$, using Newton's method. Be careful to select a suitable starting value.
7. Find the root of the equation $5(3x^4 - 6x^2 + 1) = 2(3x^5 - 5x^3)$ lying in the interval $[0, 1]$ by using Newton's method and a short program.
8. Write a complete program, using no more than six Fortran instructions, to compute and print 8 steps of Newton's method in finding a positive root of each.
   (a) $x = 2 \sin x$
   (b) $x^3 = \sin x + 7$
   (c) $x^5 + x^2 = 1 + 7x^3$ for $x \geqq 2$
   (d) $\sin x = 1 - x$
9. On a certain modern computer floating-point numbers have 48-bit fractional parts. Moreover, floating-point hardware can perform addition, subtraction, multiplication, and reciprocation, but not division. Unfortunately, the reciprocation hardware produces a result accurate to only 30 bits, whereas the other operations produce results accurate to full floating-point precision.
   (a) Show that Newton's method can be used to solve $f(x) = 1 - 1/(ax)$ for an approximation to $1/a$ which is accurate to full floating-point precision. How many iterations are required?
   (b) Show how to obtain an approximation to b/a which is accurate to full floating-point precision.

10. In the Newton method for finding a root r of $f(x) = 0$ we start with x_0 and compute the sequence $x_1, x_2, \ldots$, using the formula $x_{n+1} = x_n - f(x_n)/f'(x_n)$. To avoid computing the derivative at each step, it has been proposed to replace $f'(x_n)$ with $f'(x_0)$ in all steps. If Newton's method converges, it does so quadratically, whereas the proposed method is only linear. To obtain more rapid convergence of Newton's method, it has also been suggested that the derivative be computed every other step. This method is given by

$$\begin{cases} x_{2n+1} = x_{2n} - \dfrac{f(x_{2n})}{f'(x_{2n})} \\[2ex] x_{2n+2} = x_{2n+1} - \dfrac{f(x_{2n+1})}{f'(x_{2n})} \end{cases}$$

Numerically compare both proposed methods to Newton's method for several simple functions with known roots. Print the error of each method on every iteration in order to monitor the convergence. How do the proposed methods work?

11. Write `FUNCTION SQRTF(X)` to compute the square root of a real argument x by the following algorithm. First, reduce the range of x by finding a real number r and an integer m such that $x = 2^{2m}r$ with $\frac{1}{4} \leqq r < 1$. Next, compute x_2 by using two iterations of Newton's method given by

$$x_{n+1} = \frac{1}{2}\left(x_n + \frac{r}{x_n}\right)$$

and the special initial approximation

$$x_0 = 1.2723\ 5367 + 0.2426\ 93281r - \frac{1.0296\ 6039}{1 + r}$$

Then set $\sqrt{x} \approx 2^m x_2$. Test this algorithm on various values of x. Obtain a listing of the code for the square root function on your computer system. By reading the comment cards, try to determine what algorithm it uses.

12. Write `FUNCTION CUBERTF(X)` to compute the cube root of a real argument x by the following procedure. First, determine a real number r and an integer m such that $x = r\,2^{3m}$ with $\frac{1}{8} \leqq r < 1$. Compute x_4 using four iterations of Newton's method

$$x_{n+1} = \frac{2}{3}\left(x_n + \frac{r}{2x_n^2}\right)$$

using the special starting value

$$x_0 = 2.502926 - \frac{8.045125(r + 0.3877552)}{(r + 4.612244)(r + 0.3877552) - 0.3598496}$$

Then set $\sqrt[3]{x} \approx 2^m x_4$. Test this algorithm on a variety of x values.

3.3 Secant Method

We next consider a general-purpose procedure that converges almost as fast as Newton's method. This method mimics Newton's method but avoids the calculation of derivatives. Recall that Newton's iteration defines x_{n+1} in terms of x_n via the formula

$$x_{n+1} = x_n - \frac{f(x_n)}{f'(x_n)}$$

For the secant method, this formula is modified in an obvious way. The term $f'(x_n)$ is replaced by the approximation

$$f'(x_n) \approx \frac{f(x_n) - f(x_{n-1})}{x_n - x_{n-1}} \tag{1}$$

The name of the method is taken from the fact that the right member of Eq. (1) is the slope of a *secant* line to the graph of f. (See Fig. 3.4.) Of course, the left member is the

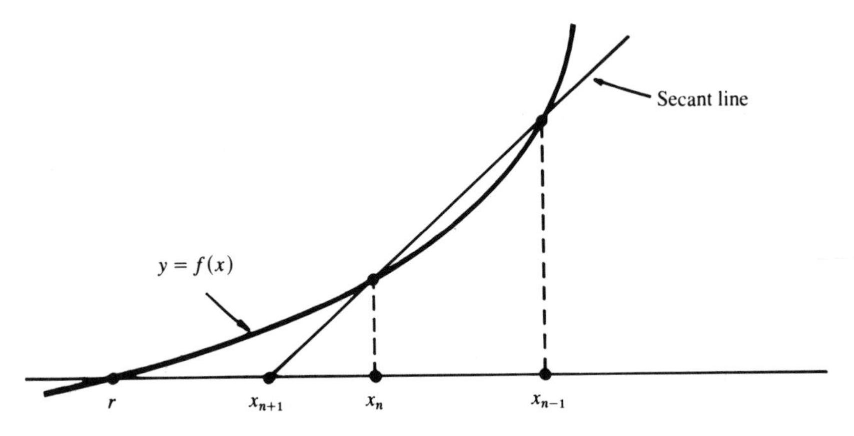

Figure 3.4 Secant Method

slope of a *tangent* line to the graph of f. When approximation (1) is used in the formula for Newton's method, the resulting formula defines the *secant method.*

$$x_{n+1} = x_n - f(x_n)\left(\frac{x_n - x_{n-1}}{f(x_n) - f(x_{n-1})}\right) \tag{2}$$

A few remarks about Eq. (2) are in order. Clearly x_{n+1} depends on *two* previous elements of the sequence. So in starting, two points (x_0 and x_1) must be provided. Formula (2) then can generate $x_2, x_3, \ldots$. In programming the secant method, it is advisable to calculate and test the quantity $f(x_n) - f(x_{n-1})$. If it is nearly zero, an overflow can occur in Eq. (2). Of course, if the method is succeeding, the points x_n will be approaching a zero of f. So $f(x_n)$ will be converging to zero. (We are assuming that f is continuous.) Also, $f(x_{n-1})$ will be converging to zero, and, *a fortiori,* $f(x_n) - f(x_{n-1})$ will approach zero. If the terms $f(x_n)$ and $f(x_{n-1})$ have the same sign, an additional cancellation of significant digits occurs in the subtraction. So we could perhaps halt the iteration when $|f(x_n) - f(x_{n-1})| \leqq \varepsilon\, |f(x_n)|$ for some specified tolerance ε, such as 10^{-6}.

The advantages of the secant method are that only one function evaluation is required per step (in contrast to Newton's iteration, which requires two) and that it is almost as rapidly convergent. It can be shown that the errors obey an equation of the form

$$e_{n+1} = e_n e_{n-1} \frac{f''(\xi_n)}{2f'(\zeta_n)} \tag{3}$$

where ξ_n and ζ_n are in the smallest interval containing r, x_n, and x_{n-1}. Thus the ratio

$e_{n+1}(e_n e_{n-1})^{-1}$ converges to $\frac{1}{2}f''(r)/f'(r)$. Rapidity of convergence of this method is, in general, between that for bisection and Newton's method.

In fact, it can be shown that for the secant method

$$e_{n+1} \leqq Ce_n^{(1+\sqrt{5})/2} \tag{4}$$

for some constant C. Since $(1 + \sqrt{5})/2 \approx 1.62$, we say that the convergence is *superlinear.* It is beyond the scope of this book to establish this relation; however, here is an indication of how it is obtained. From (3) we have

$$|e_{n+1}| \leqq c|e_n||e_{n-1}|$$

Now letting $c|e_m| = d_m$, we establish that $d_{n+1} \leqq d_n d_{n-1}$. If $d = \max(d_1, d_2) < 1$, then $d_{m+1} \leqq d^{\lambda_m}$, where the λ_m are the Fibonacci numbers given by

$$\begin{cases} \lambda_{n+1} = \lambda_n + \lambda_{n-1} \\ \lambda_0 = \lambda_1 = 1 \end{cases}$$

Moreover, it follows that

$$\lambda_n = \frac{1}{\sqrt{5}}\left[\left(\frac{1+\sqrt{5}}{2}\right)^{n+1} - \left(\frac{1-\sqrt{5}}{2}\right)^{n+1}\right] \tag{5}$$

and

$$\lim_{n\to\infty} \frac{\lambda_{n+1}}{\lambda_n} = \frac{1+\sqrt{5}}{2} \tag{6}$$

Then (4) is established by some further analysis which puts all the pieces together. (See also Problem 5 in Sec. 10.1.)

Problems 3.3

1. The formula for the secant method can also be written

$$x_{n+1} = \frac{x_{n-1}f(x_n) - x_n f(x_{n-1})}{f(x_n) - f(x_{n-1})}$$

Establish this, and explain why it is inferior to Eq. (2) in a computer program.

2. Show that if the iterates in Newton's method converge to a point r for which $f'(r) \neq 0$, then $f(r) = 0$. Establish the same assertion for the secant method. *Hint:* In the latter the Mean-Value Theorem of differential calculus is useful. This is the case $n = 0$ in Taylor's theorem.
3. Establish Eq. (3) governing the errors in the secant method.
4. A method of finding a zero of a given function f proceeds as follows. Two initial approximations x_0 and x_1 to the zero are chosen, the value of x_0 is fixed, and successive iterations are given by

$$x_{n+1} = x_n - \left(\frac{x_n - x_0}{f(x_n) - f(x_0)}\right) f(x_n)$$

This process will converge to a zero of f under certain conditions. Show that the rate of convergence to a simple zero is *linear.*

5. Try to justify Eq. (4).

Computer Problems 3.3

1. Write a simple program to compare the secant method with Newton's method for finding a root of each function.
(a) $x^3 - 3x + 1$ with starting point $x_0 = 2$
(b) $x^3 - 2 \sin x$ with starting point $x_0 = \frac{1}{2}$
Use the x_1 value from Newton's method as the second starting point for the secant method. Print out each iteration for both methods.
2. Write a simple program to find the root of $f(x) = x^3 + 2x^2 + 10x - 20$, using the secant method with starting values $x_0 = 2$ and $x_1 = 1$. Let it run at most 20 steps, but stop the iteration when $f(x_n) - f(x_{n-1})$ is "very small." Compare the number of steps needed here to the number needed in Newton's method. Is the convergence quadratic?
3. Test the secant method on the set of functions $f_k(x) = 2e^{-k}x + 1 - 3e^{-kx}$ for $k = 1, 2, 3, \ldots, 10$. Use the interval $[0, 1]$.
4. An example due to Wilkinson shows that minute alterations in the coefficients of a polynomial may have massive effects on the roots. Let

$$f(x) = (x - 1)(x - 2) \cdots (x - 20)$$

The zeros of f are, of course, the integers $1, 2, \ldots, 20$. Try to determine what happens to the zero $r = 20$ when the function is altered to $f(x) - 10^{-8}x^{19}$. *Hint:* The secant method in double precision will locate accurately a zero in the interval $[20, 21]$.
5. Test the secant method on an example in which r, $f'(r)$, $f''(r)$ are known in advance. Monitor the ratios $e_{n+1}/(e_n e_{n-1})$ to see whether they converge to $\frac{1}{2}f''(r)/f'(r)$. The function $f(x) = \text{Arctan } x$ is suitable for this experiment.
6. Use the secant method to find the zero near -0.5 of $f(x) = e^x - 3x^2$. This function also has a zero near 4. Find this positive zero by Newton's method.
7. Write `SUBROUTINE SECANT(F, X1, X2, EPSI, DELT, MAXF, X, IERR)` that uses the secant method to solve $f(x) = 0$. The input parameters are `F`—name of the given function; `X1, X2`—two initial estimates of the solution; `EPSI`—nonnegative tolerance such that the iteration stops if the difference between two consecutive iterates is smaller than this value; `DELT`—nonnegative tolerance such that the iteration stops if a function value is smaller in magnitude than this value; `MAXF`—positive integer bounding the number of evaluations of the function allowed. The output parameters are `X`—final estimate of the solution; `IERR`—an integer error flag indicating whether a tolerance test was violated. Test this routine, using the function of Problem 6. Print the final estimate of the solution and the value of the function at this point.
8. Using a function of your choice, verify numerically that the iterative method

$$x_{n+1} = x_n - \frac{f(x_n)}{\sqrt{[f'(x_n)]^2 - f(x_n)f''(x_n)}}$$

is cubically convergent at a simple root but only linearly convergent at a multiple root.

Supplementary Problems (Chapter 3)

1. What happens if the Newton iteration is applied to the function $f(x) = \text{Arctan } x$ with $x_0 = 2$?
2. What is the limit of the sequence $x_{n+1} = x_n/2 + 1/x_n$?
3. Newton's method can be interpreted as follows. Suppose that $f(x + h) = 0$. Then $f'(x) \approx [f(x + h) - f(x)]/h = -f(x)/h$. Continue this argument.
4. For what starting values will Newton's iteration converge if $f(x) = x^2/(1 + x^2)$?
5. Analyze what happens when Newton's method is applied to the function $f(x) = 2x^3 - 9x^2 + 12x + 15$, starting at $x = 3$, or $x < 3$, or $x > 3$.

6. (Continuation) Repeat Problem 5 for $f(x) = \sqrt{|x|}$, starting with $x < 0$ or $x > 0$.
7. In order to determine $x = \sqrt[3]{R}$, we can solve the equation $x^3 = R$ by Newton's method. Write the DO loop that carries out four steps of this process, starting from the initial approximation $x_0 = R$.
8. The reciprocal of a number R can be computed without division by the formula

$$x_{n+1} = x_n(2 - x_n R)$$

Establish this relation by applying Newton's method to some $f(x)$. Beginning with $x_0 = 0.2$, compute the reciprocal of 4 correct to six decimals or more by this rule. Tabulate the error at each step and observe the quadratic convergence.
9. Using Newton's method, establish the iteration formula

$$x_{n+1} = \frac{1}{2}\left[x_n + \left(\frac{R}{x_n}\right)\right]$$

for finding $\sqrt{R}$. Perform three iterations of this scheme for $R = 2$, starting with $x_0 = 1$, and of the bisection method, starting with interval $[1, 2]$. How many iterations are needed for each method in order to obtain 10^{-6} accuracy?
10. (Continuation) Newton's method for finding $\sqrt{R}$, where $R = AB$, gives this approximation.

$$\sqrt{AB} \approx \frac{A + B}{4} + \frac{AB}{A + B}$$

Show that if $x_0 = A$ or B, then two iterations of Newton's method are needed to obtain this approximation, whereas if $x_0 = \frac{1}{2}(A + B)$, then only one iteration is needed.
11. Every polynomial of degree n has n zeros in the complex plane. Does it follow that every function of the form $f(x) = \sum_{n=0}^{\infty} a_n x^n$ has infinitely many zeros?
12. Consider the algorithm of which *one* step consists of two steps of Newton's method. What is its order of convergence?
13. Show that Newton's method applied to $x^m - R$ and to $1 - (R/x^m)$ for determining $\sqrt[m]{R}$ $(R > 0,\ m \geqq 2)$ results in two similar yet different iterative formulas. Which is better and why?
14. A proposed generalization of Newton's method is

$$x_{n+1} = x_n - \omega \frac{f(x_n)}{f'(x_n)}$$

where the constant ω is an acceleration factor chosen to increase the rate of convergence. For what range of values of ω is a simple root r of $f(x)$ a *point of attraction*, that is, $|g'(r)| < 1$, where $g(x) = x - \omega f(x)/f'(x)$? This method is quadratically convergent *only* if $\omega = 1$, since $g'(r) \neq 0$ for $\omega \neq 1$.
15. Suppose that r is a double root of $f(x) = 0$, that is, $f(r) = f'(r) = 0$ but $f''(r) \neq 0$, and suppose that f and all derivatives up to and including the second are continuous in some neighborhood of r. Show that $e_{n+1} \approx \frac{1}{2}e_n$ for Newton's method and, thereby, conclude that the rate of convergence is *linear* near a double root.
16. This problem and the next three deal with the method of *functional iteration*. A *fixed point* of a function f is an x such that $f(x) = x$. The method of functional iteration is as follows. Starting with any x_0, we define $x_{n+1} = f(x_n)$ $(n = 0, 1, 2, \ldots)$. Show that if f is continuous and if the sequence $\{x_n\}$ converges, then its limit is a fixed point of f.
17. (Continuation) If f is a function defined on the whole real line whose derivative satisfies $|f'(x)| \leqq c$ with a constant c less than 1, then the method of functional iteration produces a fixed point of f. *Hint:* In establishing this, the Mean Value Theorem is helpful.
18. (Continuation) With a calculator, try the method of functional iteration with $f(x) = x/2 + 1/x$, taking $x_0 = 1$. What is the limit of the resulting sequence?

19. (Continuation) Using Problem 17, show that the equation $10 - 2x + \sin x = 0$ has a root. Locate the root approximately by drawing a graph. Starting with your approximate root, use functional iteration to obtain the root accurately by using a calculator. *Hint:* Write the equation in the form $x = 5 + \frac{1}{2}\sin x$.
20. (Simultaneous Nonlinear Equations) Using the Taylor series in two variables (x, y) of the form

$$f(x + h, y + k) = f(x, y) + hf_x(x, y) + kf_y(x, y) + \cdots$$

where $f_x = \partial f/\partial x$ and $f_y = \partial f/\partial y$, establish that Newton's method for solving the two simultaneous nonlinear equations

$$\begin{cases} f(x, y) = 0 \\ g(x, y) = 0 \end{cases}$$

can be described with the formulas

$$x_{n+1} = x_n - \frac{fg_y - gf_y}{f_x g_y - g_x f_y}$$

$$y_{n+1} = y_n - \frac{f_x g - g_x f}{f_x g_y - g_x f_y}$$

Here, the functions f, f_x, and so on are evaluated at (x_n, y_n).
21. Newton's method can be defined for the equation $f(z) = g(z) + ih(z)$, where $f(z)$ is an analytic function of the complex variable $z = x + iy$ (x and y real) and $g(z)$ and $h(z)$ are real functions for all z. The derivative $f'(z)$ is given by $f'(z) = g_x + ih_x = h_y - ig_y$, since the *Cauchy–Riemann equations* $g_x = h_y$, $h_x = -g_y$ hold. Here $g_x = \partial g/\partial x$, $g_y = \partial g/\partial y$, and so on. Show that Newton's method

$$z_{n+1} = z_n - \frac{f(z_n)}{f'(z_n)}$$

can be written in the form

$$x_{n+1} = x_n - \frac{gh_y - hg_y}{g_x h_y - g_y h_x}$$

$$y_{n+1} = y_n - \frac{hg_x - gh_x}{g_x h_y - g_y h_x}$$

Here all functions are evaluated at $z_n = x_n + iy_n$.
22. Test the following sequences for different types of convergence (i.e., linear, superlinear, or quadratic) where $n = 1, 2, 3, \ldots$.
(a) $x_n = n^{-2}$ (b) $x_n = 2^{-n}$
(c) $x_n = 2^{-a_n}$ with $a_0 = a_1 = 1$ and $a_{n+1} = a_n + a_{n-1}$ for $n \geqq 2$ (d) $x_n = 2^{-2^n}$
23. Every polynomial of degree n has n zeros in the complex plane. Does it follow that every real polynomial of degree n has n zeros on the real line?

Supplementary Computer Problems (Chapter 3)

1. Would you like to see the number 0.55887766 come out of a calculation? Take three steps in Newton's method on $10 + x^3 - 12\cos x = 0$, starting with $x = 1$.
2. Write a short program to solve for a root of the equation $e^{-x^2} = \cos x + 1$ on $[0, 4]$. What happens in Newton's method if we start with $x = 0$ or $x = 1$?
3. Find the root of the equation $\frac{1}{2}x^2 + x + 1 - e^x = 0$ by Newton's method, starting with $x_0 = 1$, and account for the slow convergence.

4. Find the zero of the function $f(x) = x - \tan x$ that is closest to 99 (radians) by both the bisection method and Newton's method. *Hint:* Extremely accurate starting values are needed for this function. Use the computer to construct a table of values of $f(x)$ around 99 to determine the nature of this function.
5. Using the bisection method, find the positive root of the equation $2x(1 + x^2)^{-1} = \text{Arctan } x$. Using the root as x_0, apply Newton's method to the function Arctan x. Interpret the results.
6. If the root r of $f(x) = 0$ is a double root, then Newton's method can be accelerated by using

$$x_{n+1} = x_n - \frac{2f(x_n)}{f'(x_n)}$$

Numerically compare the convergence of this scheme with Newton's method on a function with a known double root.

7. Test numerically whether Olver's method, given by

$$x_{n+1} = x_n - \frac{f(x_n)}{f'(x_n)} - \frac{1}{2}\left(\frac{f(x_n)^2 f''(x_n)}{[f'(x_n)]^3}\right)$$

is cubically convergent to a root of f. Try to establish that it is.

8. (Continuation) Repeat for Halley's method

$$x_{n+1} = x_n - \frac{1}{a_n}$$

with

$$a_n = \frac{f'(x_n)}{f(x_n)} - \frac{1}{2}\left(\frac{f''(x_n)}{f'(x_n)}\right)$$

9. Using Supplementary Problem 20, find a root of the nonlinear system

$$\begin{cases} 4y^2 + 4y + 52x - 19 = 0 \\ 169x^2 + 3y^2 + 111x - 10y - 10 = 0 \end{cases}$$

using Newton's method and starting at $(-0.01, -0.01)$.

10. Using Supplementary Problem 21, find a complex root of each.
(a) $z^3 - z - 1 = 0$ (b) $z^4 - 2z^3 - 2iz^2 + 4iz = 0$
(c) $2z^3 - 6(1 + i)z^2 - 6(1 - i) = 0$ (d) $z = e^z$
Hint (*d*): Use Euler's relation $e^{iy} = \cos y + i \sin y$.
11. A circular metal shaft is being used to transmit power. It is known that at a certain critical angular velocity ω any jarring of the shaft while rotating will cause it to deform or buckle. This is a dangerous situation, for the shaft might shatter under the increased centrifugal forces. To find this critical velocity ω, we must first compute a number x that satisfies the equation

$$\tan x + \tanh x = 0$$

This number is then used in a formula to obtain ω. Solve for x $(x > 0)$.

4

Numerical Integration

In electrical field theory it is proved that the magnetic field induced by a current flowing in a circular loop of wire has intensity

$$H(x) = \frac{4Ir}{r^2 - x^2} \int_0^{\pi/2} \sqrt{1 - \left(\frac{x}{r}\right)^2 \sin^2\theta}\, d\theta$$

where I is the current, r the radius of the loop, and x the distance from the center to the point where the magnetic intensity is being computed ($0 \leqq x \leqq r$). If I, r, and x are given, we have a nasty integral to evaluate. It is an *elliptic* integral and not expressible in terms of familiar functions; but H *can* be computed precisely by the methods of this chapter. For example, if $I = 15.3$, $r = 120$, and $x = 84$, we find

$$H = 1.355661135$$

accurate to nine decimals.

4.1 Definite Integral

Elementary calculus focuses largely on two important processes of mathematics: differentiation and integration. In Chapter 1 differentiation was considered briefly and that topic will be taken up again later. In this chapter the process of integration will be examined from the standpoint of numerical analysis.

It is customary to distinguish two types of integrals, the definite and the indefinite integral. The indefinite integral of a function is another *function,* or a class of functions, whereas the definite integral of a function over a fixed interval is a *number.* For example,

Indefinite integral: $\displaystyle\int x^2\, dx = \frac{1}{3}x^3 + C$

Definite integral: $\displaystyle\int_0^2 x^2\, dx = \frac{8}{3}$

Actually, a function has not just one but many indefinite integrals. These differ from each other by constants. Thus in the preceding example any constant value may be assigned to C and the result is still an indefinite integral. The concept of an indefinite integral is identical with the concept of antiderivative. An antiderivative of a function f is any function F with the property that $F' = f$.

The definite and indefinite integrals are related by the *Fundamental Theorem of Calculus,* which states that $\int_a^b f(x)\,dx$ can be computed by first finding an antiderivative F of f and then evaluating $F(b) - F(a)$. Thus using traditional notation,

$$\int_1^3 (x^2 - 2)\,dx = \left[\frac{x^3}{3} - 2x\right]_1^3 = \left(\frac{27}{3} - 6\right) - \left(\frac{1}{3} - 2\right) = \frac{14}{3}$$

As another example of the Fundamental Theorem of Calculus, we can write

$$\int_a^b F'(x)\,dx = F(b) - F(a)$$

$$\int_a^x F'(t)\,dt = F(x) - F(a)$$

If this second equation is differentiated with respect to x, the result is (and here we have put $f = F'$)

$$\frac{d}{dx}\int_a^x f(t)\,dt = f(x)$$

This last equation shows that $\int_a^x f(t)\,dt$ must be an antiderivative (indefinite integral) of f.

The foregoing technique for computing definite integrals is virtually the only one emphasized in elementary calculus. The definite integral of a function, however, has an interpretation as the area under a curve, and so the existence of a numerical value for $\int_a^b f(x)\,dx$ should not depend logically on our limited ability to find antiderivatives. Thus, for instance, $\int_0^1 e^{x^2}\,dx$ has a precise numerical value despite the fact that there is no elementary function F such that $F'(x) = e^{x^2}$. By the preceding remarks, e^{x^2} does have an antiderivative, one of which is

$$F(x) = \int_0^x e^{t^2}\,dt$$

But this form of the function F is of no help in obtaining the numerical value sought.

The existence of the definite integral of a function f on a closed interval $[a, b]$ is based on an interpretation of that integral as the area under the graph of f. The definite integral is defined by means of two concepts, the *lower sums* of f and the *upper sums* of f; these are approximations to the area under the graph.

Let P be a partition of the interval $[a, b]$, given by division points

$$a = x_1 < x_2 < x_3 < \cdots < x_{n-1} < x_n = b$$

Now denote by m_i and M_i the minimum and maximum values of f on the subinterval $[x_{i-1}, x_i]$:

$$m_i = \min_{x_{i-1} \leqq x \leqq x_i} f(x) \qquad M_i = \max_{x_{i-1} \leqq x \leqq x_i} f(x)$$

The lower and upper sums of f corresponding to the given partition P are then

$$L(f;P) = \sum_{i=2}^{n} m_i(x_i - x_{i-1})$$

$$U(f;P) = \sum_{i=2}^{n} M_i(x_i - x_{i-1})$$

If f is a positive function, these two quantities can be interpreted as estimates of the area under the curve for f. The upper sum is shown in Fig. 4.1. It is intuitively clear that

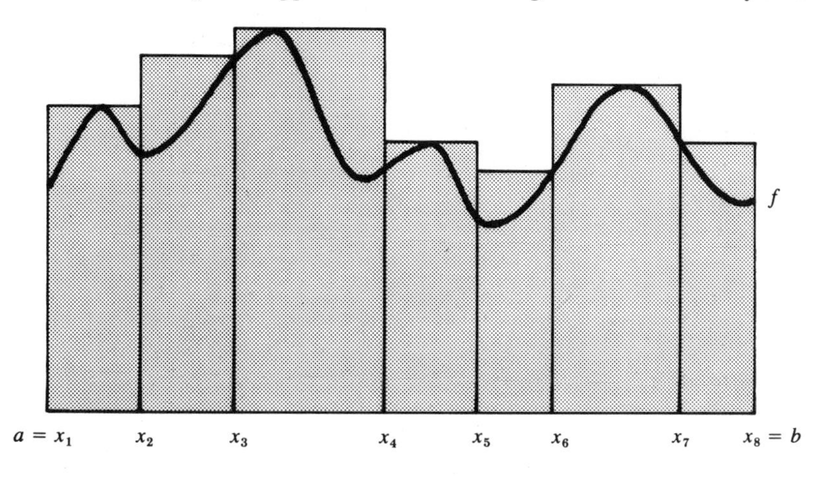

Figure 4.1

the upper sum *overestimates* the area under the curve, and the lower sum *underestimates* it. So

$$L(f;P) \leqq \int_a^b f(x)\,dx \leqq U(f;P) \tag{1}$$

The next step in defining the integral is to consider the least upper bound of the lower sums and the greatest lower bound of the upper sums when we allow P to range over *all* partitions of the interval $[a, b]$. In symbols, we consider

$$\sup_P L(f;P) \qquad \inf_P U(f;P)$$

If these two numbers are the same, then their common value is taken as the definition of $\int_a^b f(x)\,dx$; if different, then we say that f is not "Riemann integrable" or that its integral does not exist.

In calculus it is proved that the Riemann integral of a continuous function on a closed finite interval *does* exist and that it can be obtained by two limits:

$$\lim_{n\to\infty} L(f;P_n) = \int_a^b f(x)\,dx = \lim_{n\to\infty} U(f;P_n)$$

in which $P_1, P_2, P_3, \ldots$ is any sequence of partitions with the property that the length of the largest subinterval in P_n converges to zero as $n \to \infty$. Furthermore, if it is so arranged that P_{n+1} is obtained from P_n by adding new points (and not deleting points), then the lower sums converge *upward* to the integral and the upper sums converge *downward* to the integral. From the numerical standpoint this is a desirable feature of the process, for at each step an interval containing the unknown number $\int_a^b f(x)\,dx$ will be available. Moreover, these intervals shrink in width in each succeeding step.

The process just described can easily be carried out on a computer. To illustrate, we select the function $f(x) = e^{-x^2}$ and the interval $[0, 1]$. This function is of great importance in statistics but is one whose indefinite integral cannot be obtained by the elementary techniques of calculus. For partitions, we take equally spaced points in $[0, 1]$. Thus if there are to be $n - 1$ subintervals in P_n, then we define

$$P_n\colon x_1, x_2, \ldots, x_n \qquad \text{with } x_i = (i-1)h \quad \text{and} \quad h = \frac{1}{n-1}$$

Since e^{-x^2} is *decreasing* on $[0, 1]$, the least value of f on the subinterval $[x_{i-1}, x_i]$ occurs at x_i. Similarly, the greatest value occurs at x_{i-1}. Hence $m_i = f(x_i)$ and $M_i = f(x_{i-1})$. Putting this into the formulas for the upper and lower sums, we obtain for this function

$$L(f;P_n) = \sum_{i=2}^{n} hf(x_i) = h\sum_{i=2}^{n} e^{-x_i^2}$$

$$U(f;P_n) = \sum_{i=2}^{n} hf(x_{i-1}) = h\sum_{i=2}^{n} e^{-x_{i-1}^2}$$

Since these sums are almost the same, it is more economical to compute $L(f; P_n)$ by the given formula and to obtain $U(f;P_n)$ by observing that

$$U(f;P_n) = hf(x_1) + L(f;P_n) - hf(x_n) = L(f;P_n) + h(1 - e^{-1})$$

The last equation also shows that the interval defined by (1) is of width $h(1 - e^{-1})$.

Here is a program to carry out this experiment with $n = 1000$.

```
      N = 1000
      H = 1.0/FLOAT(N-1)
      SUM = 0.0
      DO 2 I = 1,N-1
      X = H*FLOAT(N-I)
    2 SUM = SUM + EXP(-X*X)
      SUMLO = H*SUM
      SUMUP = SUMLO + H*(1.0 - EXP(-1.0))
      PRINT 3,SUMLO,SUMUP
    3 FORMAT(5X,2(5X,F10.5))
      END
```

A few comments about this program may be helpful. First, a subscripted variable is not needed in the program for the points x_i. Each point is labeled X. After it is defined (fifth line of code) and used (sixth line of code), it need not be saved. Next, observe that the program has been written so that only one line of code would need to be changed if another value of N were required. Finally, the numbers $e^{-x_i^2}$ are added in order of *ascending* magnitude to reduce roundoff error. However, roundoff errors in this program are negligible compared to the error in our final estimation of the integral.

The preceding Fortran program produces as output the following values of the lower and upper sums:

```
SUMLO = 0.74651     SUMUP = 0.74714
```

The true value of the integral, computed by other means, is

$$\int_0^1 e^{-x^2}\,dx \approx 0.74682413$$

The function

$$\frac{2}{\sqrt{\pi}}\int_0^x e^{-t^2}\,dt$$

is known as the *error function* and is usually denoted by erf (x). A tabulation of this function (and many others) can be found in standard mathematical handbooks, such as Abramowitz and Stegun [1965].

At this juncture the reader is urged to program this experiment or one like it. The experiment shows how the computer can mimic the abstract definition of the Riemann integral, at least in cases where the numbers m_i and M_i can be obtained easily. Another conclusion that can be drawn from the experiment is that the direct translation of a definition into a computer algorithm may leave much to be desired in *precision*. With 999 evaluations of the function, the error is still about 0.0003 (absolute). We shall soon see that more sophisticated algorithms (such as Romberg's) improve this situation dramatically.

Problems 4.1

1. Show that if $\theta_i \geqq 0$ and $\sum_{i=1}^n \theta_i = 1$, then $\sum_{i=1}^n \theta_i a_i$ lies between the least and the greatest of the numbers a_i.
2. Justify the "midpoint rule" for estimating an integral:

$$\int_a^b f(x)\,dx \approx \sum_{i=2}^{n} (x_i - x_{i-1}) f\left(\frac{1}{2}(x_i + x_{i-1})\right)$$

   Show the relationship between the midpoint rule and the upper and lower sums.
3. (Continuation) Show that the midpoint rule for equal subintervals is given by

$$\int_a^b f(x)\,dx \approx h \sum_{i=1}^{n-1} f\left(x_i + \frac{1}{2}h\right)$$

   where $h = (b-a)/(n-1)$ and $x_i = a + (i-1)h$, $(1 \leqq i \leqq n)$.
4. Calculate an approximate value of

$$\int_0^\alpha \left(\frac{e^x - 1}{x}\right) dx$$

   for $\alpha = 10^{-4}$ correct to 14 decimal places.
5. For a decreasing function $f(x)$ over an interval $[a, b]$ with $n - 1$ uniform subintervals, show that the difference between the upper sum and the lower sum is

$$\frac{(b-a)}{n-1}[f(a) - f(b)]$$

6. (Continuation) Repeat Problem 5 for an increasing function $f(x)$.
7. If upper and lower sums are used with regularly spaced points to compute $\int_2^5 (dx/\log x)$, how many points are needed to achieve accuracy of 10^{-4}?
8. Let f be an increasing function. If $\int_0^1 f(x)\,dx$ is to be estimated by using the method of upper sums and lower sums, taking n equally spaced points, what is the worst possible error?

Computer Problems 4.1

1. Estimate the definite integral $\int_0^1 x^{-1} \sin x\,dx$ by computing the upper and lower sums, using 800 points in the interval. The function is decreasing, and this fact should be shown by calculus. (For a decreasing function f, $f' < 0$).
Cultural note: The function

$$\text{Si}(x) = \int_0^x t^{-1} \sin t\,dt$$

is an important special function known as the *sine integral.* It is represented by a Taylor series that converges for all real or complex values of x. The easiest way to obtain this series is to start with the series for sin t, divide by t, and integrate term by term.

$$\text{Si}(x) = \int_0^x t^{-1} \sin t\,dt = \int_0^x \sum_{n=0}^{\infty} (-1)^n \frac{t^{2n}}{(2n+1)!}\,dt =$$

$$= \sum_{n=0}^{\infty} (-1)^n \frac{x^{2n+1}}{(2n+1)!(2n+1)} = x - \frac{x^3}{18} + \frac{x^5}{600} - \frac{x^7}{35280} + \cdots$$

This series is rapidly convergent. For example, by using only the terms shown, Si (1) is computed to be 0.9460827 with an error of at most four units in the last digit shown.
2. Write a general-purpose subprogram to estimate integrals of decreasing functions by the method of upper and lower sums with a uniform partition. Give the subprogram the calling sequence

```
SUBROUTINE INTGRL(F, A, B, EPSI, N, SUMLO, SUMUP, VALUE)
```

where `F` is the function name, `A` and `B` are the endpoints of the interval, and `EPSI` is a tolerance. The subroutine determines `N` so that `SUMUP−SUMLO<2*EPSI`. `VALUE` is the average of `SUMUP` and `SUMLO`. Test it on the sine integral of the preceding problem, using `EPSI=1.0E−5`.
3. From calculus the length of a curve is $\int_a^b \sqrt{1 + [f'(x)]^2}\,dx$, where f is a function whose graph is the curve on the interval $a \leqq x \leqq b$. Find the length of the ellipse $y^2 + 4x^2 = 1$.

4.2 Trapezoid Rule

The next method we consider is an improvement over the coarse method of the preceding section and is, moreover, an important ingredient of the Romberg algorithm of the next section. This method is called the *trapezoid rule,* and is based on an estimation of the area beneath a curve, using trapezoids. Again, the estimation of $\int_a^b f(x)\,dx$ is approached by first dividing the interval $[a, b]$ into subintervals:

$$P: a = x_1 < x_2 < x_3 < \cdots < x_n = b$$

For each such partition of the interval (the subdivision points x_i need not be uniformly spaced), an estimation of the integral by the trapezoid rule is obtained. We denote it by $T(f;P)$. Figure 4.2 shows what the trapezoids are. A typical trapezoid has the sub-

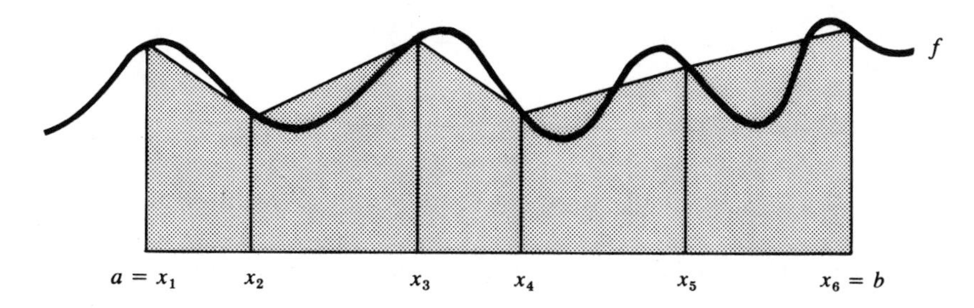

Figure 4.2

interval $[x_{i-1}, x_i]$ as its base, and the two vertical sides are $f(x_{i-1})$ and $f(x_i)$. The area is equal to the base times the average height.

$$A_i = \frac{1}{2}(x_i - x_{i-1})[f(x_{i-1}) + f(x_i)]$$

Hence, the total area approximating $\int_a^b f(x)\,dx$ is

$$T(f:P) = \frac{1}{2}\sum_{i=2}^{n}(x_i - x_{i-1})[f(x_{i-1}) + f(x_i)]$$

In practice, and in the Romberg algorithm (to be discussed in the next section), the trapezoid rule is used with a *uniform* partition of the interval. This means that

$$x_i = a + (i - 1)h \qquad h = \frac{b - a}{n - 1} \qquad (1 \leqq i \leqq n)$$

In this case, the formula for $T(f;P)$ can be given in simpler form, since $x_i - x_{i-1} = h$. Thus

$$T(f;P) = \frac{h}{2}\sum_{i=2}^{n}[f(x_i) + f(x_{i-1})] = h\sum_{i=2}^{n-1} f(x_i) + \frac{h}{2}[f(x_1) + f(x_n)] \qquad (1)$$

We now analyze the errors involved in using the trapezoid rule to estimate $\int_a^b f(x)\,dx$. A useful formula is needed for the difference

$$\int_a^b f(x)\,dx - \frac{h}{2}\sum_{i=2}^{n}[f(x_i) + f(x_{i-1})]$$

Consider first the case $n = 2$. Let F be an indefinite integral of f—specifically,

$$F(t) = \int_a^t f(x)\,dx$$

The Taylor series for F is

$$F(a + h) = F(a) + hF'(a) + \frac{1}{2}h^2F''(a) + \frac{1}{6}h^3F'''(a) + \cdots$$

By the Fundamental Theorem of Calculus, $F' = f$. Also, we observe that $F(a) = 0$ and $F(a + h) = \int_a^{a+h} f(x)\,dx$. Since $F'' = f'$, $F''' = f''$, and so on, we have

$$\int_a^{a+h} f(x)\,dx = hf(a) + \frac{1}{2}h^2 f'(a) + \frac{1}{6}h^3 f''(a) + \cdots \tag{2}$$

The Taylor series for f is

$$f(a + h) = f(a) + hf'(a) + \frac{1}{2}h^2 f''(a) + \frac{1}{6}h^3 f'''(a) + \cdots$$

To both sides of this equation, add $f(a)$ and then multiply by $\frac{1}{2}h$. The result is

$$\frac{1}{2}h\left[f(a + h) + f(a)\right] = hf(a) + \frac{1}{2}h^2 f'(a) + \frac{1}{4}h^3 f''(a) + \frac{1}{12}h^4 f'''(a) + \cdots \tag{3}$$

Next, subtract Eq. (3) from Eq. (2). The result is

$$\int_a^{a+h} f(x)\,dx - \frac{h}{2}\left[f(a + h) + f(a)\right] = -\frac{1}{12}h^3 f''(a) + \cdots$$

By making the transition from the interval $[a, a + h]$ to $[x_{i-1}, x_i]$ and by using a more complicated analysis, it can be shown that

$$\int_{x_{i-1}}^{x_i} f(x)\,dx - \frac{h}{2}\left[f(x_i) + f(x_{i-1})\right] = -\frac{1}{12}h^3 f''(\xi_i)$$

for some point ξ_i in the interval $x_{i-1} < \xi_i < x_i$. We use this result now in the general case of the trapezoid rule.

$$\int_a^b f(x)\,dx = \sum_{i=2}^{n} \int_{x_{i-1}}^{x_i} f(x)\,dx = \sum_{i=2}^{n} \left\{\frac{h}{2}\left[f(x_i) + f(x_{i-1})\right] - \frac{1}{12}h^3 f''(\xi_i)\right\}$$

The error term is analyzed as follows.

$$\frac{-h^3}{12}\sum_{i=2}^{n} f''(\xi_i) = -\frac{(n-1)h^3}{12}\left[\frac{1}{n-1}\sum_{i=2}^{n} f''(\xi_i)\right] = -\frac{b-a}{12}h^2 f''(\xi)$$

In this last step we use $(n - 1)h = b - a$. Also, the average $(n - 1)^{-1}\sum_{i=2}^{n} f''(\xi_i)$ lies between the least and greatest values of f'' on the interval $[a, b]$. By the Intermediate Value Theorem for Continuous Functions, the value $(n - 1)^{-1}\sum_{i=2}^{n} f''(\xi_i)$ is assumed by f'' at some point ξ in the interval. So the trapezoid rule and its error term obey the equation

$$\int_a^b f(x)\,dx = \frac{h}{2}\sum_{i=2}^{n}\left[f(x_i) + f(x_{i-1})\right] - \frac{b-a}{12}h^2 f''(\xi)$$

where $h = (b - a)/(n - 1)$.

How can an error formula like the one just derived be used? Our first application is in predicting how small h must be in order to attain a specified precision in the trapezoid rule. Suppose, for instance, that the trapezoid rule is to be used to compute

$$\int_0^1 \frac{\sin x}{x} dx$$

with an error of less than 10^{-4}. In this case, $f(x) = x^{-1} \sin x$, and we wish to establish a bound on $f''(x)$ for x in the range $[0, 1]$. Taking derivatives is not satisfactory, since each term contains x with a negative power, and it is difficult to find an upper bound on $|f''(x)|$. However, using Taylor series, we have

$$f(x) = 1 - \frac{x^2}{3!} + \frac{x^4}{5!} - \frac{x^6}{7!} + \frac{x^8}{9!} - \cdots$$

$$f'(x) = -\frac{2x}{3!} + \frac{4x^3}{5!} - \frac{6x^5}{7!} + \frac{8x^7}{9!} - \cdots$$

$$f''(x) = -\frac{2}{3!} + \frac{3 \times 4x^2}{5!} - \frac{5 \times 6x^4}{7!} + \frac{7 \times 8x^6}{9!} - \cdots$$

Thus, on the interval $[0, 1]$, $|f''(x)|$ cannot exceed

$$\frac{2}{3!} + \frac{3 \times 4}{5!} + \frac{5 \times 6}{7!} + \frac{7 \times 8}{9!} + \cdots < \frac{1}{2}$$

The error term $|(b - a)h^2 f''(\xi)/12|$ can therefore not exceed $(1/24)h^2$. In order that this be less than 10^{-4}, it suffices to take $h < \sqrt{24} \times 10^{-2}$ or $n - 1 > (1/\sqrt{24})10^2 = 20.4$. So for this example we need at least 21 subintervals.

In the next section we shall require a formula for the trapezoid rule when the interval $[a, b]$ is subdivided into 2^n equal parts. By formula (1), we have

$$T(f;P) = h \sum_{i=2}^{n-1} f(x_i) + \frac{h}{2}\Big[f(x_1) + f(x_n)\Big]$$

$$= h \sum_{i=1}^{n-2} f(a + ih) + \frac{h}{2}\Big[f(a) + f(b)\Big]$$

If we now replace $n - 1$ by 2^n and h by $(b - a)/2^n$, the preceding formula is

$$R(n + 1, 1) = \frac{b - a}{2^n} \sum_{i=1}^{2^n - 1} f\left(a + i\frac{(b - a)}{2^n}\right) + \frac{b - a}{2^{n+1}}\Big[f(a) + f(b)\Big] \tag{4}$$

Here we have introduced the notation that is used in the Romberg algorithm—to wit, $R(n + 1, 1)$. It denotes the result of applying the trapezoid rule with 2^n equal subintervals.

In the Romberg algorithm it will also be necessary to have a means of computing $R(n + 1, 1)$ from $R(n, 1)$ without wasting unnecessary evaluations of f. For example, the computation of $R(3, 1)$ utilizes the values of f at the five points a, $a + (b - a)/4$, $a + 2(b - a)/4$, $a + 3(b - a)/4$, b. In computing $R(4, 1)$, values of f are needed at these five points, as well as at four new points: $a + (b - a)/8$, $a + 3(b - a)/8$, $a + 5(b - a)/8$, $a + 7(b - a)/8$. The computation should take advantage of the previously computed result. The manner of doing so will now be explained. If $R(n, 1)$ is available and if $R(n+1, 1)$ is to be computed, obviously only the *difference* $R(n+1, 1) - \frac{1}{2}R(n, 1)$ need be computed, for it can be added to $\frac{1}{2}R(n, 1)$ to obtain $R(n+1, 1)$. Thus a calculation of $R(n+1, 1) - \frac{1}{2}R(n, 1)$ is required. Let C denote $(b-a)[f(a) + f(b)]/2^{n+1}$. Then by (4) we have

$$R(n+1, 1) = \frac{b-a}{2^n} \sum_{i=1}^{2^n-1} f\left(a + i\frac{b-a}{2^n}\right) + C \tag{5}$$

In the summation we consider separately the even indices $i = 2j$ and the odd indices $i = 2k - 1$. The last even index $i = 2j$ in the range $1 \leqq i \leqq 2^n - 1$ is $i = 2j = 2^n - 2$. So the last value of j is $2^{n-1} - 1$. The last odd index $i = 2k - 1$ in the range $1 \leqq i \leqq 2^n - 1$ is $i = 2k - 1 = 2^n - 1$. Thus $k = 2^{n-1}$ is the last value of k. Similarly, the first even index is $j = 1$ and the first odd index $k = 1$. Thus

$$R(n+1, 1) = \frac{b-a}{2^n}\left[\sum_{j=1}^{2^{n-1}-1} f\left(a + 2j\frac{b-a}{2^n}\right) + \right.$$
$$\left. + \sum_{k=1}^{2^{n-1}} f\left(a + (2k-1)\frac{b-a}{2^n}\right)\right] + C$$

Now consider $\frac{1}{2}R(n, 1)$, which can be obtained from (5):

$$\frac{1}{2}R(n, 1) = \frac{b-a}{2^n} \sum_{j=1}^{2^{n-1}-1} f\left(a + 2j\frac{b-a}{2^n}\right) + C$$

Comparing the two formulas, we see that

$$R(n+1, 1) = \frac{1}{2}R(n, 1) + \frac{b-a}{2^n} \sum_{k=1}^{2^{n-1}} f\left(a + (2k-1)\frac{b-a}{2^n}\right) \tag{6}$$

This formula allows us to compute a sequence of approximations to a definite integral without re-evaluating the integrand at points where it has already been evaluated.

Problems 4.2

1. Consider the function $f(x) = |x|$ on the interval $[-1, 1]$. Calculate the results of applying the following rules to approximate $\int_{-1}^{1} f(x)\,dx$. Account for the differences in the results and compare with the true solution. Use the (a) lower sums, (b) upper sums, and (c) trapezoid rule with uniform spacing $h = 2, 1, \frac{1}{2}, \frac{1}{4}$.
2. Let f be a decreasing function on $[a, b]$. Let P be a partition of the interval. Show that

$$T(f;P) = \frac{1}{2}L(f;P) + \frac{1}{2}U(f;P)$$

   where T, L, and U are the trapezoid rule, the lower sum, and the upper sum, respectively.
3. Show that for any function f and any partition P

$$L(f;P) \leqq T(f;P) \leqq U(f;P)$$

4. Give an example of a function f and a partition P for which $L(f;P)$ is a better estimate of $\int_a^b f(x)\,dx$ than is $T(f;P)$.

5. Let f be a continuous function, and let P_n ($n = 1, 2, \ldots$) be partitions of $[a, b]$ such that the width of the largest subinterval in P_n converges to zero as $n \to \infty$. Show that $T(f;P_n)$ converges to $\int_a^b f(x)\,dx$ as $n \to \infty$. *Hint:* Use Problem 3 and known facts about upper and lower sums.
6. A function is said to be *convex* if its graph lies beneath every chord drawn between two points of the graph. What is the relationship of $L(f;P)$, $U(f;P)$, $T(f;P)$, and $\int_a^b f(x)\,dx$ for such a function?
7. How large must n be if the trapezoid rule is to estimate $\int_0^2 e^{-x^2}\,dx$ with an error not exceeding 10^{-6}?
8. Show that

$$\int_a^b f(x)\,dx - \frac{b-a}{2}\Big[f(a) + f(b)\Big] = -\sum_{k=3}^{\infty} \frac{k-2}{2 \times k!}(b-a)^k f^{(k-1)}(a)$$

9. The "rectangle rule" for numerical integration is like the upper and lower sums but simpler.

$$\int_a^b f(x)\,dx \approx \sum_{i=2}^{n} (x_i - x_{i-1}) f(x_i)$$

As in Problem 5, show that the rectangle rule converges to the integral on the left as $n \to \infty$. For uniform spacing, the rule reads

$$\int_a^b f(x)\,dx \approx h \sum_{i=2}^{n} f(x_i)$$

where $x_i = a + (i-1)h$, $(1 \leqq i \leqq n)$, and $h = (b-a)/(n-1)$. Find an expression for the error involved in this latter formula.

10. How large must n be if the trapezoid rule is to estimate $\int_0^\pi \sin x\,dx$ with error $\leqq 10^{-12}$? Will the estimate be too big or too small?
11. In the trapezoid rule the spacing need not be uniform. Establish the formula

$$\int_a^b f(x)\,dx \approx \frac{1}{2}\sum_{i=2}^{n-1} (h_i + h_{i+1}) f(x_i) + \frac{1}{2}\Big[h_2 f(x_1) + h_n f(x_n)\Big]$$

where $h_i = x_i - x_{i-1}$ and $a = x_1 < x_2 < x_3 < \cdots < x_n = b$.

12. What formula results from using the trapezoid rule on $f(x) = x^2$, with interval $[0, 1]$ and n equally spaced points? Simplify your result by using the fact that $1^2 + 2^2 + 3^2 + \cdots + n^2 = \frac{1}{6}n(2n+1)(n+1)$. Show that as $n \to \infty$, the trapezoidal estimate converges to the correct value, $\frac{1}{3}$.

Computer Problems 4.2

1. Write a subroutine TRAP(F, A, B, N, T) to calculate $T = \int_a^b f(x)\,dx$, using the trapezoid rule with n equal subintervals.
2. (Continuation) Test the program written in Problem 1 on the following functions. In each case, compare to the correct answer.
   (a) $\int_0^\pi \sin x\,dx$ (b) $\int_0^1 e^x\,dx$ (c) $\int_0^1 \text{Arctan}\, x\,dx$

4.3 Romberg Algorithm

The Romberg algorithm produces a triangular array of numbers, all of which are numerical estimates of the definite integral $\int_a^b f(x)\,dx$. The array is denoted here by the notation

$$
\begin{array}{llllll}
R(1, 1) & & & & & \\
R(2, 1) & R(2, 2) & & & & \\
R(3, 1) & R(3, 2) & R(3, 3) & & & \\
R(4, 1) & R(4, 2) & R(4, 3) & R(4, 4) & & \\
\vdots & \vdots & \vdots & \vdots & \ddots & \\
R(N, 1) & R(N, 2) & R(N, 3) & R(N, 4) & \cdots & R(N, N)
\end{array}
$$

The first column of this table will contain estimates of the integral obtained by the trapezoid rule. Explicitly, $R(n + 1, 1)$ is the result of applying the trapezoid rule with 2^n equal subintervals. The first of them, $R(1, 1)$, is obtained with just one trapezoid:

$$R(1, 1) = \frac{1}{2}(b - a)\left[f(a) + f(b)\right]$$

Similarly, $R(2, 1)$ is obtained with two trapezoids:

$$
\begin{aligned}
R(2, 1) &= \frac{1}{4}(b - a)\left[f(a) + f\left(\frac{a + b}{2}\right)\right] + \frac{1}{4}(b - a)\left[f\left(\frac{a + b}{2}\right) + f(b)\right] \\
&= \frac{1}{4}(b - a)\left[f(a) + f(b)\right] + \frac{1}{2}(b - a)f\left(\frac{a + b}{2}\right) \\
&= \frac{1}{2}R(1, 1) + \frac{1}{2}(b - a)f\left(\frac{a + b}{2}\right)
\end{aligned}
$$

These formulas agree with those developed in the preceding section. In particular, note that $R(n + 1, 1)$ is obtained easily from $R(n, 1)$ if Eq. (6) in Sec. 4.2 is used.

The second and successive columns in the Romberg array are generated by the extrapolation formula

$$R(n + 1, m + 1) = R(n + 1, m) + \frac{1}{4^m - 1}\left[R(n + 1, m) - R(n, m)\right] \qquad (n \geqq 1, m \geqq 1) \tag{1}$$

It will be derived later. Here either $m = 1, 2, 3, \ldots$ and for each m we let $n = m, m + 1, m + 2, \ldots$, or $n = 1, 2, 3, \ldots$ and for each n we let $m = 1, 2, 3, \ldots, n$. In other words, the entries in the Romberg array are computed either down the columns or across the rows.

We now develop computational formulas for the Romberg algorithm. By replacing n with $i - 1$, m with $j - 1$, and $(b - a)/2^n$ with h in Eq. (1) above and Eq. (6) of Sec. 4.2, we obtain for $i \geqq 2$, $j \geqq 2$

$$R(i, j) = R(i, j - 1) + \frac{1}{4^{j-1} - 1}\left[R(i, j - 1) - R(i - 1, j - 1)\right]$$

and

$$R(i, 1) = \frac{1}{2}R(i - 1, 1) + h\sum_{k=1}^{2^{i-2}} f\left(a + (2k - 1)\,h\right)$$

Notice that the range of the summation is $1 \leqq k \leqq 2^{i-2}$ so that $1 \leqq 2k - 1 \leqq 2^{i-1} - 1$.

One way to generate the Romberg array is to compute a reasonable number of terms in the first column, $R(1, 1)$ up to $R(n, 1)$, and then use extrapolation formula (1) to construct columns 2, 3, . . . , n in order. Another way is to compute the array row by row. Observe, for example, that $R(2, 2)$ can be computed by the extrapolation formula as soon as $R(2, 1)$ and $R(1, 1)$ are available. Here is a subroutine for computing, in this way, n rows and columns of the Romberg array for a function f and a specified interval $[a, b]$.

```
      SUBROUTINE ROMBERG(F,A,B,N,R)
      DIMENSION  R(N,N)
      H = B - A
      R(1,1) = 0.5*H*(F(A) + F(B))
      PRINT 5,R(1,1)
      L = 1
      DO 4 I = 2,N
      H = 0.5*H
      L = L + L
      SUM = 0.0
      DO 2 K = 1,L-1,2
    2 SUM = SUM + F(A + H*FLOAT(K))
      R(I,1) = 0.5*R(I-1,1) + H*SUM
      M = 1
      DO 3 J = 2,I
      M = 4*M
    3 R(I,J) = R(I,J-1) + (R(I,J-1) - R(I-1,J-1))/FLOAT(M - 1)
    4 PRINT 5,(R(I,J),J = 1,I)
    5 FORMAT(5X,5E20.13)
      RETURN
      END
```

This subroutine is used with a main program and a function subprogram (for computing values of the function F). In the main program the Fortran statement **EXTERNAL F** must be present. Remember that in the Romberg algorithm as described, the number of subintervals is 2^{N-1}. Thus a modest value of N should be chosen; for example, $N = 5$. A more sophisticated program would include automatic tests to terminate the calculation as soon as the error reaches a preassigned threshold.

Now we shall explain the source of Eq. (1), which is used for constructing the successive columns of the Romberg array. We begin with a formula that expresses the error in the trapezoid rule over 2^{n-1} subintervals

$$\int_a^b f(x)\,dx = R(n, 1) + a_2h^2 + a_4h^4 + a_6h^6 + \cdots \tag{2}$$

Here $h = (b - a)/2^{n-1}$ and the coefficients a_i depend on f but not on h. This equation is one form of the Euler–Maclaurin formula and is given here without proof. In this equation $R(n, 1)$ denotes a typical element of the first column in the Romberg array; hence it is one of the trapezoidal estimates of the integral. Notice particularly that the error is expressed in powers of h^2. For our purposes, it is not necessary to know the coefficients, but, in fact, they have definite expressions in terms of f and its derivatives. In order for the theory to work smoothly, it is assumed that f possesses derivatives of all orders on the interval $[a, b]$.

Now we write down the formula for the next value of n. We replace n with $n + 1$ and h with $h/2$ and obtain from Eq. (2)

$$\int_a^b f(x)\, dx = R(n + 1, 1) + \frac{1}{4} a_2 h^2 + \frac{1}{16} a_4 h^4 + \frac{1}{64} a_6 h^6 + \cdots \tag{3}$$

Thus we have two formulas for $\int_a^b f(x)\, dx$—namely, (2) and (3). If we multiply the second one by 4, the result is

$$4 \int_a^b f(x)\, dx = 4R(n + 1, 1) + a_2 h^2 + \frac{1}{4} a_4 h^4 + \frac{1}{16} a_6 h^6 + \cdots \tag{4}$$

Now subtract Eq. (2) from Eq. (4) to obtain

$$3 \int_a^b f(x)\, dx = 4R(n + 1, 1) - R(n, 1) - \frac{3}{4} a_4 h^4 - \frac{15}{16} a_6 h^6 + \cdots$$

Thus

$$\int_a^b f(x)\, dx = \frac{4}{3} R(n + 1, 1) - \frac{1}{3} R(n, 1) - \frac{1}{4} a_4 h^4 - \frac{5}{16} a_6 h^6 - \cdots \tag{5}$$

The new combination $\frac{4}{3}R(n + 1, 1) - \frac{1}{3}R(n, 1)$ is denoted by $R(n + 1, 2)$. It should be considerably more accurate than $R(n + 1, 1)$ or $R(n, 1)$ because its error formula, Eq. (5), begins with an h^4 term:

$$\int_a^b f(x)\, dx = R(n + 1, 2) - \frac{1}{4} a_4 h^4 - \frac{5}{16} a_6 h^6 - \cdots \tag{6}$$

Note that this analysis justifies the first case ($m = 1$) of the extrapolation formula (1):

$$R(n + 1, 2) = R(n + 1, 1) + \frac{1}{3}\Big[R(n + 1, 1) - R(n, 1)\Big] \qquad (n \geqq 1)$$

Equation (6) is now taken as the starting point of a new derivation analogous to the one above. Here are the steps to do so. First, write down Eq. (6) with $n + 1$ replaced by n and with h replaced by $2h$. Then combine the two equations appropriately to eliminate the h^4 terms. The result is a new combination of elements from column 2 in the Romberg array:

$$\int_a^b f(x)\, dx = \frac{16}{15} R(n + 1, 2) - \frac{1}{15} R(n, 2) + \frac{1}{4^3} a_6 h^6 + \frac{21}{4^5} a_8 h^8 + \cdots \tag{7}$$

So we are led to define

$$R(n+1, 3) = R(n+1, 2) + \frac{1}{15}\Big[R(n+1, 2) - R(n, 2)\Big] \qquad (n \geqq 2)$$

which agrees with Eq. (1) again.

The basic assumption on which all this analysis depends is that Eq. (2) is valid for the function f being integrated. This assumption, in turn, is valid for any function possessing derivatives of all orders on the interval $[a, b]$. If the function f does not have this property (but is at least continuous), the Romberg algorithm still converges to $\int_a^b f(x)\,dx$ in the following sense. The limit of each column in the array is the integral. The convergence of the first column is easily justified (see Sec. 4.2, Problem 5). Once this has been done, the convergence of the second, third, and succeeding columns is easy (see Supplementary Problem 1).

In practice, we may not know whether the function f whose integral we seek satisfies the smoothness criterion upon which the theory depends. Then it would not be known whether Eq. (2) is valid for f. One way of testing this in the course of the Romberg algorithm is to compute the ratios

$$\frac{R(n, m) - R(n-1, m)}{R(n+1, m) - R(n, m)}$$

and to note whether they are close to 4^m. Let us verify, at least for the case $m = 1$, that this ratio is near 4 for a function that obeys Eq. (2).

If we subtract Eq. (3) from (2), the result is

$$0 = R(n, 1) - R(n+1, 1) + \frac{3}{4}a_2h^2 + \frac{15}{16}a_4h^4 + \frac{63}{64}a_6h^6 + \cdots \tag{8}$$

If we write down the same equation for the *preceding* value of n, the h of that equation is twice the value of h used in Eq. (8). Hence

$$0 = R(n-1, 1) - R(n, 1) + 3a_2h^2 + 15a_4h^4 + 63a_6h^6 + \cdots \tag{9}$$

Equations (8) and (9) are now used to express the ratio mentioned above.

$$\begin{aligned}
\frac{R(n, 1) - R(n-1, 1)}{R(n+1, 1) - R(n, 1)} &= \frac{3a_2h^2 + 15a_4h^4 + 63a_6h^6 + \cdots}{\frac{3}{4}a_2h^2 + \frac{15}{16}a_4h^4 + \frac{63}{64}a_6h^6 + \cdots} \\
&= 4 \times \frac{1 + 5\left(\frac{a_4}{a_2}\right)h^2 + 21\left(\frac{a_6}{a_2}\right)h^4 + \cdots}{1 + \frac{5}{4}\left(\frac{a_4}{a_2}\right)h^2 + \frac{21}{16}\left(\frac{a_6}{a_2}\right)h^4 + \cdots} \\
&= 4\left[1 + \frac{15}{4}\left(\frac{a_4}{a_2}\right)h^2 + \cdots\right]
\end{aligned}$$

For small values of h, this expression is close to 4.

In closing, we return to the extrapolation process that is the heart of the Romberg algorithm. The process is often termed *Richardson extrapolation, deferred approach to the limit,* or h^2 *extrapolation*. It is an example of a general dictum in numerical analysis to the effect that if anything at all is known about the errors in a process, that knowledge can be exploited to improve the process.

In order to give a general account of Richardson extrapolation, suppose that an algorithm is available to compute a certain function $\phi(h)$ for various values of h and that we want ultimately to compute $L = \lim_{h\to 0} \phi(h)$. Suppose further that some additional knowledge is available about the *errors* $e(h) = L - \phi(h)$. For example, suppose that for small values of h

$$e(h) \approx ch^p$$

In the Romberg algorithm such additional information is indeed available and is fully exploited. In this more abstract setting we imitate the analysis of the Romberg algorithm. Thus we have

$$e\left(\frac{h}{2}\right) \approx c\left(\frac{h}{2}\right)^p = 2^{-p}ch^p \approx 2^{-p}e(h)$$

And so

$$2^p e\left(\frac{h}{2}\right) - e(h) \approx 0$$

By the definition of the error, we have

$$2^p\left[L - \phi\left(\frac{h}{2}\right)\right] - \left[L - \phi(h)\right] \approx 0$$

$$(2^p - 1)L \approx 2^p\phi\left(\frac{h}{2}\right) - \phi(h)$$

$$L \approx \frac{1}{2^p - 1}\left[2^p\phi\left(\frac{h}{2}\right) - \phi(h)\right]$$

From two estimates, $\phi(h)$ and $\phi(h/2)$ of L, we can derive another estimate

$$\phi\left(\frac{h}{2}\right) + \frac{1}{2^p - 1}\left[\phi\left(\frac{h}{2}\right) - \phi(h)\right]$$

which is more accurate.

The same analysis can be carried out for any two values of h. Thus from $\phi(h_1)$ and $\phi(h_2)$ a more accurate estimate of L is given by

$$\phi(h_2) + \frac{h_2^p}{h_1^p - h_2^p}\left[\phi(h_2) - \phi(h_1)\right] \tag{10}$$

Problems 4.3

1. In calculus a technique of integration by substitution is developed. For example, if the substitution $x = z^2$ is made in the integral $\int_0^1 (e^x/\sqrt{x})\,dx$, the result is $2\int_0^1 e^{z^2}\,dz$. Verify this

and discuss the numerical aspects of this example. Which form is likely to produce a more accurate answer by the Romberg method?

2. How many evaluations of the function (integrand) are needed if the Romberg array with m rows and m columns is to be constructed?
3. Show how Richardson extrapolation works on a sequence $x_1, x_2, \ldots$ that converges to L in such a way that $L - x_n = a_1 n^{-1} + a_2 n^{-2} + a_3 n^{-3} + \cdots$. Repeat using $L - x_n = a_2 n^{-2} + a_3 n^{-3} + a_4 n^{-4} + \cdots$.
4. Assuming that the first column of the Romberg array converges to $\int_a^b f(x)\,dx$, show that the second column does also.
5. We are going to use the Romberg method to estimate $\int_0^1 \sqrt{x} \cos x\,dx$. Will the method work? Will it work well? Explain.
6. Show that the precise form of Eq. (6) is

$$\int_a^b f(x)\,dx = R(n+1, 2) - \sum_{j=1}^{\infty} \left(\frac{4^j - 1}{3 \times 4^j}\right) a_{2j+2} h^{2j+2}$$

7. Derive Eq. (7) and show that the precise form is

$$\int_a^b f(x)\,dx = R(n+2, 3) + \sum_{j=2}^{\infty} \left(\frac{4^j - 1}{3 \times 4^j}\right)\left(\frac{4^{j-1} - 1}{15 \times 4^{j-1}}\right) a_{2j+2} h^{2j+2}$$

8. Consider a function ϕ such that $\lim_{h\to 0} \phi(h) = L$ and $L - \phi(h) \approx ce^{-1/h}$ for some constant c. By combining $\phi(h)$, $\phi(h/2)$, $\phi(h/3)$, find an accurate estimate of L.
9. Using (1), fill-in the following diagram with coefficients used in the Romberg algorithm.

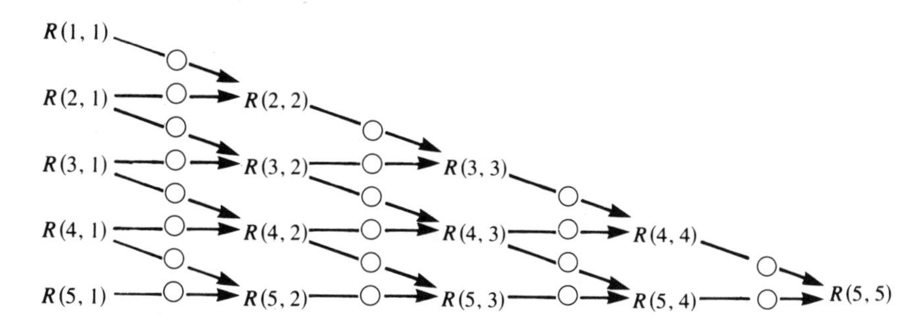

10. Show that the entries in the second column in the Romberg array are identical to those from a familiar quadrature rule over the entire interval with $2, 2^2, 2^3, \ldots, 2^n$ equal subintervals. *Hint:* See Supplementary Problem 7.
11. (Continuation) Derive the "quadrature" rule $R(3, 3)$ in terms of the function f evaluated at $a, a+h, a+2h, a+3h, b$, where $h = (b-a)/4$.
12. In the subroutine **ROMBERG** it is proposed that we replace the variable **M** with **L** and use

```
   ...
   L = 1
   DO 3 J = 2, I
   L = L + L
3  ...              .../FLOAT(L*L - 1)
   ...
```

Will this work? Notice that the value of L in the outer loop is being changed. Explain.

Computer Problems 4.3

1. Design and carry out an experiment, using the Romberg algorithm. *Suggestions:* For a function possessing many continuous derivatives on the interval, the method should work well. Try such a function first. If you choose one whose integral you can compute by other means, you will acquire a better understanding of the accuracy in the Romberg algorithm. For example, $\int (1 + x)^{-1}\, dx = \ln (1 + x)$, $\int e^x\, dx = e^x$, $\int (1 + x^2)^{-1}\, dx = \text{Arctan } x$.
2. Test the Romberg algorithm on a "bad" function, such as $\sqrt{x}$ on $[0, 1]$. Why is it bad?
3. Compute eight rows and columns in the Romberg array for the integral $\int_{1.3}^{2.19} x^{-1} \sin x\, dx$.
4. The transcendental number π is the area of a circle whose radius is 1. Show that $\pi = 8 \int_0^{1/\sqrt{2}} (\sqrt{1 - x^2} - x)\, dx$ with the help of a diagram and use this integral to approximate π by the Romberg method.
5. The *Bessel function* of order zero is defined by the equation

$$J_0(x) = \frac{1}{\pi} \int_0^{\pi} \cos (x \sin \theta)\, d\theta$$

   Calculate $J_0(1)$ by applying the Romberg algorithm to the integral.
6. Recode the Romberg subprogram so that *all* the trapezoid rule results are computed *first* and stored in the first column. Then in a separate `SUBROUTINE EXTRAP(N,R)` carry out Richardson extrapolation and store the results in the lower triangular part of the R array. What are the advantages and disadvantages of this procedure over the subroutine given in the text? Test on the two integrals $\int_0^4 dx/(1 + x)$ and $\int_{-1}^{1} e^x\, dx$, using only one run.
7. Apply the Romberg method to estimate $\int_0^{\pi} (2 + \sin 2x)^{-1}\, dx$. Observe the high precision obtained in the first column of the array—that is, by the simple trapezoidal estimates.
8. Compute $\int_0^{\pi} x \cos 3x\, dx$ by the Romberg algorithm, using $n = 8$. What is the correct answer?
9. An integral of the form $\int_a^{\infty} f(x)\, dx$ can be transformed to one on a finite interval by making a change of variable. Verify, for instance, that the substitution $x = -\ln y$ changes the integral $\int_0^{\infty} f(x)\, dx$ into $\int_0^1 y^{-1} f(-\ln y)\, dy$. Use this idea to compute $\int_0^{\infty} e^{-x}/(1 + x^2)\, dx$ by means of the Romberg algorithm, using 128 evaluations of the function.

Supplementary Problems (Chapter 4)

4, 8, 10, 16→18, 22, 23

1. A previous problem established that if $\lim_{n\to\infty} R(n, 1) = \int_a^b f(x)\, dx$, then $\lim_{n\to\infty} R(n, 2) = \int_a^b f(x)\, dx$. Show that

$$\lim_{n\to\infty} R(n, 3) = \lim_{n\to\infty} R(n, 4) = \cdots = \lim_{n\to\infty} R(n, n) = \int_a^b f(x)\, dx$$

2. Show that the first column in the Romberg array converges to the integral in such a way that the error at the nth step is bounded in magnitude by a constant times 4^{-n}.
3. Establish Eq. (10) in Sec. 4.3.
4. Let x_n be a sequence that converges to L. If $L - x_n$ is known to be of the form $a_3 n^{-3} + a_4 n^{-4} + \cdots$ (in which the coefficients are unknown), how can the convergence of the sequence be accelerated by taking combinations of x_n and x_{n+1}? Imitate the Richardson extrapolation procedure.
5. Suppose that we want to estimate $Z = \lim_{h\to 0} f(h)$ and that we calculate $f(1)$, $f(\frac{1}{2})$, $f(\frac{1}{4})$, $f(\frac{1}{8})$, . . . , $f(1/2^{10})$. Suppose also that it is known that $Z = f(h) + ah^2 + bh^4 + ch^6$. Show how to obtain an improved estimate of Z from the 11 numbers already computed. Show how Z can be determined exactly from any 4 of the 11 computed numbers.

6. A useful numerical integration scheme is "Simpson's $\frac{1}{3}$ rule" over two subintervals

$$\int_a^{a+2h} f(x)\,dx \approx \frac{h}{3}\left\{f(a) + 4f(a+h) + f(a+2h)\right\}$$

Establish the error term $-(h^5/90)f^{iv}(\xi)$, for some $\xi \in (a, a+2h)$, for this rule.

7. (Continuation) Using Problem 6, establish the general Simpson's $\frac{1}{3}$ rule over $n-1$ (even) subintervals

$$\int_a^b f(x)\,dx \approx \frac{h}{3}\left\{[f(a)+f(b)] + 4\sum_{i=1}^{(n-1)/2} f(a+(2i-1)h) + 2\sum_{i=1}^{(n-3)/2} f(a+2ih)\right\}$$

where $x_i = a + (i-1)h$, $(1 \leqq i \leqq n)$, and $h = (b-a)/(n-1)$. Derive the general error term $-((b-a)h^4/180)f^{iv}(\xi)$ for some $\xi \in (a, b)$.

8. A numerical integration scheme that is not as well known is "Simpson's $\frac{3}{8}$ rule" over three subintervals

$$\int_a^{a+3h} f(x)\,dx \approx \frac{3h}{8}\left\{f(a) + 3f(a+h) + 3f(a+2h) + f(a+3h)\right\}$$

Establish the error term for this rule and explain why this rule is overshadowed by Simpson's $\frac{1}{3}$ rule.

9. (Continuation) Using Problem 8, establish the general Simpson's $\frac{3}{8}$ rule over $n-1$ (divisible by 3) subintervals. Derive the general error term.

10. Compute $\int_0^1 (1+x^2)^{-1}\,dx$ by Simpson's $\frac{1}{3}$ rule, using the three points $x = 0, 0.5, 1$. Compare to the true solution.

11. Consider the integral $\int_0^1 \sin(\pi x^2/2)\,dx$. Suppose that we wish to integrate numerically, with an error of magnitude less than 10^{-3}.
    (a) What interval width h is needed if we wish to use the trapezoidal rule?
    (b) Simpson's $\frac{1}{3}$ rule?

12. A function f has the values shown.

x	1	1.25	1.5	1.75	2
f	10	8	7	6	5

    (a) Use Simpson's rule and the function values at $x = 1, 1.5, 2$ to approximate $\int_1^2 f(x)\,dx$.
    (b) Repeat the preceding part, using $x = 1, 1.25, 1.5, 1.75, 2$.
    (c) Use the results from parts (a) and (b) along with the error term to establish an improved approximation.

13. Find an approximate value of $\int_1^2 x^{-1}\,dx$, using Simpson's $\frac{1}{3}$ rule with $h = 0.25$. Give a bound on the error.

14. Let $f(x) = 2^x$. Approximate $\int_0^4 f(x)\,dx$ by the trapezoid rule using points 0, 2, 4. Repeat by using points 0, 1, 2, 3, 4. Now apply Romberg extrapolation to obtain a better approximation.

15. Obtain an expression for the error term of the midpoint rule for one, $(n-1)$-unequal, and $(n-1)$-uniform subintervals—that is, for

$$\int_a^{a+h} f(x)\,dx \approx hf\left(a + \frac{1}{2}h\right)$$

$$\int_a^b f(x)\,dx \approx \sum_{i=2}^{n} h_i f\left(x_{i-1} + \frac{1}{2}h_i\right)$$

$$\int_a^b f(x)\,dx \approx h\sum_{i=2}^{n} f\left[a + \left(i - \frac{3}{2}\right)h\right]$$

where $h_i = x_i - x_{i-1}$ and $h = (b-a)/(n-1)$.

16. Consider the integral $I(h) \equiv \int_a^{a+h} f(x)\,dx$. Establish an expression for the error term for each of the following rules.
(a) $I(h) \approx hf(a+h)$
(b) $I(h) \approx hf(a+h) - \frac{1}{2}h^2 f'(a)$
(c) $I(h) \approx hf(a)$
(d) $I(h) \approx hf(a) - \frac{1}{2}h^2 f'(a)$
For each, determine the corresponding general rule and error terms for the integral $\int_a^b f(x)\,dx$, where the partition is uniform—that is, $x_i = a + (i-1)h$ and $h = (b-a)/(n-1)$.
17. Construct a rule of the form

$$\int_{-1}^{1} f(x)\,dx \approx \alpha f\left(-\frac{1}{2}\right) + \beta f(0) + \gamma f\left(\frac{1}{2}\right)$$

that is exact for all polynomials of degree less than or equal to 2—that is, determine values for α, β, γ. *Hint:* Make the relation exact for 1, x, x^2 and find a solution of the resulting equations. If it is exact for these polynomials, it is exact for all polynomials of degree $\leqq 2$.
18. Establish a numerical integration formula of the form

$$\int_a^b f(x)\,dx \approx Af(a) + Bf'(b)$$

that is accurate for polynomials of as high a degree as possible.
19. Derive a formula for $\int_a^{a+h} f(x)\,dx$ in terms of $f(a)$, $f(a+h)$, and $f(a+2h)$ that is correct for polynomials of as high a degree as possible. *Hint:* Use polynomials 1, $x-a$, $(x-a)^2$, and so on.
20. Derive a formula of the form

$$\int_a^b f(x)\,dx \approx w_1 f(a) + w_2 f(b) + w_3 f'(a) + w_4 f'(b)$$

that is exact for polynomials of highest degree possible.
21. Derive the Gauss quadrature rule of the form

$$\int_{-1}^{1} f(x)\,dx \approx af(-\alpha) + bf(0) + cf(\alpha)$$

that is exact for all polynomials of as high a degree as possible—that is, determine α, a, b, and c.
22. Determine a formula of the form

$$\int_0^h f(x)\,dx \approx w_1 f(0) + w_2 f(h) + w_3 f''(0) + w_4 f''(h)$$

that is exact for polynomials of as high a degree as possible.
23. Derive a nontrivial numerical integration formula of the form

$$\int_{x_{n-1}}^{x_{n+1}} f(x)\,dx \approx Af(x_n) + Bf'(x_{n-1}) + Cf''(x_{n+1})$$

for uniformly spaced points x_{n-1}, x_n, x_{n+1} with spacing h. *Hint:* Consider

$$\int_{-h}^{h} f(x)\,dx \approx Af(0) + Bf'(-h) + Cf''(h)$$

24. Simpson's rule for calculating $\int_a^b f(x)\,dx$ is

$$S_{n-1} = \frac{h}{3}\Big[f(x_1) + 4f(x_2) + 2f(x_3) + \cdots + 4f(x_{n-1}) + f(x_n)\Big]$$

with $h = (b - a)/(n - 1)$ when $n - 1$ is even. Its error is of the form Ch^4. Show how two values of S_{n-1} can be combined to obtain a more accurate estimate of the integral.

25. In the Romberg algorithm $R(n, 1)$ denotes an estimate of $\int_a^b f(x)\,dx$ using step size $h = (b - a)/2^{n-1}$. If it were known that

$$\int_a^b f(x)\,dx = R(n, 1) + a_3h^3 + a_6h^6 + \cdots$$

how should we modify the Romberg algorithm?

26. (Adams–Bashforth–Moulton Formulas) Verify that the numerical integration formulas

(a) $$\int_t^{t+h} g(s)\,ds \approx \frac{h}{24}\Big\{55g(t) - 59g(t - h) + 37g(t - 2h) - 9g(t - 3h)\Big\}$$

(b) $$\int_t^{t+h} g(s)\,ds \approx \frac{h}{24}\Big\{9g(t + \mathrm{h}) + 19g(t) - 5g(t - h) + g(t - 2h)\Big\}$$

are exact for polynomials of third degree.
Cultural note: These two formulas can also be derived by replacing the integrand g by an interpolating polynomial for the values $\{t, t - h, t - 2h, t - 3h\}$ and $\{t + h, t, t - h, t - 2h\}$, respectively. Interpolating polynomials are discussed in Chapter 5.

27. We want to compute $X = \lim_{n\to\infty} S_n$ and have already computed the two numbers $u = S_{10}$ and $v = S_{30}$. It is known that $X = S_n + Cn^{-3}$. What is X in terms of u and v?

Supplementary Computer Problems (Chapter 4)

1. The Gaussian quadrature rule

$$\int_{-1}^{1} g(t)\,dt \approx \frac{8}{9}g(0) + \frac{5}{9}\left[g\left(-\sqrt{\frac{3}{5}}\right) + g\left(\sqrt{\frac{3}{5}}\right)\right]$$

can be used over the interval $[a, b]$ by using the transformation $t = (2x - (b + a))/(b - a)$ —that is,

$$\int_a^b f(x)\,dx = \frac{1}{2}(b - a)\int_{-1}^{1} f\left(\frac{1}{2}(b - a)t + \frac{1}{2}(b + a)\right)dt$$

Write a function subprogram with parameters f, a, b that evaluates the latter integral. Test by using $\int_0^2 e^{-\cos^2 x}\,dx$.

2. Calculate

$$\int_0^1 \frac{\sin x}{\sqrt{x}}\,dx$$

by the Romberg algorithm. *Hint:* Consider making a change of variable.

3. Calculate $\int_0^\infty e^{-x}\sqrt{1 - \sin x}\,dx$ by the Romberg algorithm.

4. Compute log 2 by using the Romberg algorithm on a suitable integral.

5. Write a subroutine to implement Simpson's rule. See Supplementary Problem 7.
6. (Continuation) Test the subroutine written in Problem 5 on some functions whose integrals are known.
7. Calculate π from an integral of the form $c \int_a^b dx/(1 + x^2)$.
8. Calculate ln 2 from an integral of the form $\int_a^b dx/x$.
9. Calculate $\int_0^\infty e^{-x^2}\, dx$ numerically.
10. Estimate $\int_0^\infty (\sin x)/x \, dx$.
11. Calculate the Fresnel integral $\int_0^\infty \sin x^2 \, dx$.

5

Interpolation and Numerical Differentiation

As a result of an expensive and laborious calculation, the following values of a function have been obtained.

x	$f(x)$
0.924	−0.00851372555933
0.928	−0.00387718871002
0.932	0.00080221229861
0.936	0.00552498098693
0.940	0.01029163044063

It is also known that f possesses derivatives of all orders and that $f'(x) > 0$. From this information, what is the probable value of ξ for which $f(\xi) = 0$?

The hypotheses on f allow us to conclude that the inverse function f^{-1} exists and can be well approximated by a polynomial p. So

$$\xi = f^{-1}(0) \approx p(0)$$

The methods of this chapter enable us to find a polynomial of degree 4 that represents f^{-1} precisely at the points in the table. Once p is known, it is a simple matter to compute $\xi \approx p(0) = 0.93131687362136$.

We pose *three* problems concerning the representation of functions to give an indication of the subject matter of Chapters 5, 7, and 12.

First, suppose that we have a table of numerical values of a function.

x	x_1	x_2	$\cdots$	x_n
y	y_1	y_2	$\cdots$	y_n

Is it possible to find a simple and convenient formula that reproduces the given points exactly?

The second problem is similar, but it is assumed that the given table of numerical values is contaminated by errors, as might occur if the table results from a physical experiment. Now we ask for a formula that represents the data (approximately) and, if possible, filters out the errors.

As a third problem, a function f is given, perhaps in the form of a computer subroutine, but it is an expensive function to evaluate. In this case, we ask for another function g that is simpler to evaluate and produces a reasonable approximation to f.

In all of these problems a simple function p can be obtained that represents or approximates the given table or function f. The representation p can always be taken to be a polynomial, although many other types of simple functions can also be used. Once a simple function p has been obtained, it can be used in place of f in many situations. For example, the integral of f could be estimated by the integral of p, and the latter would generally be simpler to evaluate.

In many situations, a polynomial solution to the problems outlined above will be unsatisfactory from a practical point of view, and other classes of functions must be considered. In this book, one other class of versatile functions is discussed: the *spline* functions (see Chapter 7). The present chapter concerns polynomials exclusively, and Chapter 12 discusses general linear families of functions, with splines and polynomials being important examples.

The obvious way in which a polynomial can *fail* as a practical solution to one of the preceding problems is that its degree may be unreasonably high. For instance, if the table previously considered contains 1000 entries, a polynomial of degree 999 may be required to represent it.

Polynomials also may have the surprising defect of being highly *oscillatory*. If the table is precisely represented by a polynomial p, then $p(x_i) = y_i$ for $1 \leqq i \leqq n$. For points other than the given x_i, however, $p(x)$ may be a very poor representation of the function from which the table arose. An example will be given in Sec. 5.2.

5.1 Polynomial Interpolation

We begin again with a table of values

x	x_1	x_2	$\cdots$	x_n
y	y_1	y_2	$\cdots$	y_n

and assume that the x_i form a set of n distinct points. The table represents n points in the Cartesian plane, and we want to find a polynomial curve that passes through the points. Thus we seek to determine the unique polynomial, defined for *all* x, that takes on the corresponding values of y_i for each of the n distinct x_i in this table. Generally a different polynomial is required for different sized tables—that is, the desired polynomial depends on n.

Consider first the simplest case, $n = 1$. Here a constant function solves the problem. That is, the polynomial p of degree zero defined by the equation $p(x) = y_1$ reproduces the two-element table.

The next simplest case is when $n = 2$. Since a straight line can be passed through two points, a linear function is capable of solving the problem. Explicitly, the polynomial p defined by

$$p(x) = y_1 + \left(\frac{y_2 - y_1}{x_2 - x_1}\right)(x - x_1)$$

is of first degree (at most) and reproduces the table. That means (in this case) that $p(x_1) = y_1$ and $p(x_2) = y_2$, as is easily verified.

Now suppose that we have succeeded in finding a polynomial p that reproduces *part* of the table. Assume, say, that $p(x_i) = y_i$ for $1 \leqq i \leqq k$. We will attempt to add to p another term that will enable the new polynomial to reproduce one more entry in the table. We consider

$$p(x) + c(x - x_1)(x - x_2) \cdots (x - x_k)$$

This is surely a polynomial. It also reproduces the first k points in the table because p itself does so, and the added portion takes the value zero at each of the points $x_1, \ldots, x_k$. (Its form is chosen for precisely this reason.) Now we adjust the parameter c so that the new polynomial takes the value y_{k+1} at x_{k+1}. Imposing this condition, we obtain

$$p(x_{k+1}) + c(x_{k+1} - x_1)(x_{k+1} - x_2) \cdots (x_{k+1} - x_k) = y_{k+1}$$

The proper value of c *can* be obtained from this equation because none of the factors $x_{k+1} - x_i$ $(1 \leqq i \leqq k)$ can be zero. Remember our original assumption that the x_i are all distinct.

This analysis is an example of inductive reasoning. We have shown that the process can be started and that it can be continued. Hence the following formal statement has been partially justified.

> If the points $x_1, \ldots, x_n$ are distinct, then for arbitrary real values $y_1, \ldots, y_n$ there is a unique polynomial p of degree $\leqq n - 1$ such that $p(x_i) = y_i$ $(1 \leqq i \leqq n)$.

Two parts of this formal statement must still be established. First, the degree of the polynomial increases by at most one in each step of the inductive argument, and at the beginning the degree was zero. So at the end the degree is at most $n - 1$.

Next, we establish the uniqueness of the polynomial p. Suppose that another polynomial q claims to accomplish what p does—that is, q is also of degree at most $n - 1$ and satisfies $q(x_i) = y_i$ for $1 \leqq i \leqq n$. Then the polynomial $p - q$ is of degree at most $n - 1$ and takes the value zero at $x_1, \ldots, x_n$. Recall, however, that a *nonzero* polynomial of degree $n - 1$ can have at most $n - 1$ roots. We conclude that $p = q$, which establishes the uniqueness of p.

A polynomial p for which $p(x_i) = y_i$ is said to *interpolate* the given table at the *nodes* x_i $(1 \leqq i \leqq n)$. Our previous discussion provides a construction of the interpolating polynomial. This construction is known as the *Newton algorithm*.

Taking a concrete example, consider this table.

x	1.3	2.5	1.8	3.9
y	24.8	13.2	11.1	17.0

In the construction four successive polynomials will appear, which we label p_0, p_1, p_2, and p_3 to indicate the degree of each polynomial. The polynomial p_0 is defined to be $p_0(x) = 24.8$. The polynomial p_1 has the form

$$p_1(x) = p_0(x) + c(x - x_1) = 24.8 + c(x - 1.3)$$

The interpolating condition placed on p_1 is $13.2 = 24.8 + c(2.5 - 1.3)$, from which we obtain $c = -29/3$. By carrying out all four steps, we obtain

$$p_2(x) = 24.8 - \frac{29}{3}(x - 1.3) + \frac{76}{3}(x - 1.3)(x - 2.5),$$

$$p_3(x) = 24.8 - \frac{29}{3}(x - 1.3) + \frac{76}{3}(x - 1.3)(x - 2.5)$$

$$- \frac{480}{49}(x - 1.3)(x - 2.5)(x - 1.8)$$

Later we shall develop a better algorithm for carrying out the construction of the Newton interpolating polynomial. Nevertheless, the method just explained is a systematic one and involves very little computation. An important feature to notice is that each new polynomial in the algorithm is obtained from its predecessor by adding a new term. Thus at the end the final polynomial exhibits all the previous polynomials as constituents.

Before continuing, let us rewrite the Newton interpolating polynomial for efficient evaluation. Using the above polynomial p_3 as an example, we arrange it as

$$p_3(x) = 24.8 + (x - 1.3)\left\{\left(-\frac{29}{3}\right) + (x - 2.5)\right.$$

$$\left.\left[\left(\frac{76}{3}\right) + (x - 1.8)\left(-\frac{480}{49}\right)\right]\right\}$$

This form is obtained by systematic factoring of the original polynomial. It is known as the *nested* form and its evaluation is by *nested multiplication*.

To describe nested multiplication in a formal way (so that it can be translated into a program), consider a general polynomial in the Newton form. It might be

$$p(x) = a_1 + a_2[(x - x_1)] + a_3[(x - x_1)(x - x_2)] + \cdots$$

$$+ a_n[(x - x_1)(x - x_2) \cdots (x - x_{n-1})]$$

This can be written succinctly as

$$p(x) = a_1 + \sum_{i=2}^{n} a_i \left[\prod_{j=1}^{i-1} (x - x_j)\right]$$

where the standard product notation has been used. The nested form of $p(x)$ is

$$p(x) = a_1 + (x - x_1)[a_2 + (x - x_2)$$

$$[a_3 + (x - x_3)[a_4 + \cdots + (x - x_{n-1})a_n]] \cdots]$$

$$= [\cdots [[a_n(x - x_{n-1}) + a_{n-1}](x - x_{n-2}) + a_{n-2}] \cdots](x - x_1) + a_1$$

In evaluating $p(t)$ for a given value of t, we naturally start with the innermost brackets, forming successively the following quantities.

$$
\begin{aligned}
v_1 &= a_n \\
v_2 &= v_1(t - x_{n-1}) + a_{n-1} \\
v_3 &= v_2(t - x_{n-2}) + a_{n-2} \\
&\cdots \\
v_n &= v_{n-1}(t - x_1) + a_1
\end{aligned}
$$

The quantity v_n is now $p(t)$. In a Fortran code a subscripted variable is not needed for v_i. Instead we can write

```
      V = A(N)
      DO 2 I = 1,N-1
    2 V = V*(T - X(N-I)) + A(N-I)
```

Here the A array contains the N coefficients of the Newton interpolating polynomial of degree at most N−1 and the X array contains the N nodes x_i.

We turn now to the problem of determining the coefficients $a_1, \ldots, a_n$ efficiently. Again we start with a table of values of a function f.

x	x_1	x_2	x_3	$\cdots$	x_n
$f(x)$	$f(x_1)$	$f(x_2)$	$f(x_3)$	$\cdots$	$f(x_n)$

The points $x_1, x_2, \ldots, x_n$ are assumed to be distinct, but no assumption about their positions on the real line is made.

Previously we established that, for each $n = 1, 2, 3, \ldots$, there exists a unique polynomial p_{n-1} such that

1. the degree of p_{n-1} is less than or equal to $n - 1$.
2. $p_{n-1}(x_i) = f(x_i)$ for $1 \leqq i \leqq n$.

It was shown that p_{n-1} can be expressed in the Newton form

$$
\begin{aligned}
p_{n-1}(x) = a_1 + a_2(x - x_1) + a_3(x - x_1)(x - x_2) + \cdots \\
+ a_n(x - x_1) \cdots (x - x_{n-1})
\end{aligned}
\tag{1}
$$

The compact form of this equation is

$$
p_{n-1}(x) = \sum_{i=1}^{n} a_i \prod_{j=1}^{i-1} (x - x_j) \tag{2}
$$

in which $\prod_{j=1}^{0} (x - x_j)$ is interpreted to be 1. A crucial observation about p_{n-1} is that the coefficients $a_1, a_2, \ldots$ do not depend on n. In other words, p_{n-1} is obtained from p_{n-2} by adding one more term, without altering the coefficients already present in p_{n-2} itself. This is because we began with the hope that p_{n-1} could be expressed in the form

$$
p_{n-1}(x) = p_{n-2}(x) + a_n(x - x_1) \cdots (x - x_{n-1})
$$

and discovered that it was indeed possible.

One way of systematically determining the unknown coefficients $a_1, a_2, \ldots$ is to set x equal in turn to $x_1, x_2, \ldots$ in (1) and to write down the resulting equations.

$$\begin{aligned} f(x_1) &= a_1 \\ f(x_2) &= a_1 + a_2(x_2 - x_1) \\ f(x_3) &= a_1 + a_2(x_3 - x_1) + a_3(x_3 - x_1)(x_3 - x_2) \\ &\text{etc.} \end{aligned} \tag{3}$$

The compact form of (3) is

$$f(x_k) = \sum_{i=1}^{k} a_i \prod_{j=1}^{i-1} (x_k - x_j) \qquad (k = 1, 2, 3, \ldots) \tag{4}$$

Equation (3) can be solved for the a_i's in turn, starting with a_1. Then we see that a_1 depends on $f(x_1)$, that a_2 depends on $f(x_1)$ and $f(x_2)$, and so on. In general, a_n depends on $f(x_1), \ldots, f(x_n)$. In other words, a_n depends on f at $x_1, \ldots, x_n$. The traditional notation is

$$a_n = f[x_1, x_2, \ldots, x_n] \tag{5}$$

This equation defines $f[x_1, \ldots, x_n]$. The a_k's are, in turn, defined by (3) or (4). The new form of Eq. (2) is

$$p_{n-1}(x) = \sum_{i=1}^{n} f[x_1, \ldots, x_i] \prod_{j=1}^{i-1} (x - x_j) \tag{6}$$

Using Eq. (3), we can write some explicit formulas.

$$f[x_1] = f(x_1)$$

$$f[x_1, x_2] = \frac{f(x_2) - f(x_1)}{x_2 - x_1}$$

$$f[x_1, x_2, x_3] = \frac{f(x_3) - f(x_1) - \dfrac{f(x_2) - f(x_1)}{x_2 - x_1}(x_3 - x_1)}{(x_3 - x_1)(x_3 - x_2)}$$

etc.

In principle, one could continue this list of formulas, but we prefer the following remarkable *recursive* formula.

$$f[x_1, \ldots, x_k] = \frac{f[x_2, \ldots, x_k] - f[x_1, \ldots, x_{k-1}]}{x_k - x_1} \tag{7}$$

We justify this formula as follows.

An alternative interpretation of $f[x_1, \ldots, x_k]$ is that it is the coefficient of x^{k-1} in the polynomial p_{k-1} of degree $\leqq k - 1$ that interpolates f at $x_1, \ldots, x_k$. A glance at the right-hand terms in (7) shows that we should consider two other polynomials: p_{k-2}, which interpolates f at $x_1, \ldots, x_{k-1}$, and q, the polynomial of degree $\leqq k - 2$ that interpolates f at $x_2, \ldots, x_k$. The relationship between these three polynomials is

$$p_{k-1}(x) = q(x) + \frac{x - x_k}{x_k - x_1}[q(x) - p_{k-2}(x)] \tag{8}$$

In order to establish (8), observe that the right side is a polynomial of degree $\leqq k - 1$. Evaluating it at x_1 gives $f(x_1)$:

$$q(x_1) + \frac{x_1 - x_k}{x_k - x_1}[q(x_1) - p_{k-2}(x_1)]$$
$$= q(x_1) - [q(x_1) - p_{k-2}(x_1)] = p_{k-2}(x_1) = f(x_1)$$

Evaluating it at x_i $(2 \leqq i \leqq k - 1)$ results in $f(x_i)$:

$$q(x_i) + \frac{x_i - x_k}{x_k - x_1}[q(x_i) - p_{k-2}(x_i)]$$
$$= f(x_i) + \frac{x_i - x_k}{x_k - x_1}[f(x_i) - f(x_i)] = f(x_i)$$

Similarly, at x_k we get $f(x_k)$:

$$q(x_k) + \frac{x_k - x_k}{x_k - x_1}[q(x_k) - p_{k-2}(x_k)] = q(x_k) = f(x_k)$$

By the uniqueness of interpolating polynomials, the right side of (8) must be $p_{k-1}(x)$, and Eq. (8) is established.

Completing the argument to justify (7), take the coefficient of x^{k-1} on both sides of Eq. (8). The result is Eq. (7). Indeed, $f[x_2, \ldots, x_k]$ is the coefficient of x^{k-2} in q, and $f[x_1, \ldots, x_{k-1}]$ is the coefficient of x^{k-2} in p_{k-2}.

Notice that $f[x_1, \ldots, x_k]$ is not changed if the nodes $x_1, \ldots, x_k$ are permuted; thus, for example, $f[x_1, x_2, x_3] = f[x_2, x_3, x_1]$. The reason is that $f[x_1, x_2, x_3]$ is the coefficient of x^2 in the quadratic polynomial interpolating f at x_1, x_2, x_3, whereas $f[x_2, x_3, x_1]$ is the coefficient of x^2 in the quadratic polynomial interpolating f at x_2, x_3, x_1. These two polynomials are, of course, the same.

Since the variables $(x_1, x_2, \ldots, x_k)$ and k are arbitrary, the recursive formula (7) can also be written

$$f[x_i, x_{i+1}, \ldots, x_{j-1}, x_j] = \frac{f[x_{i+1}, \ldots, x_j] - f[x_i, \ldots, x_{j-1}]}{x_j - x_i} \tag{9}$$

The first three divided differences are thus

$$f[x_i] = f(x_i)$$

$$f[x_i, x_{i+1}] = \frac{f[x_{i+1}] - f[x_i]}{x_{i+1} - x_i}$$

$$f[x_i, x_{i+1}, x_{i+2}] = \frac{f[x_{i+1}, x_{i+2}] - f[x_i, x_{i+1}]}{x_{i+2} - x_i}$$

Using formula (9), it is possible to construct a divided-difference table for a function f. It is customary to arrange it as follows (here $n = 4$).

$$
\begin{array}{c|c|c|c|c}
x_1 & f[x_1] & & & \\
 & & f[x_1, x_2] & & \\
x_2 & f[x_2] & & f[x_1, x_2, x_3] & \\
 & & f[x_2, x_3] & & f[x_1, x_2, x_3, x_4] \\
x_3 & f[x_3] & & f[x_2, x_3, x_4] & \\
 & & f[x_3, x_4] & & \\
x_4 & f[x_4] & & &
\end{array}
$$

Here is a concrete numerical example. The numbers on the left of the dark line are given and the divided differences on the right have been computed.

$$
\begin{array}{c|c|c|c|c}
1 & 3 & & & \\
 & & \frac{1}{2} & & \\
\frac{3}{2} & \frac{13}{4} & & \frac{1}{3} & \\
 & & \frac{1}{6} & & -2 \\
0 & 3 & & -\frac{5}{3} & \\
 & & -\frac{2}{3} & & \\
2 & \frac{5}{3} & & &
\end{array}
$$

The coefficients in the top diagonal are the ones needed to form the interpolating polynomial. Thus the data of the first two columns are reproduced by the polynomial

$$p_3(x) = 3 + \frac{1}{2}(x - 1) + \frac{1}{3}(x - 1)(x - 1.5) - 2(x - 1)(x - 1.5)(x)$$

The interpolating polynomial is obtained in different forms by reading the divided-difference table in other ways. For example, the coefficients of the *backward-difference polynomial* are found along the bottom diagonal.

$$p_3(x) = \frac{5}{3} - \frac{2}{3}(x - 2) - \frac{5}{3}(x - 2)(x) - 2(x - 2)(x)\left(x - \frac{3}{2}\right)$$

Turning next to algorithms, suppose that a table for f is given at points $x_1, \ldots, x_n$ and that all the divided differences $a_{ij} \equiv f[x_i, x_{i+1}, \ldots, x_j]$ are to be computed. The following segment of code accomplishes this.

```
      DO 2 I = 1,N
    2 A(I,I) = F(X(I))
      DO 3 J = 1,N-1
      DO 3 I = 1,N-J
    3 A(I,I+J) = (A(I+1,I+J) - A(I,I+J-1))/(X(I+J) - X(I))
```

Observe that the coefficients of the interpolating polynomial (2) are stored in the first row of the A array.

Generally, not all of the divided differences need be saved, only the coefficients in the Newton form of the interpolation polynomial. Using a one-dimensional array, the divided differences can be overwritten each time from the last storage location backward so that, finally, only the desired coefficients remain. In this case, the amount of computing is the same as in the preceding case, but storage requirements are less. (Why?) Here is a code to do this:

```
      DO 2 I = 1,N
    2 A(I) = F(X(I))
      DO 3 J = 1,N-1
      DO 3 I = 1,N-J
    3 A(N-I+1) = (A(N-I+1) - A(N-I))/(X(N-I+1) - X(N-I-J+1))
```

This algorithm is more intricate, and the reader is invited to verify it, say in the case $n = 4$.

Finally, we consider a last simplification that is possible if the original data are not needed. Suppose that the data points

x array	x_1	$\cdots$	x_n
y array	y_1	$\cdots$	y_n

are given and the Newton interpolating polynomial is sought. If the y array is not needed, our algorithm can store the divided differences a_i in the y array. Thus at the end of the process the x array is unchanged but the y array contains the coefficients denoted by a_i in the previous discussion. Here is the algorithm.

```
      DO 2 J = 1,N-1
      DO 2 I = 1,N-J
    2 Y(N-I+1) = (Y(N-I+1) - Y(N-I))/(X(N-I+1) - X(N-I-J+1))
```

The y array *now* contains the coefficients of the desired polynomial, which is

$$p_{n-1}(x) = \sum_{i=1}^{n} y_i \prod_{j=1}^{i-1} (x - x_j)$$

The interpolating polynomial can be evaluated at one particular point t, using nested multiplication. As outlined previously, this procedure involves the coefficients of the polynomial which are stored in y_i, the nodal points x_i, the degree n, and the point t.

Here are two subprograms to accomplish the process. The first is called COEF and is a subroutine. It accepts as input the original tabular values in the X and Y arrays and then generates the coefficients required in the Newton interpolating polynomial, storing them in the Y array.

```
      SUBROUTINE COEF(N,X,Y)
      DIMENSION  X(N),Y(N)
      DO 2 J = 1,N-1
      DO 2 I = 1,N-J
    2 Y(N-I+1) = (Y(N-I+1) - Y(N-I))/(X(N-I+1) - X(N-I-J+1))
      RETURN
      END
```

The second routine is a function subprogram EVAL, which accepts as input an X and Y array and a real number T. It interprets the elements of X as the nodes for interpolation and those of Y as the *coefficients of the Newton interpolating polynomial*. This polynomial is evaluated at T, and this is the value returned by the function subprogram.

```
      FUNCTION EVAL(N,X,Y,T)
      DIMENSION  X(N),Y(N)
      EVAL = Y(N)
      DO 3 I = 1,N-1
    3 EVAL = EVAL*(T - X(N-I)) + Y(N-I)
      RETURN
      END
```

Since the coefficients of the interpolating polynomial need be computed but once, we call COEF first, and then all subsequent calls for evaluating this polynomial are accomplished with EVAL. Notice that only the T parameter should be changed between successive calls to this subprogram.

As an example, suppose that a polynomial of degree 9 is wanted that approximates $\sin x$ on the interval $[0, \pi/2]$. Now $\pi/2 \approx 1.57$, and a set of ten convenient points covering this interval is $x_i = (i-1)h$ for $h = 0.1875$ and $1 \leqq i \leqq 10$. (Since $0.1875 = 2^{-3} + 2^{-5}$, it is a machine number.) The following code tests the preceding subprograms by determining the value of the Newton interpolation polynomial at T $\approx \pi/2$ and S $\approx \pi/6$ for comparison to the correct values $\sin \pi/2 = 1.0$ and $\sin \pi/6 = 0.5$.

```
      DIMENSION  X(10),Y(10)
      DATA  T,S/1.57079632679490,0.523598775598299/
      DO 2 I = 1,10
      X(I) = FLOAT(I-1)*0.1875
    2 Y(I) = SIN(X(I))
      CALL COEF(10,X,Y)
      P1 = EVAL(10,X,Y,T)
      P2 = EVAL(10,X,Y,S)
      PRINT 3,P1,P2
    3 FORMAT(5X,2F20.15)
      END
```

The results are

$$\sin\frac{\pi}{2} \approx 0.99999\ 99997\ 21723 \qquad \sin\frac{\pi}{6} \approx 0.49999\ 99999\ 89811$$

Notice that more accurate approximations are obtained near the middle of the interval of interpolation. The reader should add additional code to this test program to evaluate the polynomial at the 20 points with $h/2$. This test will show that the polynomial agrees with the function `SIN(X)`, to machine precision, at the nodal points and that at the intermediate points the accuracy varies with the location of the `T` value in the interval.

Problems 5.1

1, 4, 5, 7 Ans only for 7

1. Complete the following divided-difference table and use it to obtain a polynomial of degree 3 that interpolates the four indicated function values.

x	$f(x)$	$f[\,,\,]$	$f[\,,\,,\,]$	$f[\,,\,,\,,\,]$
-1	2			
		–		
1	-4		–	
		–		–
3	46		–	
		53.5		
4	99.5			

2. Without using a divided-difference table, derive and simplify the polynomial of least degree that assumes these values.

x	-2	-1	0	1	2
y	2	14	4	2	2

3. (Continuation) Find a polynomial that takes the values shown in Prob. 2 and has at $x = 3$ the value 10. *Hint:* Add a suitable polynomial to the $p(x)$ of the previous problem.
4. Find a polynomial of least degree that takes these values.

x	1.73	1.82	2.61	5.22	8.26
y	0	0	7.8	0	0

Hint: Rearrange the table so that the nonzero value of y is the *last* entry or think of some better way.

5. Form a divided-difference table for the following and explain what happened.

x	1	2	3	1
y	3	5	5	7

6. Verify directly that for any three distinct points x_1, x_2, x_3

$$f[x_1, x_2, x_3] = f[x_3, x_1, x_2] = f[x_2, x_3, x_1]$$

Compare this argument to the one in the text.

7. From a table of logarithms we obtain the following values of $\log x$ at the indicated tabular points.

x	1	1.5	2	3	3.5	4
$\log x$	0	0.17609	0.30103	0.47712	0.54407	0.60206

Form a divided-difference table based on these values. Interpolate for log 2.4 and log 1.2, using third-degree interpolation polynomials in Newton form.

8. There exists a unique polynomial $p(x)$ of degree 2 or less such that $p(0) = 0$, $p(1) = 1$, $p'(\alpha) = 2$ for any value of α between zero and one (inclusive) except one value of α, say α_0. Determine α_0 and give this polynomial for $\alpha \neq \alpha_0$.
9. Determine by two methods the polynomial of degree 2 or less whose graph passes through the points (0, 1.1), (1, 2), (2, 4.2). Verify that they are the same.
10. Develop the divided-difference table from the given data. Write down the interpolating polynomial and rearrange it for fast computation without simplifying.

x	0	1	3	2	5
$f(x)$	2	1	5	6	−183

Check point: $f[1, 3, 2, 5] = -7$.

11. Determine whether the following assertion is true or false. If $x_1, \ldots, x_n$ are distinct, then for arbitrary real values $y_1, \ldots, y_n$ there is a unique polynomial p of degree $\leqq n$ such that $p(x_i) = y_i$ $(1 \leqq i \leqq n)$.
12. Establish the following algorithm for determining the coefficients in the Newton interpolating polynomial. Set $a_1 = f(x_1)$. After the $a_1, \ldots, a_{k-1}$ are determined, let

$$a_k = \frac{f(x_k) - \sum_{i=1}^{k-1} a_i \prod_{j=1}^{i-1} (x_k - x_j)}{\prod_{j=1}^{k-1} (x_k - x_j)}$$

Computer Problems 5.1

1. Test the subprograms given in the text for determining the Newton interpolating polynomial. For example, consider this table.

x	1	2	3	−4	5
y	2	48	272	1182	2262

Find the interpolating polynomial and verify that $p(-1) = 12$.

2. A table of values of a function f is given at the points $x_i = i/10$ for $0 \leqq i \leqq 100$. In order to obtain a graph of f with the aid of an automatic plotter, the values of f are required at the points $z_i = i/20$, $0 \leqq i \leqq 200$. Write a subprogram to do this, using a cubic interpolating polynomial with nodes $x_i, x_{i+1}, x_{i+2}, x_{i+3}$ to compute f at $\frac{1}{2}(x_{i+1} + x_{i+2})$. For z_1 and z_{199}, use the cubic polynomial associated with z_3 and z_{197}, respectively. Compare this routine to COEF for a given function f.
3. Another form for the interpolating polynomial through n points (x_i, y_i), $i \leqq i \leqq n$, is the *Lagrange* form

$$p(x) = \sum_{i=1}^{n} y_i \prod_{\substack{j=1 \\ j \neq i}}^{n} \left(\frac{x - x_j}{x_i - x_j} \right)$$

Write a subprogram analogous to COEF, using this polynomial. Test on the example given in this section at 20 points with $h/2$. Has the Lagrange form any advantage over the Newton form? (For additional information on the Lagrange form of the interpolation polynomial see Supplementary Problems 6 to 11.)

4. Use as input data to COEF the annual rainfall in your town for each of the last 15 years. Using this routine, predict the rainfall for this year. Is the answer reasonable?
5. Recode COEF/EVAL so that a DO loop with a negative increment is used in each. Is the code simplified?
6. Write a simple program using subroutine COEF that interpolates e^x by a polynomial of degree 10 on $[0, 2]$ and then compares the polynomial to EXP at 100 points.

5.2 Errors in Polynomial Interpolation

When a function f is approximated on an interval $[a, b]$ by means of an interpolating polynomial p, the discrepancy between f and p will (theoretically) be zero at each node of interpolation. A natural expectation is that the function f will be well approximated at all intermediate points and that, as the number of nodes increases, this agreement will become better and better.

In the history of numerical analysis, a severe shock occurred when it was realized that this "expectation" was ill-founded. Of course, if the function being approximated is not required to be continuous, then there may be no agreement at all between $p(x)$ and $f(x)$ except at the nodes. As a pathological example, consider the so-called Dirichlet function f, defined to be one at each irrational point and zero at each rational point. If we choose nodes that are rational numbers, then $p(x) \equiv 0$, and $f(x) - p(x) = 0$ for all rational values of x, but $f(x) - p(x) = 1$ for all irrational values of x.

However, if the function f is well behaved, can we not assume that $f(x) - p(x)$ will be small when the number of interpolating nodes is large? The answer is still *no* even for functions possessing continuous derivatives of all orders on the interval! A specific example of this remarkable phenomenon is provided by the function $f(x) = (1 + x^2)^{-1}$ on the interval $[-5, 5]$. Let p_n be the polynomial that interpolates this function at $n + 1$ equally spaced points on the interval $[-5, 5]$, including the endpoints. Then

$$\lim_{n\to\infty} \max_{-5 \leqq x \leqq 5} |f(x) - p_n(x)| = \infty$$

Thus, the effect of requiring the agreement of f and p_n at more and more points is to *increase* the error between the nodes, and the error actually increases beyond all bounds!

The moral of this example, then, is that polynomial interpolation of high degree with many nodes is a risky operation; the resulting polynomials may be very unsatisfactory as representations of functions.

The reader can easily observe the phenomenon just described by using the programs already developed in this chapter. See Computer Problem 1 for a suggested numerical experiment. In a more advanced study of this topic it would be shown that the divergence of the polynomials can often be ascribed to the fact that the nodes are equally spaced. Again, contrary to intuition, equally-distributed nodes are a very poor choice in interpolation. A much better choice for n nodes in $[a, b]$ is the set of *Chebyshev nodes:*

$$x_i = \frac{1}{2}(a + b) + \frac{1}{2}(b - a)\cos\frac{(2i - 1)\pi}{2n} \qquad (1 \leqq i \leqq n)$$

See Computer Problem 2.

A simple graph illustrates this phenomenon best. Again consider the "Runge function" $f(x) = (1 + x^2)^{-1}$ on the interval $[-5, 5]$. First, we select nine equally spaced nodes and use subprograms **COEF** and **EVAL** with an automatic plotter to graph p_8. As shown in Fig. 5.1, the resulting curve assumes negative values which, of course, $f(x)$ does not have! Adding more equally-spaced nodes—and thereby obtaining a higher-degree polynomial—only makes matters worse with wider oscillations. In Fig. 5.2 nine Chebyshev nodes are used, and the resulting polynomial curve is smoother. However, cubic splines (which are discussed in Chapter 7) will produce an even better curve fit.

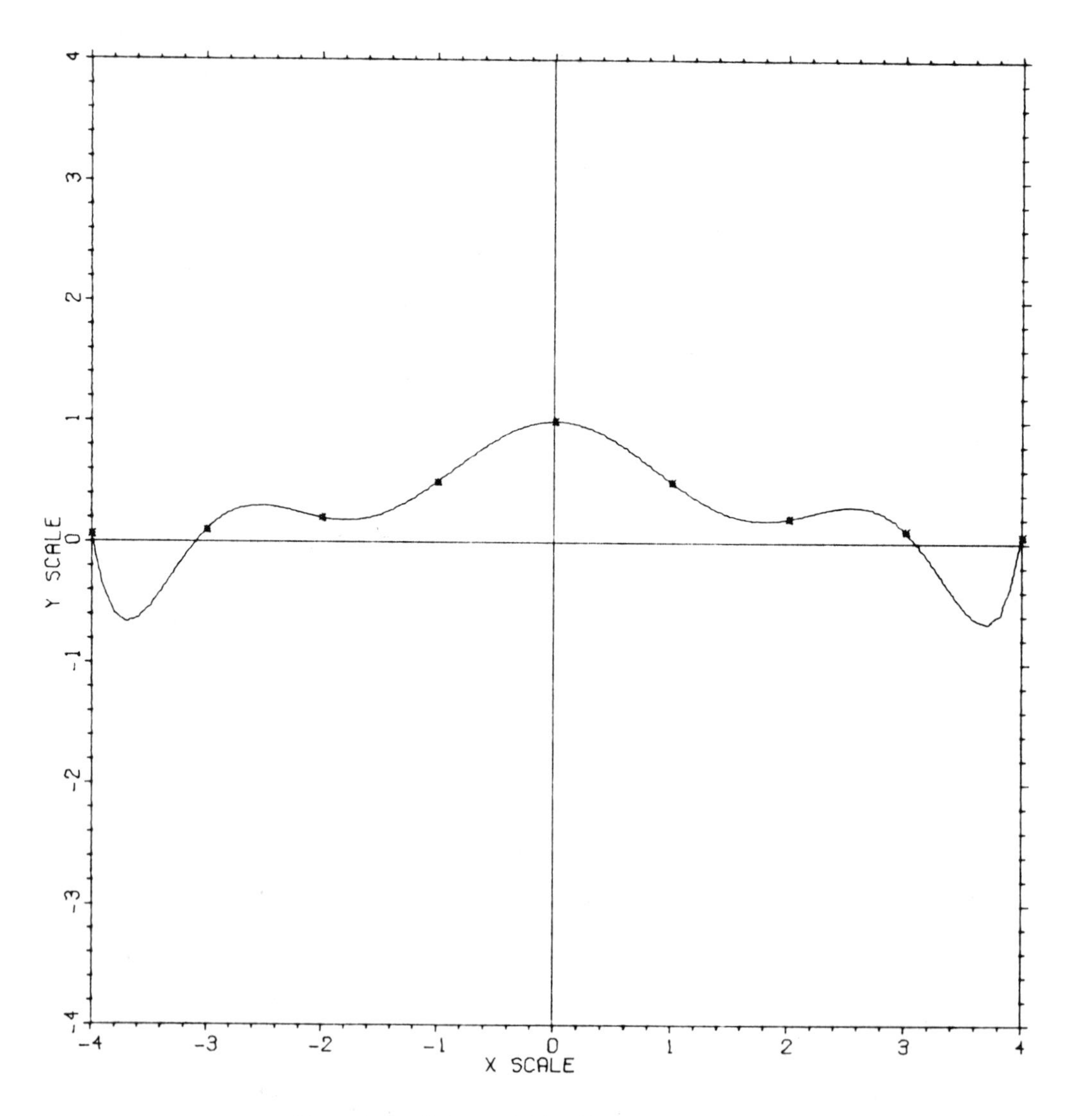

Figure 5.1 Polynomial fit with nine equally spaced nodes.

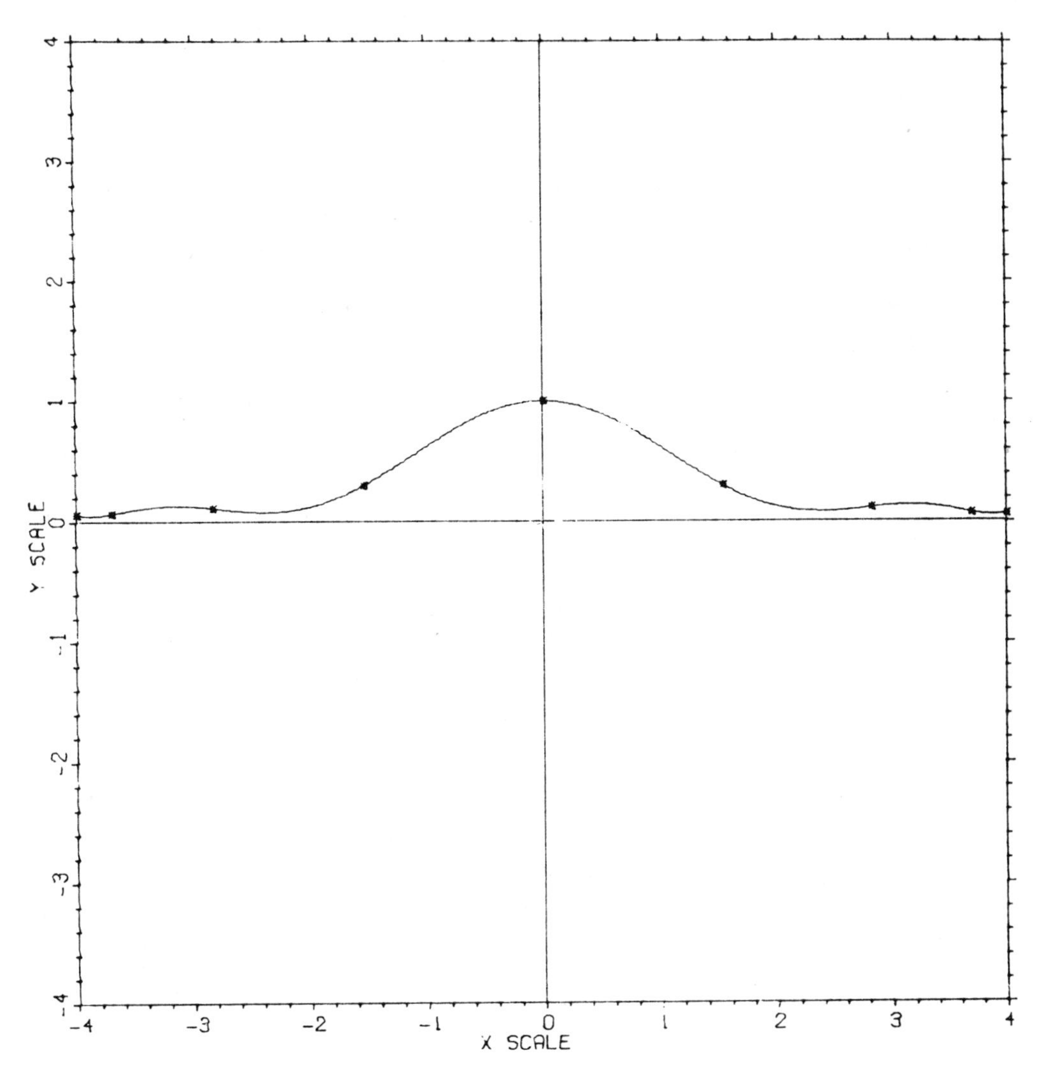

Figure 5.2 Polynomial fit with nine Chebyshev nodes.

It is possible to assess the errors of interpolation by means of a formula involving the nth derivative of the function being interpolated. Here is the formal statement.

> If p is the polynomial of degree at most $n - 1$ that interpolates f at the n distinct nodes $x_1, \ldots, x_n$ belonging to an interval $[a, b]$ and if $f^{(n)}$ is continuous, then for each x in $[a, b]$ there is a ξ in $[a, b]$ for which
>
> $$f(x) - p(x) = \frac{1}{n!}(x - x_1)(x - x_2) \cdots (x - x_n) f^{(n)}(\xi) \tag{10}$$

The justification of this formula proceeds as follows. Observe first that it is obviously valid if x is one of the nodes, for then both sides of the equation reduce to zero.

If x is not a node, let it be fixed in the remainder of the discussion, and define

$$w(t) = (t - x_1)(t - x_2) \cdots (t - x_n) \qquad \text{(polynomial in the variable } t\text{)}$$

$$c = \frac{f(x) - p(x)}{w(x)} \qquad \text{(constant)}$$

$$\phi(t) = f(t) - p(t) - cw(t) \qquad \text{(function in the variable } t\text{)}$$

Observe that c is well-defined because $w(x) \neq 0$ (x is not a node). Note also that ϕ takes the value zero at the $n + 1$ points, $x_1, x_2, \ldots, x_n$, and x. Now invoke *Rolle's Theorem,* which states that between any two roots of ϕ there must occur a root of ϕ'. Thus ϕ' has at least n roots. By similar reasoning, ϕ'' has at least $n - 1$ roots, ϕ''' has at least $n - 2$ roots, and so on. Finally, it can be inferred that $\phi^{(n)}$ must have at least one root. Let ξ be a root of $\phi^{(n)}$. All the roots being counted in this argument are in (a, b). Thus

$$0 = \phi^{(n)}(\xi) = f^{(n)}(\xi) - p^{(n)}(\xi) - cw^{(n)}(\xi)$$

In this equation, $p^{(n)}(\xi) = 0$ because p is a polynomial of degree $\leqq n - 1$. Also, $w^{(n)}(\xi) = n!$ because $w(t) = t^n +$ (lower-order terms). Thus, we have

$$0 = f^{(n)}(\xi) - cn! = f^{(n)}(\xi) - \frac{n!}{w(x)}[f(x) - p(x)]$$

This equation is a rearrangement of Eq. (10). Here is the theorem cited above.

ROLLE'S THEOREM. Let f be a function that is continuous on $[a, b]$ and differentiable on (a, b). If $f(a) = f(b) = 0$, then $f'(c) = 0$ for some point c in (a, b).

To illustrate how Eq. (10) can be used, consider the interpolating polynomial of degree 9 constructed previously for the function $f(x) = \sin x$ on the interval $[0, \pi/2]$. The tenth derivative of $\sin x$ is $-\sin x$ and so $|f^{(n)}(\xi)| \leqq 1$ for this function. The term $|(x - x_1)(x - x_2) \cdots (x - x_{10})|$ cannot exceed

$$\left(\frac{h}{2}\right)\left(\frac{h}{2}\right)\left(2h\right)\left(3h\right) \cdots \left(9h\right) = \frac{1}{4} 9! h^{10}$$

The bound is obtained by considering that x lies in one of the subintervals created by the nodes, say $[x_{i-1}, x_i]$. For the quadratic function $|(x - x_{i-1})(x - x_i)|$, the maximum occurs when x is the midpoint of this interval. For this value of x, the distance to the adjacent nodes x_{i-1} and x_i is $h/2$. For any x in this interval, the distance to the next nearest node cannot exceed $2h$ and so on. Clearly the largest possible bound is obtained when x is within the subinterval $[x_1, x_2]$ or $[x_9, x_{10}]$, since nine factorial would be obtained. Now when the error formula (10) is evaluated with $h = 0.1875$, the result is

$$\left| f(x) - p(x) \right| \leqq \frac{1}{10!}\left(\frac{1}{4} 9! h^{10}\right) = \frac{1}{40} h^{10} = 1.3 \times 10^{-9}$$

Thus the polynomial represents $\sin x$ on the interval $[0, \pi/2]$ with an error of at most two units in the ninth decimal place. (A careful check on a computer would reveal that the polynomial is, in fact, accurate to more decimal places. Why?)

Problems 5.2

1. Show that the maximum error associated with linear interpolation is bounded by $\frac{1}{8}(x_2 - x_1)^2M$, where $M = \max_{x_1 \leqq x \leqq x_2} |f''(x)|$.
2. Let the function $f(x) = \ln x$ be approximated by an interpolation polynomial of degree 9 with ten nodes uniformly distributed in the interval $[1, 2]$. What bound can be placed on the error?
3. In the statement governing the error in interpolation show that if $x_1 < x_2 < \cdots < x_n$ and $x_1 < x < x_n$, then $x_1 < \xi < x_n$.
4. (Continuation) In the same statement, considering ξ as a function of x, show that $f^{(n)}(\xi(x))$ is a continuous function of x. Note that $\xi(x)$ need not be a continuous function of x.
5. Cos x is to be approximated by an interpolating polynomial of degree $n - 1$, using n equally spaced nodes in the interval $[0, 1]$. How accurate is the approximation? (Express your answer in terms of n.) How accurate is the approximation when $n = 10$? For what values of n is the error less than 10^{-7}?
6. Let n nodes x_i be uniformly distributed in $[a, b]$ including endpoints. Show that if $a \leqq x \leqq b$ then

$$\prod_{i=1}^{n} |x - x_i| \leqq \frac{1}{4} h^n (n - 1)!$$

   where $h = (b - a)/(n - 1)$.
7. If p is a polynomial of degree $< n$ that interpolates f at nodes $a = x_1 < \cdots < x_n = b$ in the interval $[a, b]$, then for all x

$$|f(x) - p(x)| \leqq \frac{Mh^n}{4n}$$

   where M is any number satisfying $|f^{(n)}(x)| \leqq M$ for all x in $[a, b]$ and $h = \max_{2 \leqq i \leqq n} (x_i - x_{i-1})$.
8. In interpolating with n equally spaced nodes on an interval, we could use $x_i = a + (2i - 1)h/2$, $1 \leqq i \leqq n$, with $h = (b - a)/n$. What bound can be given now for $\prod_{i=1}^{n} |x - x_i|$ when $a \leqq x \leqq b$? Note that we are not requiring the endpoints to be nodes.

Computer Problems 5.2

1. Using 21 equally spaced nodes on the interval $[-5, 5]$, find the interpolating polynomial p of degree 20 for the function $f(x) = (x^2 + 1)^{-1}$. Print the values of $f(x)$ and $p(x)$ at 41 equally spaced points, including the nodes. Observe the large discrepancy between $f(x)$ and $p(x)$.
2. (Continuation) Perform the experiment above, using the Chebyshev nodes $x_i = 5 \cos(i\pi/20)$, $0 \leqq i \leqq 20$, or $x_i = 5 \cos[(2i + 1)\pi/42]$, $0 \leqq i \leqq 20$.
3. Using subprograms in the text, find a polynomial of degree 13 that interpolates $f(x) = \text{Arctan } x$ on the interval $[-1, 1]$. Test numerically, by taking 100 points, to determine how accurate the polynomial approximation is.
4. (Continuation) Write a subprogram for Arctan x that uses the polynomial of Prob. 3. If x is not in the interval $[-1, 1]$, use the formula $1/\tan\theta = \cot\theta = \tan(\pi/2 - \theta)$.
5. Approximate $f(x) = \text{Arcsin } x$ on the interval $[-1/\sqrt{2}, 1/\sqrt{2}]$ by an interpolating polynomial of degree 15. Determine how accurate the approximation is by numerical tests.
6. (Continuation) Write a function subprogram for Arcsin x, using the polynomial of Prob. 5. If x is in the interval $|x| > 1/\sqrt{2}$, use equation $\sin(\pi/2 - \theta) = \cos\theta = \sqrt{1 - \sin^2\theta}$.
7. Let $f(x) = \max\{0, 1 - x\}$. Sketch the function f. Then find interpolating polynomials p of degrees 2, 4, 8, 16, 32 to f on the interval $[-4, 4]$, using equally spaced nodes. Print out the discrepancy $f(x) - p(x)$ at 128 equally spaced points. Then redo the problem, using Chebyshev nodes.

8. Using COEF/EVAL and an automatic plotter, fit a polynomial through the following data.

x	0.0	0.6	1.5	1.7	1.9	2.1	2.3	2.6	2.8	3.0
y	−0.8	−0.34	0.59	0.59	0.23	0.1	0.28	1.03	1.5	1.44

Does the resulting curve look like a good fit? Explain.

9. Why are the Chebyshev nodes generally better than equally spaced nodes in polynomial interpolation? The answer lies in the term $(x - x_1) \cdots (x - x_n)$ that occurs in the error formula. If $x_1, \ldots, x_n$ are the Chebyshev nodes on $[-1, 1]$, then

$$|(x - x_1) \cdots (x - x_n)| \leqq 2^{1-n}$$

for all x in $[-1, 1]$. Devise a numerical experiment to test this assertion for $n = 2, 4, 8, 16$.

5.3 Estimating Derivatives

A numerical experiment outlined in Chapter 1 showed that determining the derivative of a function f at a point x is not a trivial numerical problem. Specifically, if $f(x)$ can be computed with only n digits of precision, it is difficult to calculate $f'(x)$ numerically with n digits of precision. This difficulty can be traced to the subtraction of quantities that are nearly equal. In this section several alternatives are offered for the numerical computation of $f'(x)$.

First, consider again the obvious method based on the definition of $f'(x)$. It consists in selecting one or more small values of h and writing

$$f'(x) \approx \frac{f(x + h) - f(x)}{h} \tag{1}$$

What error is involved in this formula? In order to find out, expand f in a Taylor series.

$$f(x + h) = f(x) + hf'(x) + \frac{1}{2}h^2 f''(\xi) \tag{2}$$

Here ξ is in the interval whose endpoints are x and $x + h$. Rearranging Eq. (2) gives

$$\frac{f(x + h) - f(x)}{h} = f'(x) + \frac{1}{2}hf''(\xi) \tag{3}$$

This equation shows that, in general, as $h \to 0$, the difference between $f'(x)$ and the estimate $h^{-1}[f(x + h) - f(x)]$ approaches zero at the same rate that h does. Of course, if $f''(x) = 0$, then $\frac{1}{2}hf''(\xi) \approx 0$ and the error term will be $\frac{1}{6}h^2 f'''(\gamma)$, which converges to zero somewhat faster. But usually $f''(x)$ is not zero.

Equation (3) gives the *truncation* error for this numerical procedure—namely, $\frac{1}{2}hf''(\xi)$. It is an error that will be present even if the calculations are performed with *infinite* precision and is due to our imitating the mathematical limit process by means of an approximate formula. Additional (and worse) errors must be expected when calculations are performed on a computer with finite word length.

As we saw in Newton's method (Chapter 3) and in the Romberg method (Chapter 4), it is advantageous to have convergence of numerical processes occur with higher powers of some quantity approaching zero. In the present situation, we want an approximation to $f'(x)$ in which the error behaves like h^2. One such method is easily obtained with the aid of the Taylor series.

$$f(x + h) = f(x) + hf'(x) + \frac{1}{2}h^2f''(x) + \frac{1}{6}h^3f'''(x) + \cdots$$

$$f(x - h) = f(x) - hf'(x) + \frac{1}{2}h^2f''(x) - \frac{1}{6}h^3f'''(x) + \cdots \tag{4}$$

By subtraction, we obtain

$$f(x + h) - f(x - h) = 2hf'(x) + \frac{1}{3}h^3f'''(x) + \cdots$$

This leads to a very important formula.

$$f'(x) = \frac{f(x + h) - f(x - h)}{2h} - \frac{h^2}{6}f'''(x) - \cdots \tag{5}$$

This formula will be useful in the numerical solution of certain types of differential equations (as we shall see in Chapters 14 and 15).

Since the error terms $-(h^2/6)f'''(x) - \cdots$ are on the order of h^2, the process of Richardson extrapolation can be used. In order to do so, put

$$\phi(h) = \frac{f(x + h) - f(x - h)}{2h}$$

Of course, f and x are fixed. We know that as $h \to 0$, $\phi(h) \to f'(x)$ and the error behaves like h^2. So, as in Chapter 4, the combination

$$\phi\left(\frac{h}{2}\right) + \frac{1}{3}\left[\phi\left(\frac{h}{2}\right) - \phi(h)\right]$$

should be a more accurate estimate (of order h^4) for $f'(x)$ than $\phi(h)$ or $\phi(h/2)$. By carrying more terms in the Taylor expansions of Eq. (5) we could establish that

$$f'(x) = \phi(h) + a_2h^2 + a_4h^4 + a_6h^6 + \cdots$$

So just as in developing the Romberg algorithm, we can carry out Richardson extrapolation repeatedly. If this is done, a triangular array can be computed in which the first column gives $\phi(h)$, $\phi(h/2)$, $\phi(h/4)$, . . . , and the remaining columns result from Richardson extrapolation. Such a program is suggested as a problem.

An important general strategem can be used to approximate derivatives (as well as integrals and other quantities). The function f is first approximated by a polynomial p so that $f \approx p$. Then we simply proceed to $f'(x) \approx p'(x)$ as a consequence.

In practice, the approximating polynomial p is often determined by interpolation at a few points. For example, suppose that p is the polynomial of degree $\leqq 1$ that interpolates f at two nodes, x_1 and x_2. Then from Eq. (6) in Sec. 5.1,

$$p_1(x) = f(x_1) + f[x_1, x_2](x - x_1)$$

Consequently,

$$f'(x) \approx p_1'(x) = f[x_1, x_2] = \frac{f(x_2) - f(x_1)}{x_2 - x_1} \tag{6}$$

If $x_1 = x$ and $x_2 = x + h$, this formula is one previously considered—namely, Eq. (1).

$$f'(x) \approx \frac{f(x + h) - f(x)}{h} \tag{7}$$

If $x_1 = x - h$ and $x_2 = x + h$, the resulting formula is Eq. (5).

$$f'(x) \approx \frac{f(x + h) - f(x - h)}{2h} \tag{8}$$

Now consider interpolation with three nodes, x_1, x_2, and x_3. The interpolating polynomial is obtained from Eq. (6) in Sec. 5.1:

$$p_2(x) = f(x_1) + f[x_1, x_2](x - x_1) + f[x_1, x_2, x_3](x - x_1)(x - x_2)$$

and its derivative is

$$p_2'(x) = f[x_1, x_2] + f[x_1, x_2, x_3](2x - x_1 - x_2) \tag{9}$$

Here the right-hand side consists of two terms: the first is the previous estimate in Eq. (6), and the second is a refinement or correction. If Eq. (9) is used to evaluate $f'(x)$ when $x = \frac{1}{2}(x_1 + x_2)$ [as in Eq. (8)], then the correction term in Eq. (9) is zero. Thus the first term in this case must be more accurate than in other cases, for the correction term adds nothing. This is why Eq. (8) is more accurate than (7).

An analysis of the errors in this general procedure goes as follows. Suppose that p_{n-1} is the polynomial of least degree that interpolates f at the nodes $x_1, \ldots, x_n$. Then according to a result established in the previous section,

$$f(x) - p_{n-1}(x) = \frac{1}{n!} f^{(n)}(\xi) w(x) \tag{10}$$

with ξ dependent on x and $w(x) = (x - x_1)(x - x_2) \cdots (x - x_n)$. Differentiating gives

$$f'(x) - p_{n-1}'(x) = \frac{1}{n!} w(x) \frac{d}{dx} f^{(n)}(\xi) + \frac{1}{n!} f^{(n)}(\xi) w'(x) \tag{11}$$

Here we had to assume that $f^{(n)}(\xi)$ is differentiable as a function of x, a fact that is known if $f^{(n+1)}$ exists and is continuous.

The first observation to make about the error formula in Eq. (11) is that $w(x)$ vanishes at each node; so if the evaluation is at a node x_i, the resulting equation is simpler:

$$f'(x_i) = p_{n-1}'(x_i) + \frac{1}{n!} f^{(n)}(\xi) w'(x_i) \tag{12}$$

For example, taking just two points x_1 and x_2, we obtain

$$f'(x_1) = f[x_1, x_2] + \frac{1}{2} f''(\xi) \frac{d}{dx} [(x - x_1)(x - x_2)] \bigg|_{x=x_1}$$

$$= f[x_1, x_2] + \frac{1}{2} f''(\xi)(x_1 - x_2) \tag{13}$$

This is Eq. (3) in disguise.

The second observation to make about Eq. (11) is that it becomes simpler if x is chosen as a point where $w'(x) = 0$. For instance, if $n = 2$, then w is a quadratic function that vanishes at the two nodes x_1 and x_2. Because a parabola is symmetric about its axis, $w'((x_1 + x_2)/2) = 0$. The resulting formula is

$$f'\left(\frac{x_1 + x_2}{2}\right) = f[x_1, x_2] - \frac{1}{8}(x_2 - x_1)^2 \frac{d}{dx} f''(\xi) \tag{14}$$

As a final example, consider four interpolation points, x_1, x_2, x_3, x_4. The interpolating polynomial is

$$p_3(x) = f(x_1) + f[x_1, x_2](x - x_1) + f[x_1, x_2, x_3](x - x_1)(x - x_2) + f[x_1, x_2, x_3, x_4](x - x_1)(x - x_2)(x - x_3)$$

Its derivative is

$$p_3'(x) = f[x_1, x_2] + f[x_1, x_2, x_3](2x - x_1 - x_2) + f[x_1, x_2, x_3, x_4](3x^2 - 2x_1x - 2x_2x - 2x_3x + x_1x_2 + x_1x_3 + x_2x_3) \tag{15}$$

A useful special case occurs if $x_1 = x - h$, $x_2 = x + h$, $x_3 = x - 2h$, and $x_4 = x + 2h$. The resulting formula is

$$f'(x) \approx \frac{f(x + h) - f(x - h)}{2h} - \frac{f(x + 2h) - 2f(x + h) + 2f(x - h) - f(x - 2h)}{12h} \tag{16}$$

which has been arranged in the form in which it probably should be computed: a principal term plus a correction or refining term.

In the numerical solution of differential equations it is often necessary to approximate second derivatives. We shall derive the most important formula for accomplishing this. Simply add the two equations in Eq. (4). After rearrangement, the result is

$$\frac{f(x + h) - 2f(x) + f(x - h)}{h^2} = f''(x) + \frac{2h^2}{4!} f^{(4)}(x) + \frac{2h^4}{6!} f^{(6)}(x) + \cdots \tag{17}$$

Hence, we have

$$f''(x) \approx \frac{f(x + h) - 2f(x) + f(x - h)}{h^2} \tag{18}$$

with error of order h^2.

Problems 5.3

1. In approximating an operation on a function f, one stratagem is first to approximate f by a polynomial p and then apply the operation to p. Numerical differentiation provides examples of this. What other examples of this stratagem can you find in previous chapters?
2. Derive the approximate formula

$$f'(x) \approx \frac{1}{2h}[4f(x+h) - 3f(x) - f(x+2h)]$$

and show that its error is of the form $\frac{1}{3}h^2 f'''(\xi)$.
3. Let $w(x) = (x - x_1)(x - x_2) \cdots (x - x_n)$. Show that

$$w'(x) = \sum_{i=1}^{n} \prod_{\substack{j=1 \\ j \neq i}}^{n} (x - x_j) \quad \text{and} \quad w'(x_i) = \prod_{\substack{j=1 \\ j \neq i}}^{n} (x_i - x_j).$$

4. Carry out the derivation of Eq. (16) from (15).
5. Show that

$$f'(x) = \frac{1}{2h}[f(x+h) - f(x-h)] - \frac{1}{2}\sum_{k=1}^{\infty} \frac{1}{(2k+1)!} h^{2k} f^{(2k+1)}(x)$$

Hint: Use full Taylor series in Eqs. (4) and (5).
6. Carry out the details in deriving Eq. (17).
7. A certain scientific calculation requires an approximate formula for $f'(x) + f''(x)$. How well does the expression

$$\left(\frac{2+h}{h^2}\right) f(x+h) - \left(\frac{2}{h^2}\right) f(x) + \left(\frac{2-h}{h^2}\right) f(x-h)$$

serve? Derive this approximation and its error term.
8. Using the unevenly spaced points $x_1 < x_2 < x_3$, where $x_2 - x_1 = h$ and $x_3 - x_2 = \alpha h$, establish the formula

$$f''(x) \approx \frac{2}{h^2}\left(\frac{f(x_1)}{(1+\alpha)} - \frac{f(x_2)}{\alpha} + \frac{f(x_3)}{\alpha(\alpha+1)}\right)$$

in the following two ways. Notice that this formula reduces to the standard central difference formula (18) when $\alpha = 1$.
   (a) Approximate $f(x)$ by the Newton interpolating polynomial of degree 2.
   (b) Calculate the undetermined coefficients A, B, C in the expression $f''(x) \approx Af(x_1) + Bf(x_2) + Cf(x_3)$ by making it exact for the three polynomials 1, $x - x_2$, $(x - x_2)^2$ and thus exact for all polynomials of degree $\leqq 2$.
9. (Continuation) Using Taylor series, show that

$$f'(x_2) = \frac{f(x_3) - f(x_1)}{x_3 - x_1} + (\alpha - 1)\frac{h}{2} f''(x_2) + \mathcal{O}(h^2)$$

Consequently, establish that the error for approximating $f'(x_2)$ by $[f(x_3) - f(x_1)]/(x_3 - x_1)$ is $\mathcal{O}(h^2)$ when x_2 is midway between x_1 and x_3 but only $\mathcal{O}(h)$ otherwise.
10. Using Taylor series, derive the formula

$$f'''(x_0) \approx -\frac{1}{h^3}[f(x_1) - f(x_{-1})] + \frac{1}{2h^3}\Big[f(x_2) - f(x_{-2})\Big]$$

and determine the error term. Here $x_{-2} < x_{-1} < x_0 < x_1 < x_2$ and $h = x_i - x_{i-1}$.

11. Use Taylor series to derive the error term for the derivative approximation

$$f^{iv}(x_0) \approx \frac{1}{h^4}[f(x_{-2}) - 4f(x_{-1}) + 6f(x_0) - 4f(x_1) + f(x_2)]$$

for evenly-spaced points $x_{-2} < x_{-1} < x_0 < x_1 < x_2$ with $h = x_i - x_{i-1}$.

12. Criticize the following analysis. By Taylor's formula, we have

$$f(x+h) - f(x) = hf'(x) + \frac{h^2}{2}f''(x) + \frac{h^3}{6}f'''(\xi_1)$$

$$f(x-h) - f(x) = -hf'(x) + \frac{h^2}{2}f''(x) - \frac{h^3}{6}f'''(\xi_2)$$

Therefore

$$\frac{f(x+h) - 2f(x) + f(x-h)}{h^2} = f''(x) + \frac{h}{6}[f'''(\xi_1) - f'''(\xi_2)]$$

The error in the approximate formula for f'' is thus $\mathcal{O}(h)$.

Computer Problems 5.3

1. Test a subroutine called `DERIV(F, X, H, N, D)` that computes the derivative of the function f at the point x. Use Richardson extrapolation to create and print a triangular array D with n rows and columns. In the `(I, 1)` position of the array, compute $[f(x + h/2^{i-1}) - f(x - h/2^{i-1})]/(h/2^i)$. *Hint:* Use the Romberg algorithm of Chapter 4 as a guide.
2. Carry out a numerical experiment to compare the accuracy of formulas (13), (16), and (17) on a function f whose derivatives can be computed precisely. Take a sequence of values of h, such as 4^{-n} with $1 \leqq n \leqq 24$.
3. Find $F'(0.25)$ as accurately as possible, using only the single-precision subroutine for F below and a method for numerical differentiation.

```
      FUNCTION F(X)
      A = 1.0
      B = COS(X)
      DO 2 N = 1,5
      D = B
      B = SQRT(A*B)
    2 A = (A + D)/2.0
      F = 2.0*ATAN(1.0)/A
      RETURN
      END
```

Supplementary Problems (Chapter 5)

1. Use the divided-difference method to obtain a polynomial of least degree that fits the values shown.

(a)

x	0	1	2	-1	3
y	-1	-1	-1	-7	5

(b)

x	1	3	-2	4	5
y	2	6	-1	-4	2

2. Count the number of multiplications, divisions, and additions/subtractions in generating the divided-difference table having n points.
3. Find the interpolating polynomial for these data.

x	1.0	2.0	2.5	3.0	4.0
$f(x)$	−1.5	−0.5	0.0	0.5	1.5

4. It is suspected that the table

x	−2	−1	0	1	2	3
y	1	4	11	16	13	−4

comes from a cubic polynomial. How can this be tested? Explain.

5. The polynomial $p(x) = x^4 - x^3 + x^2 - x + 1$ has the values shown.

x	−2	−1	0	1	2	3
$p(x)$	31	5	1	1	11	61

Find a polynomial q that takes these values:

x	−2	−1	0	1	2	3
$q(x)$	31	5	1	1	11	30

Hint: It can be done with little work.

6. (*Lagrange* form of the interpolating polynomial) For each i, define a polynomial ℓ_i by writing

$$\ell_i(x) = \prod_{\substack{j=1 \\ j \neq i}}^{n} \left(\frac{x - x_j}{x_i - x_j} \right)$$

Verify that ℓ_i is of degree $n - 1$, that $\ell_i(x_i) = 1$, and that $\ell_i(x_j) = 0$ if $j \neq i$.

7. (Continuation) Show that the interpolating polynomial for f with nodes $x_1, \ldots, x_n$ is

$$\sum_{i=1}^{n} f(x_i)\ell_i(x)$$

8. (Continuation) Use the uniqueness of the interpolating polynomial to verify that

$$\sum_{i=1}^{n} f(x_i)\ell_i(x) = \sum_{i=1}^{n} f[x_1, \ldots, x_i] \prod_{j=1}^{i-1} (x - x_j)$$

9. (Continuation) Show that the following explicit formula is valid for divided differences.

$$f[x_1, \ldots, x_n] = \sum_{i=1}^{n} f(x_i) \prod_{\substack{j=1 \\ j \neq i}}^{n} (x_i - x_j)^{-1}$$

Hint: If two polynomials are equal, the coefficients of x^{n-1} in each are equal.

10. Verify directly that

$$\sum_{i=1}^{n} \ell_i(x) = 1$$

for the case n =2. Then establish the result for arbitrary values of n.

11. For the four interpolation nodes -1, $+1$, 3, 4, what are the ℓ_i functions required in the Lagrange interpolation procedure? Draw the graphs of these four functions to show their essential properties.
12. If f is a polynomial of degree n, show that in a divided-difference table for f the nth column has a single constant value. This is the column containing entries $f[x_i, x_{i+1}, \ldots, x_{i+n}]$.
13. Use the Lagrange interpolation process to obtain a polynomial of least degree that assumes these values.

x	0	2	3	4
y	7	11	28	63

14. (Continuation) Rearrange the points in the table of Prob. 13 and find the Newton interpolating polynomial. Show that the polynomials obtained are identical, although their forms may differ.
15. Complete the following divided-difference table and use it to obtain a polynomial of degree 3 that interpolates the function values indicated.

x	$f(x)$	$f[\,,\,]$	$f[\,,\,,\,]$	$f[\,,\,,\,,\,]$
-1	2			
		–		
1	-4		2	
		–		–
3	6		–	
		2		
5	10			

Write the final polynomial in a form most efficient for computing.
16. Find an interpolating polynomial for this table.

x	1	2	2.5	3	4
y	-1	$-\frac{1}{3}$	$\frac{3}{32}$	$\frac{4}{3}$	25

17. A table of values of $\sin x$ is required so that linear interpolation will yield five-decimal-place accuracy for any value of x in $[0, \pi]$. Assuming that the tabular values are equally spaced, what is the minimum number of entries needed in this table?
18. Does every polynomial p of degree $\leqq n - 1$ obey this equation?

$$p(x) = \sum_{i=1}^{n} p[x_1, \ldots, x_i] \prod_{j=1}^{i-1} (x - x_j)$$

Hint: Use the uniqueness of the interpolating polynomial.
19. Show that f is a polynomial of degree $\leqq n - 1$ if and only if $f[x_1, x_2, \ldots, x_i] = 0$ for $i \geqq n + 1$ and for arbitrary points x_i. *Hint:* Use the preceding problem and Prob. 22.
20. Use a difference table and Prob. 19 to show that the following data can be represented by a polynomial of degree 3.

x	-2	-1	0	1	2	3
y	1	4	11	16	13	-4

21. Show that if a function g interpolates the function f at $x_1, \ldots, x_{n-1}$ and h interpolates f at $x_2, \ldots, x_n$, then

$$g(x) + \frac{x_1 - x}{x_n - x_1}[g(x) - h(x)]$$

interpolates f at $x_1, \ldots, x_n$.

22. Let p be the polynomial of degree $\leqq n - 1$ that interpolates the function f at nodes $x_1, \ldots, x_n$. Then the discrepancy between f and p at any point x is

$$f(x) - p(x) = f[x_1, \ldots, x_n, x] \prod_{j=1}^{n} (x - x_j)$$

Establish this fact. *Hint:* Let q be the polynomial that interpolates f at $x_1, \ldots, x_n, t$. Then

$$q(x) = p(x) + f[x_1, \ldots, x_n, t] \prod_{j=1}^{n} (x - x_j)$$

Next write $f(t) - p(t) = q(t) - p(t)$ and so on and put $t = x$.

23. Let p be a polynomial of degree n. What is $p[x_1, x_2, \ldots, x_{n+2}]$?
24. Show that if f is continuously differentiable on the interval $[x_1, x_2]$, then $f[x_1, x_2] = f'(c)$ for some c in (x_1, x_2).
25. If $x_1, \ldots, x_{n+1}$ are distinct points and if $f^{(n)}$ exists, show that $f[x_1, \ldots, x_{n+1}] = [(1/n!) \times f^{(n)}(\xi)]$ for some ξ in the interval spanned by the points $x_1, \ldots, x_{n+1}$. *Hint:* There are two distinct assessments of the error in interpolation, one in the text and another in Prob. 22. Set them equal to each other.
26. Simple polynomial interpolation in two dimensions is not always possible. For example, suppose that the following data are to be represented by a polynomial of first degree in x and y, $p(t) = a + bx + cy$ where $t = (x, y)$.

t	(1, 1)	(3, 2)	(5, 3)
$f(t)$	3	2	6

Show that it is not possible.

27. Consider a function $f(x)$ such that $f(2) = 1.5713$, $f(3) = 1.5719$, $f(5) = 1.5738$, $f(6) = 1.5751$. Estimate $f(4)$, using a second-degree interpolating polynomial and a third-degree polynomial. Round the final results off to four decimal places. Is there any advantage here in using a third-degree polynomial?
28. A process called ***inverse interpolation*** is often used to solve equations of the form $f(x) = 0$. Suppose that values y_i of $f(x_i)$ have been computed at $x_1, \ldots, x_n$. We use an interpolating polynomial of degree $n - 1$ to estimate the entry denoted by z in the table such that $f(z) = 0$.

f	y_1	y_2	$\cdots$	y_n	0
x	x_1	x_2	$\cdots$	x_n	z

Write an algorithm to do so.

29. Find a polynomial $p(x)$ of degree at most 3 such that $p(0) = 1$, $p(1) = 0$, $p'(0) = 0$, $p'(1) = -1$.
30. An interpolating polynomial of degree 20 is to be used to approximate e^{-x} on the interval $[0, 2]$. How accurate will it be?
31. How accurately can we determine sin x by linear interpolation, given a table of sin x to ten decimal places, for x in $[0, 2]$ with $h = 0.01$?
32. Given the data

x	sin x	cos x
0.70	0.64421 76872	0.76484 21872
0.71	0.65183 37710	0.75836 18759

find approximate values of sin (0.705) and cos (0.702) by linear interpolation. What is the maximum error involved?

33. Show that there exist coefficients $w_1, w_2, \ldots, w_n$ such that

$$\int_a^b p(x)\,dx = \sum_{i=1}^{n} w_i p(x_i)$$

for all polynomials p of degree $\leqq n-1$, any a and b, and any distinct points $x_1, x_2, \ldots, x_n$. *Hint:* Use the Lagrange form of the interpolating polynomial.

34. Which are the most accurate and why?

$$f'(x) \approx \frac{1}{h}[f(x+h) - f(x)]$$

$$f'(x) \approx \frac{1}{h}[f(x) - f(x-h)]$$

$$f'(x) \approx \frac{1}{2h}[f(x+2h) - f(x)]$$

$$f'(x) \approx \frac{1}{2h}[f(x) - f(x-2h)]$$

35. Averaging the forward-difference formula $f'(x) \approx [f(x+h) - f(x)]/h$ and the backward-difference formula $f'(x) \approx [f(x) - f(x-h)]/h$, each with error term $\mathcal{O}(h)$, results in the central-difference formula $f'(x) \approx [f(x+h) - f(x-h)]/(2h)$ with error $\mathcal{O}(h^2)$. Show why. *Hint:* Determine at least the first term in the error series for each formula.

36. Derive the two formulas

(a) $$f'(x) \approx \frac{1}{4h}[f(x+2h) - f(x-2h)]$$

(b) $$f''(x) \approx \frac{1}{4h^2}[f(x+2h) - 2f(x) + f(x+2h)]$$

and establish formulas for the errors in using them.

37. Derive the following rules for estimating derivatives

(a) $$f'''(x) \approx \frac{1}{2h^3}[f(x+2h) - 2f(x+h) + 2f(x-h) - f(x-2h)]$$

(b) $$f^{iv}(x) \approx \frac{3}{h^4}[f(x+2h) - 4f(x+h) + 6f(x) - 4f(x-h) + f(x+2h)]$$

and their error terms. Which is more accurate? *Hint:* Consider the Taylor series for $D(h) \equiv f(x+h) - f(x-h)$ and $S(h) \equiv f(x+h) + f(x-h)$.

38. (Vandermonde Determinants) Using $f_i = f(x_i)$, show that

(a) $$f[x_1, x_2] = \frac{\begin{vmatrix} 1 & f_1 \\ 1 & f_2 \end{vmatrix}}{\begin{vmatrix} 1 & x_1 \\ 1 & x_2 \end{vmatrix}}$$ and (b) $$f[x_1, x_2, x_3] = \frac{\begin{vmatrix} 1 & x_1 & f_1 \\ 1 & x_2 & f_2 \\ 1 & x_3 & f_3 \end{vmatrix}}{\begin{vmatrix} 1 & x_1 & x_1^2 \\ 1 & x_2 & x_2^2 \\ 1 & x_3 & x_3^2 \end{vmatrix}}$$

39. Write the Lagrange form of the interpolating polynomial of degree $\leqq 2$ that interpolates $f(x)$ at x_1, x_2, x_3, where $x_1 < x_2 < x_3$.
40. (Continuation) Write the Newton form of the interpolating polynomial $p_2(x)$ and show that it is equivalent to the Lagrange form.
41. (Continuation) Show directly that

$$p_2''(x) = 2f[x_1, x_2, x_3]$$

42. (Continuation) Show directly for uniform spacing $h = x_2 - x_1 = x_3 - x_2$ that

$$f[x_1, x_2] = \frac{\Delta f_1}{h} \quad \text{and} \quad f[x_1, x_2, x_3] = \frac{\Delta^2 f_1}{2h^2}$$

where $\Delta f_i = f_{i+1} - f_i$, $\Delta^2 f_i = \Delta f_{i+1} - \Delta f_i$, and $f_i = f(x_i)$.
43. (Continuation) Establish the *Newton forward-difference interpolating polynomial* for uniform spacing

$$p_2(x) = f_1 + \binom{s}{1} \Delta f_1 + \binom{s}{2} \Delta^2 f_1$$

where $x = x_1 + sh$. Here $\binom{s}{m}$ is the binomial coefficient $[s!]/[(s - m)!m!] = [s(s - 1) \times (s - 2) \cdots (s - m + 1)]/[m(m - 1)(m - 2) \cdots 3 \times 2 \times 1]$.
44. (Continuation) The *Newton backward-difference interpolating polynomial* for uniform spacing is given by $x = x_3 + sh$ and

$$p_2(x) = f_3 - \binom{-s}{1} \Delta f_2 + \binom{-s}{2} \Delta^2 f_1$$

Show that forward- and backward-difference polynomials $p_2(x_1 + sh)$ and $p_2(x_3 + sh)$ are equivalent.
45. (Continuation) From the following table of values of ln x interpolate ln 2.352 and ln 2.387, using both the forward- and backward-difference forms of Newton's interpolating polynomial.

x	$f(x)$	Δf	$\Delta^2 f$
2.35	0.85442		
		0.00424	
2.36	0.85866		−0.00001
		0.00423	
2.37	0.86289		−0.00002
		0.00421	
2.38	0.86710		−0.00002
		0.00419	
2.39	0.87129		

Using the exact values of ln 2.352 ≈ 0.85527, ln 2.387 ≈ 0.87004, show that the forward-difference formula is more exact for values near the top of the table and that the backward-difference formula is more exact for values near the bottom.
46. Consider *Stirling's polynomial* of second degree that interpolates $f(x)$ at uniformly spaced points $x_{-1} < x_0 < x_1$, where $x_i = x_0 + ih$, and $x = x_0 + sh$.

$$p_2(x) = f_0 + \frac{1}{2}\binom{s}{1}\left[\Delta f_0 + \Delta f_{-1}\right] + \frac{1}{2}\left[\binom{s+1}{2} + \binom{s}{2}\right] \Delta^2 f_{-1}$$

(a) By approximating $f(x)$ with this polynomial, determine a numerical differentiation formula for $f'(x_{-1})$ in terms of f_{-1}, f_0, f_1 and x_{-1}, x_0, x_1. Here $f_i = f(x_i)$.
(b) Use Taylor series to determine the error term.

Hint: $p_2'(x) = \dfrac{1}{h}\dfrac{d}{ds} p(x_0 + sh)$.

47. Using the backward-difference form of the Newton interpolating polynomial at $x_{-2} < x_{-1} < x_0$ —that is,

$$p(x) = f_0 - \binom{-s}{1} \Delta f_{-1} + \binom{-s}{2} \Delta^2 f_{-2}$$

where

$$x = x_0 + sh, \quad f_i = f(x_i), \quad x_i = x_0 + ih, \quad h = x_0 - x_{-1} = x_{-1} - x_{-2}$$

determine a numerical integration formula in terms of f_0, f_{-1}, f_{-2} for evaluating

$$\int_{x_{-2}}^{x_0} f(x)\, dx$$

Hint: Approximate the integrand with this polynomial and use

$$\int_{x_{-2}}^{x_0} p(x)\, dx = h \int_{-2}^{0} p(x_0 + sh)\, ds$$

48. Criticize the following analysis. By Taylor's formula, we have

$$f(x + h) - f(x) = hf'(x) + \frac{h^2}{2} f''(x) + \frac{h^3}{6} f'''(\xi)$$

$$f(x - h) - f(x) = -hf'(x) + \frac{h^2}{2} f''(x) - \frac{h^3}{6} f'''(\xi)$$

So by adding, we obtain an *exact* expression for $f''(x)$:

$$f(x + h) + f(x - h) - 2f(x) = h^2 f''(x)$$

49. Show how Richardson extrapolation would work on formula (17).
50. Let $f(x) = x^3 + 2x^2 + x + 1$. Find the polynomial of degree 4 that interpolates the values of f at $x = -2, -1, 0, 1, 2$. Find the polynomial of degree 2 interpolating the values of f at $x = -1$, 0, 1.
51. The values of a function f are given at three points, x_1, x_2, and x_3. If a quadratic interpolating polynomial is used to estimate $f'(x)$ at $x = \frac{1}{2}(x_1 + x_2)$, what formula will result?

6

Systems of Linear Equations

A simple electrical network contains a number of resistances and a single source of electromotive force (a battery) as shown in Fig. 6.1.

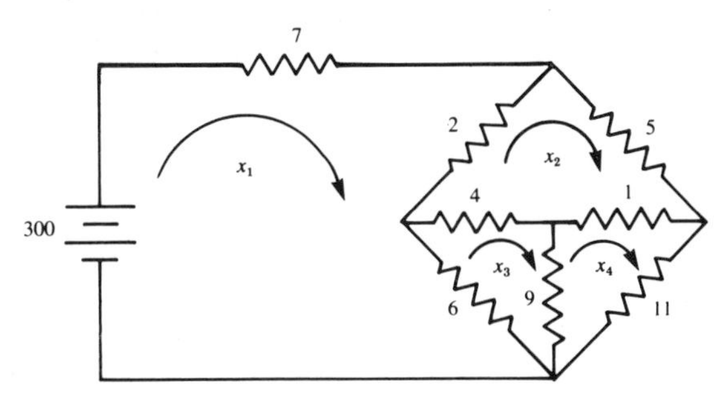

Figure 6.1 Electrical network.

Using Kirchhoff's laws and Ohm's law, a system of equations can be written governing this circuit. If x_1, x_2, x_3, and x_4 are the loop currents as shown, the equations are

$$\begin{cases} 15x_1 - 2x_2 - 6x_3 = 300 \\ -2x_1 + 12x_2 - 4x_3 - x_4 = 0 \\ -6x_1 - 4x_2 + 19x_3 - 9x_4 = 0 \\ -x_2 - 9x_3 + 21x_4 = 0 \end{cases}$$

Systems of equations like this, even containing hundreds of unknowns, can be solved by using the methods developed in this chapter. The solution to the preceding system, for instance, is $x_1 = 26.549157854$, $x_2 = 9.353701528$, $x_3 = 13.254994125$, $x_4 = 6.126126126$.

6.1 Naive Gaussian Elimination

Our objective in this chapter is to develop a good program for solving a system of n linear equations in n unknowns:

$$\begin{cases} a_{11}x_1 + a_{12}x_2 + \cdots + a_{1n}x_n = b_1 \\ a_{21}x_1 + a_{22}x_2 + \cdots + a_{2n}x_n = b_2 \\ \vdots \\ a_{n1}x_1 + a_{n2}x_2 + \cdots + a_{nn}x_n = b_n \end{cases} \tag{1}$$

In compact form this system can be written

$$\sum_{j=1}^{n} a_{ij}x_j = b_i \qquad (1 \leqq i \leqq n) \tag{2}$$

In these equations a_{ij} and b_i are prescribed real numbers (data) and the unknowns x_j are to be determined. Subscripts on the letter a are separated by a comma only if necessary: a_{ij}, $a_{32,75}$, and so on.

In this section the simplest form of Gaussian elimination will be explained. The adjective "naive" applies because it is not suitable for automatic computation unless essential modifications are made.

We illustrate with a specific example.

$$\begin{cases} 3x_1 + 2x_2 - x_3 = 7 \\ 5x_1 + 3x_2 + 2x_3 = 4 \\ -x_1 + x_2 - 3x_3 = -1 \end{cases} \tag{3}$$

In the first step of the elimination procedure certain multiples of the first equation are subtracted from the second and third equations so as to create zero coefficients in place of the 5 and -1. A moment's reflection shows that we should subtract $\frac{5}{3}$ times the first equation from the second equation, for then the new coefficient of x_1 in the second equation will be $5 - \frac{5}{3} \cdot 3 = 0$. By the same reasoning, we should subtract $-\frac{1}{3}$ times the first equation from the third equation. (At this point it might seem that adding equations occasionally rather than always subtracting them would be simpler. However, we must be consistent in order to develop an algorithm suitable for a computer.) The result is the following new system.

$$\begin{cases} 3x_1 + 2x_2 - x_3 = 7 \\ -\dfrac{1}{3}x_2 + \dfrac{11}{3}x_3 = -\dfrac{23}{3} \\ \dfrac{5}{3}x_2 - \dfrac{10}{3}x_3 = \dfrac{4}{3} \end{cases} \tag{4}$$

Before continuing, observe that systems (3) and (4) are *equivalent*. In other words, if (x_1, x_2, x_3) is a triple of numbers satisfying system (3), then it will also satisfy system (4) and vice versa. This result follows at once from the fact that "if equal quantities are added to (or subtracted from) two equal quantities, the resulting quantities are equal."

In the next step certain operations are performed on system (4). A multiple of the second equation is subtracted from the third in order to produce a zero coefficient where $\frac{5}{3}$ stands. The correct multiple is obviously -5. Afterward the result is

$$\begin{cases} 3x_1 + 2x_2 - \quad x_3 = 7 \\ \quad -\dfrac{1}{3}x_2 + \dfrac{11}{3}x_3 = -\dfrac{23}{3} \\ \qquad\qquad 15x_3 = -37 \end{cases} \tag{5}$$

The numbers $\frac{5}{3}$ and $-\frac{1}{3}$ for step one and -5 for step two are called the ***multipliers***. System (5) is equivalent to system (4) and hence also to system (3). This completes the first phase of the algorithm, the so-called ***forward elimination***. Notice that the coefficients form a two-dimensional upper-triangular array at this stage. Thus the kth row in the coefficient array begins with a string of $k - 1$ zeros.

In the second phase of the algorithm the unknowns are obtained one by one, starting at the bottom in system (5). Thus, the third equation is solved to yield $x_3 = -37/15$. If this value of x_3 is substituted in the second equation, the result is

$$-\frac{1}{3}x_2 + \left(\frac{11}{3}\right)\left(\frac{-37}{15}\right) = -\frac{23}{3}$$

from which we obtain $x_2 = -62/15$. Finally, the numerical values of x_2 and x_3 are substituted into the first equation, yielding

$$3x_1 + 2\left(\frac{-62}{15}\right) - \left(\frac{-37}{15}\right) = 7$$

from which we obtain $x_1 = 64/15$. So the solution to the original system (3) is

$$\begin{cases} x_1 = \dfrac{64}{15} \approx 4.267 \\ x_2 = -\dfrac{62}{15} \approx -4.133 \\ x_3 = -\dfrac{37}{15} \approx -2.467 \end{cases}$$

This phase is termed ***back substitution***.

Before continuing, we introduce the conventions of matrix algebra, which can often simplify notation and enhance understanding. (This material is *not* essential to what follows.)

A ***matrix*** is any rectangular array of numbers. For example, here are three matrices.

$$\begin{bmatrix} \frac{1}{5} & \frac{2}{7} & -1 \\ 3 & 2 & \frac{1}{8} \\ -\frac{5}{6} & \frac{2}{5} & 3 \end{bmatrix}, \quad \begin{bmatrix} 1, & 6, & \frac{9}{8}, & -5 \end{bmatrix}, \quad \begin{bmatrix} \frac{11}{2} & \frac{4}{9} \\ \frac{2}{3} & -\frac{7}{8} \\ \pi & e \\ \frac{1}{\pi} & \frac{1}{e} \end{bmatrix}$$

The shape of a matrix is described by giving first the number of rows and next the number of columns. (The columns are the vertical subarrays and the rows are the horizontal subarrays.) The three examples shown are, respectively, 3×3, 1×4, and 4×2. An $n \times 1$ matrix is called a *vector* (or *column vector*), and a $1 \times n$ matrix is called a *row vector*. The numbers used to describe the shape of a matrix are its *dimensions* or *order*. The numbers that constitute a matrix are its *elements, components,* or *entries*.

If $\mathbf{A}$ is a matrix, the notation a_{ij} or $A(i, j)$ is used to denote the element standing in the ith row and jth column of $\mathbf{A}$. For instance, if $\mathbf{A}$ denotes the first of the three matrices shown on p. 115, then $a_{11} = \frac{1}{5}$, $a_{12} = \frac{2}{7}$, $a_{23} = \frac{1}{8}$, and so on. The *transpose* of a matrix $\mathbf{A}$ is denoted $\mathbf{A}^T$. The rows and columns of $\mathbf{A}$ are interchanged in $\mathbf{A}^T$—that is, $\mathbf{A}^T = (a_{ji})$. As examples, note that

$$[1 \quad 2 \quad 3 \quad 4]^T = \begin{bmatrix} 1 \\ 2 \\ 3 \\ 4 \end{bmatrix}$$

$$\begin{bmatrix} 2 & 4 & 9 \\ 5 & 7 & 3 \\ 10 & 6 & 2 \end{bmatrix}^T = \begin{bmatrix} 2 & 5 & 10 \\ 4 & 7 & 6 \\ 9 & 3 & 2 \end{bmatrix}$$

If $\mathbf{A} = \mathbf{A}^T$, then $\mathbf{A}$ is said to be *symmetric*.

Equality of matrices is defined as follows. Two matrices $\mathbf{A}$ and $\mathbf{B}$ are equal if they have the same dimensions and the same elements. More formally, if $\mathbf{A}$ is an $n \times m$ matrix and $\mathbf{B}$ is a $p \times q$ matrix, the equation $\mathbf{A} = \mathbf{B}$ will mean

1. $n = p$
2. $m = q$
3. $a_{ij} = b_{ij}$ for all $i \in \{1, \ldots, n\}$ and all $j \in \{1, \ldots, m\}$.

We *add* or *subtract matrices* of the same dimensions by adding or subtracting the corresponding elements; that is, $\mathbf{C} = \mathbf{A} \pm \mathbf{B}$ means $c_{ij} = a_{ij} \pm b_{ij}$.

Only a few further definitions are needed at the moment. The product $\mathbf{C} = \mathbf{AB}$ of an $n \times m$ matrix $\mathbf{A}$ and a $p \times q$ matrix $\mathbf{B}$ is defined when $m = p$. The elements of the product are given by the rule

$$c_{ij} = \sum_{k=1}^{m} a_{ik} b_{kj} \qquad (1 \leqq i \leqq n,\ 1 \leqq j \leqq q)$$

The product matrix $\mathbf{C}$ is of dimensions $n \times q$. Thus an $n \times m$ matrix times an $m \times q$ matrix is an $n \times q$ matrix. Matrix multiplication is not a commutative operation; that is, we must not expect $\mathbf{AB}$ to equal $\mathbf{BA}$. For instance, if $\mathbf{v}$ is an n-component vector, then $\mathbf{v}^T\mathbf{v}$ is a scalar (a 1×1 matrix), but $\mathbf{v}\mathbf{v}^T$ is a symmetric matrix—that is, $\mathbf{v}^T\mathbf{v} = \sum_{i=1}^{n} v_i^2$ but $\mathbf{v}\mathbf{v}^T = (v_i v_j)_{n \times n}$.

If two square matrices are multiplied together and the resulting matrix is the unit diagonal matrix, then each is the *inverse* of the other. Such a diagonal matrix is the *identity* matrix denoted by **I**. A matrix that does not have an inverse is *singular*. If **A** is a square matrix, its inverse is denoted by $\mathbf{A}^{-1}$. Moreover, **A** is singular if and only if the determinant of **A** is zero; that is, $\mathbf{A}^{-1}$ exists if and only if $\det(\mathbf{A}) \neq 0$.

As examples of products of matrices, we have

$$[1, \quad 1, \quad 1, \quad 1] \begin{bmatrix} 1 \\ 1 \\ 1 \\ 1 \end{bmatrix} = [4]$$

$$\begin{bmatrix} 3 & 2 & -1 \\ 5 & 3 & 2 \\ -1 & 1 & -3 \end{bmatrix} \begin{bmatrix} x_1 \\ x_2 \\ x_3 \end{bmatrix} = \begin{bmatrix} 3x_1 + 2x_2 - x_3 \\ 5x_1 + 3x_2 + 2x_3 \\ -x_1 + x_2 - 3x_3 \end{bmatrix}$$

$$\begin{bmatrix} 1 & 0 & 0 \\ -\frac{5}{3} & 1 & 0 \\ -8 & 5 & 1 \end{bmatrix} \begin{bmatrix} 3 & 2 & -1 \\ 5 & 3 & 2 \\ -1 & 1 & -3 \end{bmatrix} = \begin{bmatrix} 3 & 2 & -1 \\ 0 & -\frac{1}{3} & \frac{11}{3} \\ 0 & 0 & 15 \end{bmatrix}$$

The reader should verify them and note how the second and third relate to the example of Gaussian elimination given earlier. In particular, we see that the system of equations (3) can be written succinctly in matrix form as

$$\mathbf{Ax} = \mathbf{b} \tag{6}$$

where

$$\mathbf{A} = \begin{bmatrix} 3 & 2 & -1 \\ 5 & 3 & 2 \\ -1 & 1 & -3 \end{bmatrix} \qquad \mathbf{x} = \begin{bmatrix} x_1 \\ x_2 \\ x_3 \end{bmatrix} \qquad \mathbf{b} = \begin{bmatrix} 7 \\ 4 \\ -1 \end{bmatrix}$$

Furthermore, the operations that led from system (3) to system (5) could be effected by an appropriate matrix multiplication. The forward part of the Gaussian elimination process (in the naive form considered above) can be interpreted as starting from $\mathbf{Ax} = \mathbf{b}$ and proceeding to $\mathbf{MAx} = \mathbf{Mb}$, where **M** is an appropriate matrix chosen so that **MA** is upper triangular.

The first step of the naive Gaussian elimination is equivalent to multiplying (6) by a lower-triangular matrix $\mathbf{M}_1$; that is, system (4) is

$$\mathbf{M}_1\mathbf{Ax} = \mathbf{M}_1\mathbf{b}$$

where

$$\mathbf{M}_1 = \begin{bmatrix} 1 & 0 & 0 \\ -\frac{5}{3} & 1 & 0 \\ \frac{1}{3} & 0 & 1 \end{bmatrix}$$

Notice the special form of $\mathbf{M}_1$. The diagonal elements are all ones and the only other nonzero elements are in the first column. These numbers are the negative of the multipliers in step one of the forward elimination. Continuing, step 2 resulted in system (5), which is equivalent to

$$\mathbf{M}_2\mathbf{M}_1\mathbf{A}\mathbf{x} = \mathbf{M}_2\mathbf{M}_1\mathbf{b}$$

where

$$\mathbf{M}_2 = \begin{bmatrix} 1 & 0 & 0 \\ 0 & 1 & 0 \\ 0 & 5 & 1 \end{bmatrix}$$

Again $\mathbf{M}_2$ differs from an identity matrix by the presence of the negative of the multipliers in the second column from the diagonal down. Now the forward elimination is complete, and system (5) has an upper-triangular coefficient matrix $\mathbf{U}$ such that

$$\mathbf{M}_2\mathbf{M}_1\mathbf{A} = \begin{bmatrix} 3 & 2 & -1 \\ 0 & -\frac{1}{3} & \frac{11}{3} \\ 0 & 0 & 15 \end{bmatrix} \equiv \mathbf{U}$$

Multiplying this equation by $\mathbf{M}_2^{-1}$ and then by $\mathbf{M}_1^{-1}$ gives

$$\mathbf{A} = \mathbf{M}_1^{-1}\mathbf{M}_2^{-1}\mathbf{U} = \mathbf{L}\mathbf{U} \tag{7}$$

where

$$\mathbf{L} \equiv \mathbf{M}_1^{-1}\mathbf{M}_2^{-1} = \begin{bmatrix} 1 & 0 & 0 \\ \frac{5}{3} & 1 & 0 \\ -\frac{1}{3} & 0 & 1 \end{bmatrix} \begin{bmatrix} 1 & 0 & 0 \\ 0 & 1 & 0 \\ 0 & -5 & 1 \end{bmatrix} = \begin{bmatrix} 1 & 0 & 0 \\ \frac{5}{3} & 1 & 0 \\ -\frac{1}{3} & -5 & 1 \end{bmatrix}$$

Since $\mathbf{M}_i$ is of such a special form, its inverse is easy to obtain by changing signs of the multipliers. By (7), we see that $\mathbf{A}$ is *factored* or *decomposed* into a unit lower-triangular matrix $\mathbf{L}$ and an upper-triangular matrix $\mathbf{U}$. The matrix $\mathbf{L}$ consists of the multipliers located in the position of the element they annihilated from $\mathbf{A}$, of unit diagonal elements, and of zero upper-triangular elements. The matrix $\mathbf{U}$ is upper triangular (not necessarily unit diagonal) and is the final coefficient matrix after the forward elimination. Notice that we do not form $\mathbf{M} = \mathbf{M}_2\mathbf{M}_1$ and then determine $\mathbf{M}^{-1} = \mathbf{L}$. (Why?)

The naive Gaussian elimination procedure can be regarded as solving

$$\mathbf{L}\mathbf{y} = \mathbf{b} \tag{8}$$

for $\mathbf{y}$ and solving

$$\mathbf{U}\mathbf{x} = \mathbf{y} \tag{9}$$

for $\mathbf{x}$. This is particularly useful for problems involving the same coefficient matrix $\mathbf{A}$ and many different right-hand vectors $\mathbf{b}$. In this case, it would be advantageous to save the multipliers so that the vector **y** could be easily determined for each **b** by Eq. (8) and then (9) solved for each **x**. What we called "back substitution" is the process of solving system (9).

Problems 6.1

1. Solve each of the following systems, using naive Gaussian elimination—that is, forward elimination and back substitution. Carry four significant figures.

(a) $$\begin{cases} 3x_1 + 4x_2 + 3x_3 = 16 \\ x_1 + 5x_2 - x_3 = -12 \\ 6x_1 + 3x_2 + 7x_3 = 102 \end{cases}$$

(b) $$\begin{cases} 3x + 2y - 5z = 4 \\ 2x - 3y + z = 8 \\ x + 4y - z = -3 \end{cases}$$

(c) $$\begin{bmatrix} 1 & -1 & 2 & 1 \\ 3 & 2 & 1 & 4 \\ 5 & 8 & 6 & 3 \\ 4 & 2 & 5 & 3 \end{bmatrix} \begin{bmatrix} x_1 \\ x_2 \\ x_3 \\ x_4 \end{bmatrix} = \begin{bmatrix} 1 \\ 1 \\ 1 \\ -1 \end{bmatrix}$$

2. Form the following products and observe that matrix multiplication is not commutative.

$$\begin{bmatrix} 3 & 1 & 7 \\ 2 & 4 & -5 \\ 1 & -3 & 2 \end{bmatrix} \begin{bmatrix} -1 & -3 & 2 \\ 1 & 1 & 1 \\ -3 & -2 & 1 \end{bmatrix}$$

$$\begin{bmatrix} -1 & -3 & 2 \\ 1 & 1 & 1 \\ -3 & -2 & 1 \end{bmatrix} \begin{bmatrix} 3 & 1 & 7 \\ 2 & 4 & -5 \\ 1 & -3 & 2 \end{bmatrix}$$

3. Does this example "prove" that matrix multiplication is commutative? Verify that **AB** = **BA**, where

$$\mathbf{A} = \begin{bmatrix} 1 & -3 & 2 \\ 1 & 1 & 1 \\ -3 & -2 & 1 \end{bmatrix} \qquad \mathbf{B} = \begin{bmatrix} -8 & -10 & 1 \\ -1 & -4 & 4 \\ -8 & 5 & -7 \end{bmatrix}$$

4. Compute the products shown and relate them to the example of Gaussian elimination in the text.

(a) $$\begin{bmatrix} 1 & 0 & 0 \\ -\frac{5}{3} & 1 & 0 \\ -8 & 5 & 1 \end{bmatrix} \begin{bmatrix} 7 \\ 4 \\ -1 \end{bmatrix}$$

(b) $\begin{bmatrix} \frac{1}{3} & 2 & -\frac{7}{15} \\ 0 & -3 & \frac{11}{15} \\ 0 & 0 & \frac{1}{15} \end{bmatrix} \quad \begin{bmatrix} 3 & 2 & -1 \\ 0 & -\frac{1}{3} & \frac{11}{3} \\ 0 & 0 & 15 \end{bmatrix}$

(c) $\begin{bmatrix} \frac{1}{3} & 2 & -\frac{7}{15} \\ 0 & -3 & \frac{11}{15} \\ 0 & 0 & \frac{1}{15} \end{bmatrix} \quad \begin{bmatrix} 7 \\ -\frac{23}{3} \\ -37 \end{bmatrix}$

5. Apply naive Gaussian elimination to these examples and account for the failures. Solve the systems by other means if possible.

(a) $\begin{cases} 3x_1 + 2x_2 = 4 \\ -x_1 - \frac{2}{3}x_2 = 1 \end{cases}$ (b) $\begin{cases} 6x_1 - 3x_2 = 6 \\ -2x_1 + x_2 = -2 \end{cases}$

(c) $\begin{cases} 0x_1 + 2x_2 = 4 \\ x_1 - x_2 = 5 \end{cases}$ (d) $\begin{cases} x_1 + x_2 + 2x_3 = 4 \\ x_1 + x_2 + 0x_3 = 2 \\ 0x_1 + x_2 + x_3 = 0 \end{cases}$

6. As an application of Gaussian elimination, the matrix $\mathbf{A}$ can be factored or decomposed such that $\mathbf{A} = \mathbf{LU}$, where $\mathbf{L}$ is a unit lower-triangular matrix and $\mathbf{U}$ an upper-triangular matrix. Find the $\mathbf{LU}$ decomposition for

$$\mathbf{A} = \begin{bmatrix} 1 & 0 & 0 & 1 \\ 1 & 1 & 0 & -1 \\ -1 & 1 & 1 & 1 \\ 1 & -1 & 1 & -1 \end{bmatrix}$$

7. Consider

$$\mathbf{A} = \begin{bmatrix} 2 & -1 & 2 \\ 2 & -3 & 3 \\ 6 & -1 & 8 \end{bmatrix}$$

(a) Find the matrix factoring $\mathbf{A} = \mathbf{LDU}'$ where $\mathbf{L}$ is unit lower triangular, $\mathbf{D}$ is diagonal, and $\mathbf{U}'$ is unit upper triangular.

(b) Use this decomposition of $\mathbf{A}$ to solve $\mathbf{Ax} = \mathbf{b}$, where $\mathbf{b} = (-2, -5, 0)^T$.

8. (Continuation) Repeat Problem 7 for

$$\mathbf{A} = \begin{bmatrix} -2 & 1 & -2 \\ -4 & 3 & -3 \\ 2 & 2 & 4 \end{bmatrix} \qquad \mathbf{b} = \begin{bmatrix} 1 \\ 4 \\ 4 \end{bmatrix}$$

9. Consider

$$\mathbf{A} = \begin{bmatrix} 0.780 & 0.563 \\ 0.913 & 0.659 \end{bmatrix}, \quad \mathbf{b} = \begin{bmatrix} 0.217 \\ 0.254 \end{bmatrix}, \quad \mathbf{x} = \begin{bmatrix} 0.999 \\ -1.001 \end{bmatrix}, \quad \hat{\mathbf{x}} = \begin{bmatrix} 0.341 \\ -0.087 \end{bmatrix}$$

Compute *residual vectors* $\mathbf{r} = \mathbf{Ax} - \mathbf{b}, \hat{\mathbf{r}} = \mathbf{A}\hat{\mathbf{x}} - \mathbf{b}$ and decide which of $\mathbf{x}$ and $\hat{\mathbf{x}}$ is the solution vector. Now compute the *error vectors* $\mathbf{e} = \mathbf{x} - \bar{\mathbf{x}}, \hat{\mathbf{e}} = \hat{\mathbf{x}} - \bar{\mathbf{x}}$, where $\bar{\mathbf{x}} = (1, -1)^T$ is the exact solution. Discuss the importance of this example.

10. Show that if $\mathbf{E} = (\mathbf{b} - \mathbf{Ax})\mathbf{x}^T/(\mathbf{x}^T\mathbf{x})$ where $\mathbf{x}^T\mathbf{x} \neq 0$, then $(\mathbf{A} + \mathbf{E})\mathbf{x} = \mathbf{b}$.
11. Let $\mathbf{x} = (1, 1, 1, 1)^T$. Find a vector $\mathbf{u} = (u_1, u_2, u_3, u_4)^T$ such that $\mathbf{u}^T\mathbf{u} = 1$ and $(\mathbf{I} - 2\mathbf{uu}^T)\mathbf{x} = (0, 2, 0, 0)^T$.
12. Given

$$A = \begin{bmatrix} \alpha & \beta \\ 0 & \alpha \end{bmatrix}$$

for constants $\alpha \neq 0$ and β, determine the general form of $\mathbf{A}^{\pm m}$ for any positive integer m.

Computer Problems 6.1

1. Write and test a subroutine that carries out the addition of two $n \times m$ matrices $\mathbf{A}$ and $\mathbf{B}$ and stores the result in $\mathbf{C}$.
2. (Continuation) Repeat Problem 1 for multiplication, $\mathbf{C} = \mathbf{AB}$, where $\mathbf{A}$ is $n \times k$, $\mathbf{B}$ is $k \times m$, and $\mathbf{C}$ is $n \times m$.
3. Given a vector $\mathbf{v} = (v_1, \ldots, v_n)$ and a permutation $p = (p_1, \ldots, p_n)$ of integers $1, \ldots, n$, can we form the vector $\tilde{\mathbf{v}} = (v_{p_1}, v_{p_2}, \ldots, v_{p_n})$ by overwriting $\mathbf{v}$ and not involving another array in memory? If so, write and test the code for doing it. If not, use another array and test.
4. Using one of the coefficient matrices $\mathbf{A}$ in this section, write a program to verify numerically that the following four lines of code perform naive Gaussian elimination on $\mathbf{A}$.

```
  DO 2 K = 1,N-1
  DO 2 I = K+1,N
  DO 2 J = K+1,N
2 A(I,J) = A(I,J) - (A(I,K)/A(K,K))*A(K,J)
```

Is it necessary to compute the zeros? Modify code to do so.

6.2 Gaussian Elimination with Pivoting

To see why direct programming of the naive Gaussian elimination algorithm is unsatisfactory, consider the system

$$\begin{cases} 0x_1 + x_2 = 1 \\ x_1 + x_2 = 2 \end{cases} \tag{1}$$

A program coded mechanically along the lines described in Sec. 6.1 would attempt to subtract some multiple of the first equation from the second in order to produce a zero coefficient of x_1 in the second equation. This, of course, is impossible. So the algorithm fails if $a_{11} = 0$.

If a numerical procedure actually fails for some values of the data, then the procedure is probably untrustworthy for values of the data *near* the failing values. To test this dictum, consider the system

$$\begin{cases} \varepsilon x_1 + x_2 = 1 \\ x_1 + x_2 = 2 \end{cases} \tag{2}$$

in which ε is a small number near zero yet different from zero. Now the naive algorithm of Sec. 6.1 works and produces first the system

$$\begin{cases} \varepsilon x_1 + \qquad\qquad x_2 = 1 \\ \left(1 - \dfrac{1}{\varepsilon}\right)x_2 = 2 - \dfrac{1}{\varepsilon} \end{cases} \tag{3}$$

In back substitution the arithmetic is as follows.

$$x_2 = \frac{2 - \dfrac{1}{\varepsilon}}{1 - \dfrac{1}{\varepsilon}}$$

$$x_1 = \frac{1 - x_2}{\varepsilon}$$

If this calculation is performed by a computer having a fixed word length, then, for small values of ε, $(2 - 1/\varepsilon)$ and $(1 - 1/\varepsilon)$ could both be computed as $-1/\varepsilon$. For example, in an 8-digit decimal machine with a 16-digit accumulator, when $\varepsilon = 10^{-9}$ it follows that $1/\varepsilon = 10^9$. In order to subtract, the computer must interpret the numbers as

$$\frac{1}{\varepsilon} = 10^9 = 0.1000\ 0000 \times 10^{10} = 0.1000\ 0000\ 0000\ 0000 \times 10^{10}$$

$$2 = 0.2000\ 0000 \times 10^1 = 0.0000\ 0000\ 0200\ 0000 \times 10^{10}$$

Thus $1/\varepsilon - 2$ is computed initially as $0.0999\ 9999\ 9800\ 0000 \times 10^{10}$ and then rounded to $0.1000\ 0000 \times 10^{10} = 1/\varepsilon$.

We conclude that, for values of ε sufficiently close to zero, the computer calculates x_2 as 1 and then x_1 as 0. Since the correct solution is

$$x_1 = \frac{1}{1 - \varepsilon} \approx 1 \qquad x_2 = \frac{1 - 2\varepsilon}{1 - \varepsilon} \approx 1$$

the relative error in the computed solution for x_1 is extremely large: 100%.

Actually, the algorithm outlined previously works well on examples (1) and (2) if the equations are first permuted:

$$\begin{cases} x_1 + x_2 = 2 \\ \qquad x_2 = 1 \end{cases} \quad \text{or} \quad \begin{cases} x_1 + x_2 = 2 \\ \varepsilon x_1 + x_2 = 1 \end{cases}$$

Indeed, the second of these systems become

$$\begin{cases} x_1 + \qquad x_2 = 2 \\ \qquad (1 - \varepsilon)x_2 = 1 - 2\varepsilon \end{cases}$$

after the forward elimination. Then from the back substitution the solution is computed as

$$\begin{cases} x_2 = \dfrac{1 - 2\varepsilon}{1 - \varepsilon} \approx 1 \\ x_1 = 2 - x_2 \approx 1 \end{cases}$$

The difficulty in system (2) is not due simply to ε being small but rather to its being small relative to other coefficients in the same row. To verify this, consider

$$\begin{cases} x_1 + \dfrac{1}{\varepsilon}x_2 = \dfrac{1}{\varepsilon} \\ x_1 + \quad x_2 = 2 \end{cases} \tag{4}$$

System (4) is mathematically equivalent to (2). The naive algorithm fails here, too, for it produces first the triangular system

$$\begin{cases} x_1 + \qquad \dfrac{1}{\varepsilon}x_2 = \dfrac{1}{\varepsilon} \\ \qquad \left(1 - \dfrac{1}{\varepsilon}\right)x_2 = 2 - \dfrac{1}{\varepsilon} \end{cases}$$

and then in the back substitution it produces the erroneous result

$$x_2 = \frac{2 - \dfrac{1}{\varepsilon}}{1 - \dfrac{1}{\varepsilon}} \approx 1$$

$$x_1 = \frac{1}{\varepsilon} - \frac{1}{\varepsilon}x_2 \approx 0$$

These simple examples should make it clear that the *order* in which we treat the equations significantly affects the accuracy of the elimination algorithm in the computer. In the naive algorithm we use the first equation to eliminate x_1 from following ones. Then we use the second equation to eliminate x_2 from the following ones, and so on. The order in which the equations are used as "operating" or "pivot" equations is the *natural* order $1, 2, 3, \ldots, n$. Note that the last equation (equation number n) is *not* used as an operating equation with the natural ordering. At no time are multiples of it added to other equations in the naive algorithm.

The algorithm now to be described employs the equations in an order that is determined by the actual system being solved. For instance, if the algorithm were asked to solve system (1) or (2), the order in which the equations would be used as pivot equations would not be the natural order (1, 2) but rather (2, 1). This order is automatically determined by the computer program.

The order in which the equations are employed is denoted by $(\ell_1, \ell_2, \ldots, \ell_n)$, with ℓ_n not actually being used in the forward elimination phase. Here the ℓ_i are integers from 1 to n in a possibly different order. The array $(\ell_1, \ldots, \ell_n)$ is called the *index array*. The strategy for determining the index array is termed *scaled partial pivoting*.

At the beginning a scale factor must be computed for each row in the system. Referring to the notation in Sec. 6.1, we define

$$s_i = \max_{1 \leqq j \leqq n} |a_{ij}| \qquad (1 \leqq i \leqq n)$$

These n numbers are recorded in the *scale array*.

In starting the forward elimination process, we do not arbitrarily use the first row as the operating or *pivot row*. Instead we use the row for which the ratio $|a_{i1}|/s_i$ is greatest. Therefore ℓ_1 is the index for which this ratio is greatest. Having determined ℓ_1, appropriate multiples of row ℓ_1 are subtracted from the other rows in order to create zeros as coefficients of each x_1 except in the pivot row.

The best way of keeping track of the indices is as follows. At the beginning define $(\ell_1, \ldots, \ell_n) = (1, \ldots, n)$. Now determine the index j for which $|a_{\ell_j 1}|/s_{\ell_j}$ is a maximum and interchange ℓ_j with ℓ_1 in $\boldsymbol{\ell}$. Next, $a_{\ell_i 1}/a_{\ell_1 1}$ times row ℓ_1 is subtracted from row ℓ_i for $2 \leqq i \leqq n$. It is important to note that only entries in $\boldsymbol{\ell}$ are being interchanged and *not* those in $\mathbf{A}$.

In the second step the numbers $|a_{\ell_i 2}|/s_{\ell_i}$ are scanned for $2 \leqq i \leqq n$. If j is the index for which this ratio is a maximum, interchange ℓ_j with ℓ_2 in $\boldsymbol{\ell}$. Then subtract $a_{\ell_i 2}/a_{\ell_2 2}$ times row ℓ_2 from row ℓ_i, for $3 \leqq i \leqq n$.

We are not quite ready to write a program for this, but let us consider what has been outlined so far in a concrete example. Consider this system.

$$\begin{cases} 2x_1 + 3x_2 - 6x_3 = -1 \\ x_1 - 6x_2 + 8x_3 = \dfrac{47}{39} \\ 3x_1 - 2x_2 + x_3 = 2 \end{cases}$$

The index array is (1, 2, 3) at the beginning. The scale array is (6, 8, 3) and does not change. To determine the first pivot row, we need to look at the three ratios $|a_{\ell_1 1}|/s_{\ell_1} = \frac{2}{6}$, $|a_{\ell_2 1}|/s_{\ell_2} = \frac{1}{8}$, and $|a_{\ell_3 1}|/s_{\ell_3} = \frac{3}{3}$. The largest of these occurs for the index $j = 3$. So row 3 is the first pivot row. In the index array ℓ_1 and ℓ_3 are interchanged, producing (3, 2, 1) as the new index array. Thus $\ell_1 = 3$. Now appropriate multiples of row 3 are subtracted from rows 2 and 1 so as to create zeros in $\mathbf{A}$. Explicitly, $\frac{1}{3}$ times row 3 is subtracted from row 2 and $\frac{2}{3}$ times row 3 is subtracted from row 1. The result is

$$\begin{cases} \dfrac{13}{3}x_2 - \dfrac{20}{3}x_3 = -\dfrac{7}{3} \\ -\dfrac{16}{3}x_2 + \dfrac{23}{3}x_3 = \dfrac{7}{13} \\ 3x_1 - 2x_2 + x_3 = 2 \end{cases}$$

In the next step we scan the ratios $|a_{\ell_i 2}|/s_{\ell_i}$ for $i = 2, 3$ and look for the largest value. The ratios are $(16/3)/8$ and $(13/3)/6$. The larger is associated with $j = 3$ and so ℓ_3 is to be interchanged with ℓ_2 in the index array. Thus the index array becomes (3, 1, 2). Next, a multiple of row 1 is subtracted from row 2. The appropriate multiple is $-(16/3)/(13/3) = -16/13$. The result is

$$\begin{cases} \quad\quad\quad \dfrac{13}{3}x_2 - \dfrac{20}{3}x_3 = -\dfrac{7}{3} \\ \quad\quad\quad\quad\quad\quad -\dfrac{7}{13}x_3 = -\dfrac{7}{3} \\ 3x_1 - 2x_2 + x_3 = 2 \end{cases}$$

So ends the forward elimination phase. The solution is obtained through back substitution, using the equations in the order 2, 1, 3. The index array (3, 1, 2) gives the order in which the equations were selected as pivot equations. Now reading the entries in the index array from the last to the first, we have the order in which the back substitution is to be performed. Namely, the second equation gives

$$x_3 = \frac{13}{3} \approx 4.333$$

the first equation gives

$$x_2 = \frac{3}{13}\left[-\frac{7}{3} + \left(\frac{20}{3}\right)\left(\frac{13}{3}\right)\right] = \frac{239}{39} \approx 18.385$$

and, finally, the third equation gives

$$x_1 = \frac{1}{3}\left[2 - \left(\frac{13}{3}\right) + 2\left(\frac{239}{39}\right)\right] = \frac{387}{117} \approx 3.308$$

The algorithm, as it will be programmed, carries out the forward elimination phase on the coefficient array **A** only. The right-hand side **B** is treated in the next phase. This method is adopted because it is more efficient if several systems must be solved with the same array **A** but differing **B**'s. Because we wish to treat **B** later, it is necessary to store not only the index array but also the various multipliers that are used. For example, in the first step, multiples of row ℓ_1 are subtracted from the other rows. Explicitly, $a_{\ell_i 1}/a_{\ell_1 1}$ times row ℓ_1 is subtracted from row ℓ_i $(2 \leqq i \leqq n)$. These multipliers are conveniently stored in the array **A** in the positions where the zero entries would have been created.

The subroutine that carries out Gaussian elimination with scaled partial pivoting on the square array **A** is called subroutine GAUSS. Its calling sequence is (A,IA,N,L,S), where **A** is the $n \times n$ coefficient array or matrix, IA is the row dimension of array **A**, N is, of course, n, L is the $\boldsymbol{\ell}$ array or index array, and S is the **s** array or scale array. Although the scale array is not required as output, it must occur in the calling sequence because it has a variable dimension.

```
      SUBROUTINE GAUSS(A,IA,N,L,S)
      DIMENSION  A(IA,N),L(N),S(N)
      DO 2 I = 1,N
      L(I) = I
      S(I) = 0.0
      DO 2 J = 1,N
    2 S(I) = AMAX1(S(I),ABS(A(I,J)))
      DO 4 K = 1,N-1
      RMAX = 0.0
      DO 3 I = K,N
      R = ABS(A(L(I),K))/S(L(I))
      IF(R .LE. RMAX)  GO TO 3
      J = I
      RMAX = R
    3 CONTINUE
      LK = L(J)
      L(J) = L(K)
      L(K) = LK
      DO 4 I = K+1,N
      XMULT = A(L(I),K)/A(LK,K)
      A(L(I),K) = XMULT
      DO 4 J = K+1,N
    4 A(L(I),J) = A(L(I),J) - XMULT*A(LK,J)
      RETURN
      END
```

A detailed explanation of the above program is now presented. The parameter **IA** is the row dimension of the array **A**—that is, the row value used in the **DIMENSION** statement for **A** in the calling program. This parameter allows the subroutine to be called several times within a single run with different-sized systems using the same array **A**. In the DO loop ending at statement 2 the initial form of the index array is being established—namely, $\ell_i = i$. Also, the scale array S is being computed. The Fortran function **AMAX1** computes the larger of its two real arguments.

The statement DO 4 K=1,N-1 initiates the principal outer loop. The index k is the subscript of the variable whose coefficients will be made 0 in the array **A**. That is, k is the index of the column in which new zeros are to be created. Remember that the zeros in the array **A** do not actually appear because those storage locations are used for the multipliers. This fact can be seen in the line of the subprogram where **XMULT** is stored in the array **A**.

Once k has been set, the first task is to select the correct pivot row, which is done by computing $|a_{\ell_i k}|/s_{\ell_i}$ for $i = k, k+1, \ldots, n$. The next set of lines in the program are calculating this greatest ratio, called **RMAX** in the program, and are calculating the index j where it occurs. The next group of lines is performing the interchange of ℓ_k and ℓ_j in the array L.

The arithmetic modifications in the array **A** due to subtracting multiples of row ℓ_k from rows $\ell_{k+1}, \ldots, \ell_n$ all occur in the final lines of code. First the multiplier is computed and stored. Then the subtraction occurs in a DO loop.

As noted, the subroutine called GAUSS performs the forward elimination phase on the square array of coefficients for the system of linear equations. The processing of the array B is done in a separate routine called SOLVE. In this subroutine the array B is first subjected to the forward elimination process, using, of course, the array L, which is obtained from subroutine GAUSS, and the multipliers that are stored in the array A. (Caution: Values in array A that result as *output* from subroutine GAUSS are not the same as those in array A at *input*. If the original array must be retained, we can store a duplicate of it in another array.)

After the array B has been processed in the forward elimination, the back substitution is carried out. It begins by solving the equation

$$a_{\ell_n, n} x_n = b_{\ell_n} \tag{5}$$

whence $x_n = b_{\ell_n} / a_{\ell_n}$. After $x_n, x_{n-1}, \ldots, x_{n-i+1}$ are obtained, x_{n-i} is found from

$$a_{\ell_{n-i}, n-i} x_{n-i} + a_{\ell_{n-i}, n-i+1} x_{n-i+1} + \cdots + a_{\ell_{n-i}, n} x_n = b_{\ell_{n-i}}$$

whose solution is

$$x_{n-i} = \frac{b_{\ell_{n-i}} - a_{\ell_{n-i}, n-i+1} x_{n-i+1} - \cdots - a_{\ell_{n-i}, n} x_n}{a_{\ell_{n-i}, n-i}} \tag{6}$$

The subprogram for processing the array B and performing back substitution is given next.

```
      SUBROUTINE SOLVE(A, IA, N, L, B, X)
      DIMENSION  A(IA, N), L(N), B(N), X(N)
      DO 2 J = 1, N-1
      DO 2 I = J+1, N
    2 B(L(I)) = B(L(I)) - A(L(I), J)*B(L(J))
      X(N) = B(L(N))/A(L(N), N)
      DO 4 I = 1, N-1
      SUM = B(L(N-I))
      DO 3 J = N-I+1, N
    3 SUM = SUM - A(L(N-I), J)*X(J)
    4 X(N-I) = SUM/A(L(N-I), N-I)
      RETURN
      END
```

Here the DO loop ending at statement 2 carries out the forward elimination process on array B, using arrays A and L that result from subroutine GAUSS. At the next line we see the Fortran solution of Eq. (5). The DO loop ending at statement 4 carries out Eq. (6). The variable SUM is a temporary variable for accumulating the numerator in Eq. (6).

As with most programs in this book, those in this chapter contain only the basic ingredients for mathematical software. They are not suitable as "production" codes, for various reasons. For example, several standard procedures for optimizing code are ignored. Furthermore, the subroutines do not give warnings of difficulties that may be encountered, such as division by zero! Finally, the programs tacitly assume that $n \geqq 2$. *General* subprograms should be *robust*—that is, anticipate every possible situation and deal with each in a prescribed way. (See Computer Problem 8.)

Solving large systems of linear equations can be expensive on a computer. To understand why, let us perform an operation count on the two algorithms whose codes have been given. We count only multiplications and divisions (long operations), for they are more time consuming. Furthermore, we lump multiplications and divisions together even though division is slower than multiplication. For instance, the execution times in a modern large computer are as follows (a microsecond is 10^{-6} second).

Multiplication:	1.0 microsecond
Division :	2.9 microseconds
Addition :	0.4 microsecond
Subtraction :	0.4 microsecond

The actual time for addition and subtraction may be twice as long as indicated, since a normalization instruction usually follows each. These numbers are machine and compiler dependent and simply illustrate the relative times for the four basic arithmetic operations on a particular computer.

Consider first subroutine GAUSS. In step 1 the choice of a pivot element requires the calculation of n ratios—that is, n divisions. Then for rows $\ell_2, \ell_3, \ldots, \ell_n$ we first compute a multiplier and then subtract from row ℓ_i that multiplier times row ℓ_1. The zero that is being created in this process is *not* computed. So the elimination requires $n - 1$ multiplications per row. If we include the calculation of the multiplier, there are n long operations (divisions or multiplications) per row. There are $n - 1$ rows to be processed for a total of $n(n - 1)$ operations. If we add the cost of computing the pivot, a total of n^2 operations is needed for step 1.

The next step is like step 1 except that row ℓ_1 is not affected; neither is the column of zeros created in step 1. So step 2 will require $(n - 1)^2$ multiplications or divisions, since it operates on a system without row ℓ_1 and without column 1. Continuing this reasoning, we conclude that the total number of long operations for subroutine GAUSS is

$$n^2 + (n-1)^2 + (n-2)^2 + \cdots + 4^2 + 3^2 + 2^2 = \frac{n}{6}(n+1)(2n+1) - 1$$

(The derivation of this formula is outlined in Problem 7.) Note that the number of long operations in this subroutine grows like $n^3/3$, the dominant term.

Now consider subroutine SOLVE. The forward processing of the array B involves $n - 1$ steps. The first step contains $n - 1$ multiplications, the second $n - 2$ multiplications, and so on. The total of the forward processing of array B is thus

$$(n-1) + (n-2) + \cdots + 3 + 2 + 1 = \frac{n}{2}(n-1)$$

(See Problem 6.) In back substitution one long operation is involved in the first step, two in the second step, and so on. The total is

$$1 + 2 + 3 + \cdots + n = \frac{n}{2}(n+1)$$

Thus subroutine SOLVE involves altogether n^2 long operations. To summarize,

The forward elimination phase of the Gaussian elimination algorithm with scaled partial pivoting, if applied only to the $n \times n$ coefficient array, involves

$$\frac{n}{6}(n + 1)(2n + 1) - 1 \approx \frac{n^3}{3}$$

long operations (multiplications or divisions). The back substitution phase entails n^2 long operations.

Problems 6.2

b) 1. Show how Gaussian elimination with scaled partial pivoting works on the systems given in Problem 1 of Sec. 6.1. What are the contents of the index array at each step?

2. Show how Gaussian elimination with scaled partial pivoting works on the following matrix **A**.

$$\begin{bmatrix} 2 & 3 & -4 & 1 \\ 1 & -1 & 0 & -2 \\ 3 & 3 & 4 & 3 \\ 4 & 1 & 0 & 4 \end{bmatrix}$$

3. Count the number of operations in the following code.

```
      DO 2 I = 1,N
      DO 2 J = 1,I
    2 Z = Z + A(I,J)*X(I,J)
```

4. Count the number of divisions in the GAUSS subroutine. Count the number of multiplications. Count the number of additions or subtractions. Using execution times given in the text, write a function of n that represents the time used in these arithmetic operations.

5. Considering long operations only and assuming 1-microsecond execution time for all long operations, give the approximate execution times and costs for subroutine GAUSS when $n = 10$, 10^2, 10^3, 10^4. Use only the dominant term in the operation count. Estimate costs at $500 per hour.

6. Derive the formula

$$\sum_{k=1}^{n} k = \frac{n}{2}(n + 1)$$

Hint: Write

$$\begin{aligned} 2s &= (1 + 2 + \cdots + n) + [n + (n - 1) + \cdots + 2 + 1] \\ &= (n + 1) + (n + 1) + \cdots \end{aligned}$$

Or use induction.

7. Derive the formula

$$\sum_{k=1}^{n} k^2 = \frac{n}{6}(n + 1)(2n + 1)$$

Hint: Induction is probably easiest.

8. Determine the number of multiplications and additions required to multiply an $m \times n$ matrix with an n-component vector.

9. Confirm that the following code carries out the forward phase of the naive version of Gaussian elimination. Here array B has been incorporated into array A as the $(n + 1)$st column.

```
    DO 2 K = 1,N-1
    DO 2 I = K+1,N
    DO 2 J = K+1,N+1
  2 A(I,J) = A(I,J) - (A(I,K)/A(K,K))*A(K,J)
```

10. (Continuation) Make the code in Problem 9 more efficient by computing the multiplier a_{ik}/a_{kk} outside the J loop. It can be stored in the a_{ik} location.
11. (Continuation) Show that if pivoting is carried out with an index aray $(\ell_1, \ell_2, \ldots, \ell_n)$, then the code of Problem 9 must be modified to read

```
    DO 2 K = 1,N-1
    DO 2 I = K+1,N
    A(L(I),K) = A(L(I),K)/A(L(K),K)
    DO 2 J = K+1,N+1
  2 A(L(I),J) = A(L(I),J) - A(L(I),K)*A(L(K),J)
```

12. What modifications would make subroutine GAUSS more efficient if division were *very* slow compared to multiplication?

Computer Problems 6.2

1. The *Hilbert matrix* of order n is the $n \times n$ matrix (two-dimensional array) defined by $a_{ij} = (i + j - 1)^{-1}$ $(1 \leqq i, j \leqq n)$. It is often used for test purposes because of its *ill-conditioned* nature. Define $b_i = \sum_{j=1}^{n} a_{ij}$. Then the solution of the system of equations $\sum_{j=1}^{n} a_{ij}x_j = b_i$ $(1 \leqq i \leqq n)$ is $\mathbf{x} = (1, 1, \ldots, 1)^T$. Verify this. Select some values of n in the range $2 \leqq n \leqq 12$, solve the system of equations for $\mathbf{x}$ using subroutine GAUSS and SOLVE, and see whether the result is as predicted. Do the case $n = 2$ by hand to see what difficulties will occur in the computer.
2. If an $n \times n$ matrix $\mathbf{A}$ has an inverse, it is an $n \times n$ matrix $\mathbf{X}$ with the property $\mathbf{AX} = \mathbf{I}$, where $\mathbf{I}$ is the identity matrix.

$$\mathbf{I} = \begin{bmatrix} 1 & 0 & 0 & \cdots & 0 \\ 0 & 1 & 0 & \cdots & 0 \\ 0 & 0 & 1 & \cdots & 0 \\ \vdots & \vdots & \vdots & & \vdots \\ 0 & 0 & 0 & \cdots & 1 \end{bmatrix}$$

If $\mathbf{X}^{(j)}$ denotes the jth column of $\mathbf{X}$ and $\mathbf{I}^{(j)}$ denotes the jth column of $\mathbf{I}$, show that $\mathbf{AX}^{(j)} = \mathbf{I}^{(j)}$. Use this idea to write a subroutine for determining $\mathbf{X}$. Employ subroutine GAUSS once and subroutine SOLVE n times. Test your program on the following matrix of order 20.

$$\mathbf{A} = \begin{bmatrix} -2 & 1 & & & & & \\ 1 & -2 & 1 & & & 0 & \\ & 1 & -2 & 1 & & & \\ & & \ddots & \ddots & \ddots & & \\ & 0 & & 1 & -2 & 1 \\ & & & & 1 & -2 \end{bmatrix}$$

3. Define an $n \times n$ array by $a_{ij} = -1 + 2 \min \{i, j\}$. Then set up the array b in such a way that the solution of the system $\sum_{j=1}^{n} a_{ij}x_j = b_i$ $(1 \leqq i \leqq n)$ is $x_j = 1$ $(1 \leqq j \leqq n)$. Test subroutines GAUSS and SOLVE on this system for a moderate value of n, say $n = 15$.
4. Consider the system

$$\begin{bmatrix} 0.4096 & 0.1234 & 0.3678 & 0.2943 \\ 0.2246 & 0.3872 & 0.4015 & 0.1129 \\ 0.3645 & 0.1920 & 0.3781 & 0.0643 \\ 0.1784 & 0.4002 & 0.2786 & 0.3927 \end{bmatrix} \begin{bmatrix} x_1 \\ x_2 \\ x_3 \\ x_4 \end{bmatrix} = \begin{bmatrix} 0.4043 \\ 0.1550 \\ 0.4240 \\ 0.2557 \end{bmatrix}$$

   Solve it by Gaussian elimination with scaled partial pivoting, using routines GAUSS and SOLVE. Use a DATA statement to input values.
5. (Continuation) Assume that an error was made when the coefficient matrix was typed and that a single digit was mistyped—namely, change 0.3645 to 0.3345. Solve this system and notice the effect of this small change. Explain.
6. Without changing the parameter list, rewrite and test subroutine GAUSS so that it does both forward elimination and back substitution. Increase the size of array A and store the right-hand side B in the $(n + 1)$st column of A. Also, return the solution in this column.
7. Computer memory can be minimized by using a different storage mode when the coefficient matrix is symmetric. An $n \times n$ symmetric matrix $\mathbf{A} = (a_{ij})$ has the property that $a_{ij} = a_{ji}$ so that only the elements on and below the main diagonal need be stored in a vector of length $n(n + 1)/2$. The elements of the matrix $\mathbf{A}$ are placed in a vector $\mathbf{V} = (v_k)$ in this order: a_{11}, a_{21}, a_{22}, a_{31}, a_{32}, a_{33}, . . . , $a_{n,n}$. Storing a matrix in this way is known as *symmetric storage mode* and effects a savings of $n(n - 1)/2$ memory locations. Here $a_{ij} = v_k$, where $k = \frac{1}{2}i(i - 1) + j$ for $i \geqq j$. Verify these statements.

   Write and test SUBROUTINE GAUSYM(N,V,L,S) and SUBROUTINE SOLSYM(N,V,L,B), which are analogous to routines GAUSS and SOLVE except that the coefficient matrix is stored in symmetric storage mode in a one-dimensional array V and the solution is returned in array B.
8. Modify subroutine GAUSS and SOLVE so that they are more like production software. Some suggested changes are as follows.

   (a) Allow 1×1 linear systems. (b) Skip elimination if A(L(I),K)=0.

   (c) Add an error parameter IERR to the parameter list and perform error checking (e.g., on division by zero or a row of zeros). Test the modified code on linear systems of varying sizes.
9. The determinant of a square matrix $\mathbf{A}$ can be defined as follows. By interchanging rows and adding multiples of one row to another, get the matrix into upper triangular form $\mathbf{U}$. The product of the elements on the diagonal of $\mathbf{U}$ is the determinant of $\mathbf{A}$. As in Gaussian elimination, it is not necessary actually to interchange rows but only to keep track of the indices. Write a program to compute the determinant of a matrix. Test it on the example $a_{ij} = (i + j - 1)^{-1}$, $n = 13$.

6.3 Band Systems

In many applications, including several considered later on, linear systems having a *band* structure are encountered. Band matrices often occur in solving ordinary and partial differential equations.

Of practical importance is the *tridiagonal* matrix. Here all the nonzero elements must be on the main diagonal or on the two diagonals just above and below the main diagonal (usually called *superdiagonal* and *subdiagonal,* respectively).

$$\mathbf{A} = \begin{bmatrix} d_1 & c_1 & & & & & \\ a_1 & d_2 & c_2 & & & \mathbf{0} & \\ & a_2 & d_3 & c_3 & & & \\ & & a_3 & d_4 & c_4 & & \\ & & & a_4 & d_5 & \ddots & \\ & \mathbf{0} & & & \ddots & \ddots & c_{n-1} \\ & & & & & a_{n-1} & d_n \end{bmatrix} \tag{1}$$

(All elements not displayed are zero—denoted by large zeros.)

The tridiagonal matrix is characterized by saying that $a_{ij} = 0$ if $|i - j| \geqq 2$. In general, a matrix is said to have *band structure* if there is an integer k (less than n) such that $a_{ij} = 0$ whenever $|i - j| \geqq k$.

The storage requirements for a band matrix are less than for a general matrix of the same size. Thus an $n \times n$ diagonal matrix requires only n memory locations in the computer, and a tridiagonal matrix requires only $3n - 2$. This factor is important if band matrices of very large order are being used.

For band matrices, the Gaussian elimination algorithm can be made more efficient and further savings are possible if it is known beforehand that pivoting is unnecessary. This situation occurs often enough to justify special programs being written for them. Here we develop a program for the tridiagonal matrix and give a listing for the "quint-diagonal" matrix ($a_{ij} = 0$ if $|i - j| > 2$).

The program to be described now is called subroutine TRI. It is designed to solve a system of n linear equations in n unknowns under the assumption that the coefficient matrix is as shown in Eq. (1). Both the forward elimination phase and the back substitution phase are incorporated in the program, and no pivoting is employed. Thus naive Gaussian elimination is used. In order to see how to proceed, we suppose that the forward elimination phase is applied to the array

$$\begin{bmatrix} d_1 & c_1 & & & & \mathbf{0} & b_1 \\ a_1 & d_2 & c_2 & & & & b_2 \\ & a_2 & d_3 & c_3 & & & b_3 \\ & & \ddots & \ddots & \ddots & & \vdots \\ & \mathbf{0} & & a_{n-2} & d_{n-1} & c_{n-1} & b_{n-1} \\ & & & & a_{n-1} & d_n & b_n \end{bmatrix}$$

In step 1 we subtract a_1/d_1 times row 1 from row 2, thus creating a zero in the a_1 position. Only the entries b_2 and d_2 are altered. Observe that c_2 is *not* altered. In step 2 the process is repeated, using the new row 2 as the operating row. Here is how the d's and b's are altered in each step.

$$\begin{cases} d_2 \leftarrow d_2 - c_1\left(\dfrac{a_1}{d_1}\right) \\ b_2 \leftarrow b_2 - b_1\left(\dfrac{a_1}{d_1}\right) \end{cases} \qquad \begin{cases} d_3 \leftarrow d_3 - c_2\left(\dfrac{a_2}{d_2}\right) \\ b_3 \leftarrow b_3 - b_2\left(\dfrac{a_2}{d_2}\right) \end{cases}$$

etc.

In the subroutine the first DO loop carries out these calculations. At the end of the forward elimination phase the *form* of the array is as follows.

$$\left[\begin{array}{cccccc|c} d_1 & c_1 & & & & & b_1 \\ & d_2 & c_2 & & \mathbf{0} & & b_2 \\ & & d_3 & c_3 & & & b_3 \\ & & & \ddots & \ddots & & \vdots \\ & \mathbf{0} & & & d_{n-1} & c_{n-1} & b_{n-1} \\ & & & & & d_n & b_n \end{array}\right]$$

Of course, these b's and d's are not the same as at the beginning. The back substitution phase solves for $x_n, x_{n-1}, \ldots$ as follows.

$$x_n \leftarrow \frac{b_n}{d_n}$$

$$x_{n-1} \leftarrow \frac{b_{n-1} - c_{n-1}x_n}{d_{n-1}}$$

$$x_{n-2} \leftarrow \frac{b_{n-2} - c_{n-2}x_{n-1}}{d_{n-2}}$$

etc.

In the code the solution is put into array B.

```
      SUBROUTINE TRI(N, A, D, C, B)
      DIMENSION  A(N), D(N), C(N), B(N)
      DO 2 I = 2, N
      XMULT = A(I-1)/D(I-1)
      D(I) = D(I) - XMULT*C(I-1)
    2 B(I) = B(I) - XMULT*B(I-1)
      B(N) = B(N)/D(N)
      DO 3 I = 1, N-1
    3 B(N-I) = (B(N-I) - C(N-I)*B(N-I+1))/D(N-I)
      RETURN
      END
```

Since this program does not involve pivoting, it is natural to ask whether it is likely to fail. Simple examples can be given to illustrate the possibility of failure because of attempted division by zero even though the matrix is nonsingular. On the other hand, it is not easy to give the weakest possible conditions on the matrix to guarantee the success of the subroutine. We content ourselves with one property that is easily checked and commonly encountered. If the tridiagonal matrix is "diagonally dominant," then subroutine TRI will not encounter zero divisors. For a general matrix **A**, *diagonal dominance* is the condition

$$|a_{ii}| > \sum_{\substack{j=1 \\ j \neq i}}^{n} |a_{ij}| \quad \text{for all } i$$

In the case of the tridiagonal matrix of Eq. (1), diagonal dominance means simply that $|d_i| > |c_i| + |a_{i-1}|$ for all i.

Let us verify that the forward elimination phase in subroutine TRI preserves diagonal dominance. The new matrix produced by Gaussian elimination has zero elements where the a's originally stood and new diagonal elements are determined recursively by

$$\begin{aligned} d_1' &= d_1 \\ d_i' &= d_i - \frac{c_{i-1}a_{i-1}}{d_{i-1}'} \qquad (2 \leqq i \leqq n) \end{aligned}$$

where d' denotes a new diagonal element. The c elements are unaltered. Now we assume that $|d_i| > |a_{i-1}| + |c_i|$, and we want to be sure that $|d_i'| > |c_i|$. Obviously this is true for $i = 1$, since $d_1' = d_1$. If it is true for index $i - 1$, then it is true for index i, since

$$\begin{aligned} |d_i'| &= \left| d_i - \frac{c_{i-1}a_{i-1}}{d_{i-1}'} \right| \\ &\geqq |d_i| - |a_{i-1}| \frac{|c_{i-1}|}{|d_{i-1}'|} \\ &> |a_{i-1}| + |c_i| - |a_{i-1}| = |c_i| \end{aligned}$$

A symmetric tridiagonal system arises in the cubic spline development of Chapter 7 and elsewhere. A general symmetric tridiagonal system has the form

$$\begin{cases} d_1x_1 + c_1x_2 & = b_1 \\ c_1x_1 + d_2x_2 + c_2x_3 & = b_2 \\ \qquad c_2x_2 + d_3x_3 + c_3x_4 & = b_3 \\ \qquad\qquad \ddots & \vdots \\ \qquad\qquad c_{n-1}x_{n-1} + d_nx_n & = b_n \end{cases} \tag{2}$$

Even less storage is required for this system, since only arrays D, C, and B are needed. A special subroutine to solve system (2) is not necessary, because subroutine TRI can be used with A = C:

```
CALL TRI(N,C,D,C,B)
```

The principles illustrated by subroutine TRI can be applied in the case of matrices having wider bands of nonzero elements. A subroutine called QUINT is given here to solve the five-diagonal system. The form of the augmented matrix $[\mathbf{A}\ \mathbf{b}]$ is

$$\begin{bmatrix} d_1 & c_1 & f_1 & & & & & & & b_1 \\ a_1 & d_2 & c_2 & f_2 & & & \mathbf{0} & & & b_2 \\ e_1 & a_2 & d_3 & c_3 & f_3 & & & & & b_3 \\ & e_2 & a_3 & d_4 & c_4 & f_4 & & & & b_4 \\ & & & \cdot & \cdot & \cdot & \cdot & \cdot & & \cdot \\ & & & & \cdot & \cdot & \cdot & \cdot & \cdot & \cdot \\ & \mathbf{0} & & & & e_{n-3} & a_{n-2} & d_{n-1} & c_{n-1} & b_{n-1} \\ & & & & & & e_{n-2} & a_{n-1} & d_n & b_n \end{bmatrix}$$

In the program the solution vector is placed in array B. Also, the user should not use this routine if $n \leqq 2$. (Why?)

```
      SUBROUTINE QUINT(N, E, A, D, C, F, B)
      DIMENSION  E(N),A(N),D(N),C(N),F(N),B(N)
      DO 2 I = 2,N-1
      XMULT = A(I-1)/D(I-1)
      D(I) = D(I) - XMULT*C(I-1)
      C(I) = C(I) - XMULT*F(I-1)
      B(I) = B(I) - XMULT*B(I-1)
      XMULT = E(I-1)/D(I-1)
      A(I) = A(I) - XMULT*C(I-1)
      D(I+1) = D(I+1) - XMULT*F(I-1)
    2 B(I+1) = B(I+1) - XMULT*B(I-1)
      XMULT = A(N-1)/D(N-1)
      D(N) = D(N) - XMULT*C(N-1)
      B(N) = (B(N) - XMULT*B(N-1))/D(N)
      B(N-1) = (B(N-1) - C(N-1)*B(N))/D(N-1)
      DO 3 I = 2,N-1
    3 B(N-I) = (B(N-I) - F(N-I)*B(N-I+2) - C(N-I)*B(N-I+1))/D(N-I)
      RETURN
      END
```

Problems 6.3

1. Count the arithmetic operations involved in subroutine TRI.
2. How many storage locations are needed for a system of n linear equations if the coefficient matrix has band structure in which $a_{ij} = 0$ for $|i - j| \geqq k$?
3. Give an example of a system of linear equations in tridiagonal form that cannot be solved without pivoting.

4. What is the appearance of a matrix if its elements satisfy $a_{ij} = 0$ when $j < i - 2$? When $j > i + 1$?
5. Consider a diagonally dominant matrix $\mathbf{A}$ whose elements satisfy $a_{ij} = 0$ when $i > j + 1$. Does Gaussian elimination without pivoting preserve the diagonal dominance?
6. Given a real tridiagonal matrix $\mathbf{A}$ of form (1) such that $d_{i+1}a_i d_i c_i \neq 0$ $(1 \leqq i \leqq n - 1)$, find a diagonal matrix $\mathbf{D}$ such that $\mathbf{D}^{-1}\mathbf{AD}$ is symmetric.
7. Consider the matrix

$$\begin{bmatrix} a_{11} & a_{12} & 0 & 0 \\ a_{21} & a_{22} & a_{23} & 0 \\ 0 & a_{32} & a_{33} & a_{34} \\ 0 & 0 & a_{43} & a_{44} \end{bmatrix}$$

where $a_{ii} \neq 0$ for $i = 1,2,3,4$.

(a) Establish the procedure

$$\ell_{11} \leftarrow a_{11}$$

$$\left[\begin{array}{l} \text{for } i = 2,3,4 \\ \ell_{i,i-1} \leftarrow a_{i,i-1} \\ u_{i-1,i} \leftarrow a_{i-1,i}/\ell_{i-1,i-1} \\ \quad \ell_{i,i} \leftarrow a_{i,i} - \ell_{i,i-1}u_{i-1,i} \end{array}\right.$$

for determining the elements of a lower-tridiagonal matrix $\mathbf{L} = (\ell_{ij})$ and a *unit* upper-tridiagonal matrix $\mathbf{U} = (u_{ij})$ such that $\mathbf{A} = \mathbf{LU}$.

(b) Establish the procedure

$$u_{11} \leftarrow a_{11}$$

$$\left[\begin{array}{l} \text{for } i = 2,3,4 \\ u_{i-1,i} \leftarrow a_{i-1,i} \\ \ell_{i,i-1} \leftarrow a_{i,i-1}/u_{i-1,i-1} \\ \quad u_{i,i} \leftarrow a_{i,i} - \ell_{i,i-1}u_{i-1,i} \end{array}\right.$$

for determining the elements of a *unit* lower-tridiagonal matrix $\mathbf{L} = (\ell_{ij})$ and an upper-tridiagonal matrix $\mathbf{U} = (u_{ij})$ such that $\mathbf{A} = \mathbf{LU}$.

By extending the loops, these algorithms can be generalized to $n \times n$ tridiagonal matrices.

Computer Problems 6.3

1. Write a special subroutine to solve the tridiagonal system in which $d_i = 1$ and $a_i = c_i = \lambda$ for all i. Here λ is a given constant.
2. Use subroutine `TRI` to solve the following system of 100 equations. Compare the numerical solution to the obvious exact solution.

$$\begin{cases} x_1 + 0.5x_2 = 1.5 \\ 0.5x_{i-1} + x_i + 0.5x_{i+1} = 2.0 \qquad (2 \leqq i \leqq 99) \\ 0.5x_{99} + x_{100} = 1.5 \end{cases}$$

What are the entries d_i after applying the algorithm?

3. (Normalized Tridiagonal Algorithm) Construct an algorithm for handling tridiagonal systems in which the "normalized" Gaussian elimination procedure without pivoting is used. In this proc-

ess each pivot row is divided by the diagonal element before a multiple of the row is subtracted from the successive rows. Write the equations involved in the forward elimination phase and store the upper-diagonal entries back in array C and the right-hand side entries back in array B. Write the equations for the back substitution phase, storing the solution in array B. Code and test this procedure. What are its advantages and/or disadvantages?

4. Using the ideas illustrated in QUINT, write a subroutine for solving seven-diagonal systems. Test it on several such systems.
5. An $n \times n$ band coefficient matrix with ℓ subdiagonals and m superdiagonals can be stored in ***band storage*** mode in an $n \times (\ell + m + 1)$ array. The matrix is stored with the row and diagonal structure preserved with almost all zero elements unstored. If the original $n \times n$ band matrix had the form shown in the Fig. 6.2, then the $n \times (\ell + m + 1)$ array in band storage mode

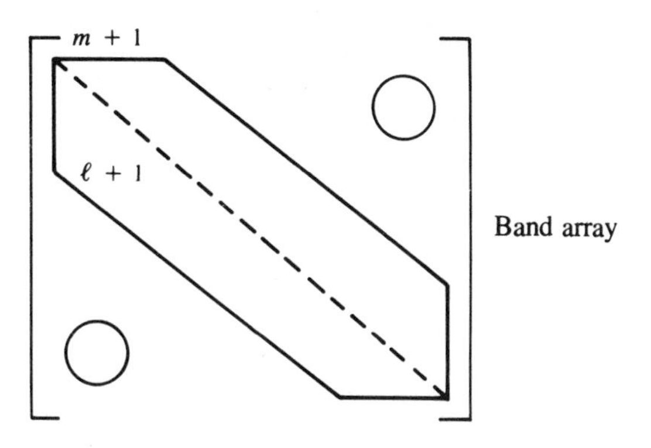

Figure 6.2

would have the form displayed in Fig. 6.3. The main diagonal would be the $\ell + 1$ column of the new array.

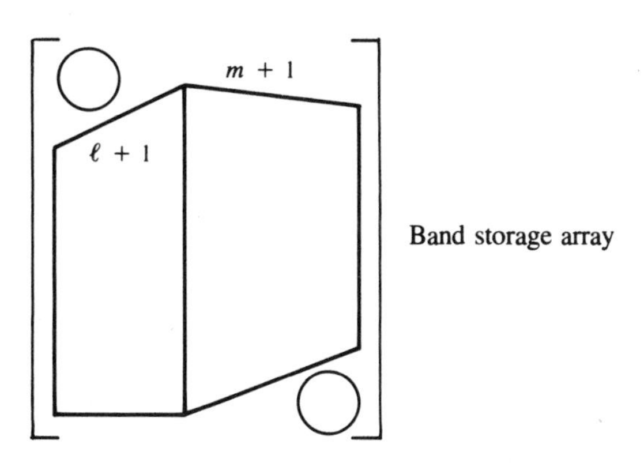

Figure 6.3

Write and test a subroutine that is analogous to subroutine QUINT except that the coefficient matrix is stored in band storage mode in array A.

6. An $n \times n$ symmetric band coefficient matrix with m subdiagonals and m superdiagonals can be stored in ***symmetric band storage*** mode in an $n \times (m + 1)$ array. Only the main diagonal and subdiagonals are stored so that the main diagonal is the last column in the new array, shown in Fig. 6.4.

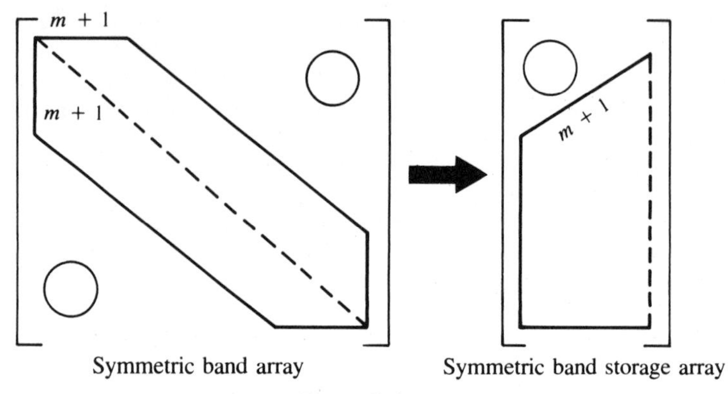

Figure 6.4

Write and test a subroutine that is analogous to subroutine QUINT except that the coefficient matrix is stored in symmetric band storage mode.

7. For a $2n \times 2n$ tridiagonal system, write and test a subroutine that proceeds as follows. In the forward elimination phase the routine simultaneously eliminates the elements in the subdiagonal from the top to the middle and in the superdiagonal from the bottom to the middle. In the back substitution phase the unknowns are determined two at a time from the middle outward.
8. (Continuation) Recode and test the subroutine of Problem 7 for a general $n \times n$ tridiagonal matrix.
9. Rewrite and test subroutine TRI so that it performs Gaussian elimination with scaled partial pivoting. *Hint:* Additional temporary storage arrays may be needed.
10. Solve the system

$$\begin{cases} 4x_1 - x_2 = -20 \\ x_{j-1} - 4x_j + x_{j+1} = 40 \qquad (2 \leqq j \leqq n-1) \\ -x_{n-1} + 4x_n = -20 \end{cases}$$

using subroutine TRI for $n = 100$.

11. Let **A** be the 50×50 tridiagonal matrix

$$\begin{bmatrix} 5 & -1 & & & & \\ -1 & 5 & -1 & & \mathbf{0} & \\ & -1 & 5 & -1 & & \\ & & \ddots & \ddots & \ddots & \\ & \mathbf{0} & & -1 & 5 & -1 \\ & & & & -1 & 5 \end{bmatrix}$$

Consider the problem $\mathbf{Ax} = \mathbf{b}$ for 50 different vectors $\mathbf{b}$ of the form $(1, 2, 3, \ldots, 49, 50)^T$, $(2, 3, 4, \ldots, 50, 1)^T$, $(3, 4, 5, \ldots, 50, 1, 2)^T$, and so on. Write and test efficient code for solving this problem. *Hint:* Rewrite TRI.

12. Rewrite and test QUINT so that it does Gaussian elimination with scaled partial pivoting. Is this worthwhile?

Supplementary Problems (Chapter 6)

1. Count the arithmetic operations involved in subroutine `QUINT`.
2. What are the final contents of the array `A` after subroutine `GAUSS` has processed the following array?

$$\begin{bmatrix} 1 & 3 & 2 & 1 \\ 4 & 2 & 1 & 2 \\ 2 & 1 & 2 & 3 \\ 1 & 2 & 4 & 1 \end{bmatrix}$$

3. Consider the matrix

$$\begin{bmatrix} -0.0013 & 56.4972 & 123.4567 & 987.6543 \\ 0. & -0.0145 & 8.8990 & 833.3333 \\ 0. & 102.7513 & -7.6543 & 69.6869 \\ 0. & -1.3131 & -9876.5432 & 100.0001 \end{bmatrix}$$

Identify the entry that will be used as the next pivot element of naive Gaussian elimination, of Gaussian elimination with partial pivoting, and of Gaussian elimination with scaled partial pivoting. The scale vector is $(987.6543, 46.79, 256.29, 1.096)^T$.

4. Consider the system

$$\begin{cases} 10^{-4}x_1 + x_2 = b_1 \\ x_1 + x_2 = b_2 \end{cases}$$

where $b_1 \neq 0$ and $b_2 \neq 0$. Its exact solution is

$$x_1 = \frac{-b_1 + b_2}{1 - 10^{-4}} \quad \text{and} \quad x_2 = \frac{b_1 - 10^{-4}b_2}{1 - 10^{-4}}$$

(a) Let $b_1 = 1$ and $b_2 = 2$. Solve this system, using naive Gaussian elimination with three-digit (rounded) arithmetic and compare to the exact solution $x_1 = 1.00010\ldots$ and $x_2 = 0.999899\ldots$.

(b) Repeat the preceding part, using Gaussian elimination with scaled partial pivoting.

(c) Find values of b_1 and b_2 so that naive Gaussian elimination does not give poor answers.

5. Using Gaussian elimination with scaled partial pivoting, reduce the matrix

$$\begin{bmatrix} 1 & 0 & 3 & 0 \\ 0 & 1 & 3 & -1 \\ 3 & -3 & 0 & 6 \\ 0 & 2 & 4 & -6 \end{bmatrix}$$

Show intermediate matrices.

6. Solve the following system using Gaussian elimination with scaled partial pivoting.

$$\begin{bmatrix} 1 & -1 & 2 \\ -2 & 1 & -1 \\ 4 & -1 & 2 \end{bmatrix} \begin{bmatrix} x_1 \\ x_2 \\ x_3 \end{bmatrix} = \begin{bmatrix} -2 \\ 2 \\ -1 \end{bmatrix}$$

Show intermediate matrices at each step.

7. A linear system is row *equilibrated* if it is scaled so that the maximum element in every row is unity.
   (a) Solve the system of equations

$$\begin{bmatrix} 1 & 1 & 2 \times 10^9 \\ 2 & -1 & 10^9 \\ 1 & 2 & 0 \end{bmatrix} \begin{bmatrix} x_1 \\ x_2 \\ x_3 \end{bmatrix} = \begin{bmatrix} 1 \\ 1 \\ 1 \end{bmatrix}$$

   by Gaussian elimination with scaled partial pivoting.
   (b) Solve by using row-equilibrated naive Gaussian elimination. Are the answers the same?
8. Show that $\mathbf{AX} = \mathbf{B}$ can be solved by Gaussian elimination with scaled partial pivoting in $(n^3/3) + mn^2 + \mathcal{O}(n^2)$ multiplications and divisions, where $\mathbf{A}$, $\mathbf{X}$, and $\mathbf{B}$ are matrices of order $n \times n$, $n \times m$, and $n \times m$, respectively. Thus if $\mathbf{B}$ is $n \times n$, then the $n \times n$ solution matrix $\mathbf{X}$ can be found by Gaussian elimination with scaled partial pivoting in $\frac{4}{3}n^3 + \mathcal{O}(n^2)$ multiplications and divisions. *Hint:* If $\mathbf{X}^{(j)}$ and $\mathbf{B}^{(j)}$ are the jth columns of $\mathbf{X}$ and $\mathbf{B}$, respectively, then $\mathbf{AX}^{(j)} = \mathbf{B}^{(j)}$.
9. Let $\mathcal{X}$ be a square matrix having the form

$$\mathcal{X} = \begin{bmatrix} \mathbf{A} & \mathbf{B} \\ \mathbf{C} & \mathbf{D} \end{bmatrix}$$

   where $\mathbf{A}$, $\mathbf{D}$ are square matrices and $\mathbf{A}^{-1}$ exists. It is known that $\mathcal{X}^{-1}$ exists if and only if $(\mathbf{D} - \mathbf{CA}^{-1}\mathbf{B})^{-1}$ exists. Verify that $\mathcal{X}^{-1}$ is given by

$$\mathcal{X}^{-1} = \begin{bmatrix} \mathbf{I} & -\mathbf{A}^{-1}\mathbf{B} \\ 0 & \mathbf{I} \end{bmatrix} \begin{bmatrix} \mathbf{A}^{-1} & 0 \\ 0 & (\mathbf{D} - \mathbf{CA}^{-1}\mathbf{B})^{-1} \end{bmatrix} \begin{bmatrix} \mathbf{I} & 0 \\ -\mathbf{CA}^{-1} & \mathbf{I} \end{bmatrix}$$

   As an application, compute the inverse of

$$\text{(a)} \quad \mathcal{X} = \left[\begin{array}{cc:cc} 1 & 0 & 0 & 1 \\ 0 & 1 & 1 & 0 \\ \hdashline 1 & 0 & 1 & 2 \\ 0 & 0 & 0 & 1 \end{array}\right] \quad \text{and} \quad \text{(b)} \quad \mathcal{X} = \left[\begin{array}{ccc:c} 1 & 0 & 0 & 1 \\ 0 & 1 & 0 & 1 \\ 0 & 0 & 1 & 1 \\ \hdashline 1 & 1 & 1 & 2 \end{array}\right]$$

10. Let $\mathbf{A}$ be an $n \times n$ complex matrix such that $\mathbf{A}^{-1}$ exists. Verify that

$$\begin{bmatrix} \mathbf{A} & \bar{\mathbf{A}} \\ -i\mathbf{A} & -i\bar{\mathbf{A}} \end{bmatrix}^{-1} = \frac{1}{2} \begin{bmatrix} \mathbf{A}^{-1} & i\mathbf{A}^{-1} \\ \bar{\mathbf{A}}^{-1} & -i\bar{\mathbf{A}}^{-1} \end{bmatrix}$$

   where $\bar{\mathbf{A}}$ denotes the complex conjugate of $\mathbf{A}$; that is, if $\mathbf{A} = (a_{ij})$, then $\bar{\mathbf{A}} = (\bar{a}_{ij})$. Recall that if a complex number z is of form $a + bi$, for real a and b, then $\bar{z} = a - bi$.
11. After processing a matrix $\mathbf{A}$ by subroutine GAUSS, how can the results be used to solve a system of equations of form $\mathbf{A}^T\mathbf{x} = \mathbf{b}$?
12. Show that the system of equations

$$\begin{cases} x_1 + 4x_2 + \alpha x_3 = 6 \\ 2x_1 - x_2 + 2\alpha x_3 = 3 \\ \alpha x_1 + 3x_2 + x_3 = 5 \end{cases}$$

   possesses a unique solution when $\alpha = 0$, no solution when $\alpha = -1$, and infinitely many solutions when $\alpha = 1$. Also, investigate the corresponding situation when the right-hand side is replaced by zeros.

13. Consider the matrix

$$\mathbf{A} = \begin{bmatrix} 25 & 0 & 0 & 0 & 1 \\ 0 & 27 & 4 & 3 & 2 \\ 0 & 54 & 58 & 0 & 0 \\ 0 & 108 & 116 & 0 & 0 \\ 100 & 0 & 0 & 0 & 24 \end{bmatrix}$$

(a) Determine the unit lower-triangular matrix $\mathbf{M}$ and the upper-triangular matrix $\mathbf{U}$ such that $\mathbf{MA} = \mathbf{U}$.

(b) Determine $\mathbf{M}^{-1} = \mathbf{L}$ such that $\mathbf{A} = \mathbf{LU}$.

14. Consider the matrix

$$\mathbf{A} = \begin{bmatrix} 2 & 2 & 1 \\ 1 & 1 & 1 \\ 3 & 2 & 1 \end{bmatrix}$$

(a) Show that $\mathbf{A}$ *cannot* be factored into the product of a unit lower-triangular matrix and an upper-triangular matrix.

(b) Interchange the rows of $\mathbf{A}$ so that this can be done.

15. Consider the matrix

$$\mathbf{A} = \begin{bmatrix} a & 0 & 0 & z \\ 0 & b & 0 & 0 \\ 0 & x & c & 0 \\ w & 0 & y & d \end{bmatrix}$$

(a) Determine a unit lower-triangular matrix $\mathbf{M}$ and an upper-triangular matrix $\mathbf{U}$ such that $\mathbf{MA} = \mathbf{U}$.

(b) Determine a lower-triangular matrix $\mathbf{L}'$ and a unit upper-triangular matrix $\mathbf{U}'$ such that $\mathbf{A} = \mathbf{L}'\mathbf{U}'$.

16. Consider the matrix

$$\mathbf{A} = \begin{bmatrix} 4 & -1 & -1 & 0 \\ -1 & 4 & 0 & -1 \\ -1 & 0 & 4 & -1 \\ 0 & -1 & -1 & 4 \end{bmatrix}$$

Factor $\mathbf{A}$ in the following ways.

(a) $\mathbf{A} = \mathbf{LU}$, where $\mathbf{L}$ is unit lower triangular and $\mathbf{U}$ is upper triangular.

(b) $\mathbf{A} = \mathbf{LDU}'$, where $\mathbf{L}$ is unit lower triangular, $\mathbf{D}$ is diagonal, and $\mathbf{U}'$ is unit upper triangular.

(c) $\mathbf{A} = \mathbf{L}'\mathbf{U}'$, where $\mathbf{L}'$ is lower triangular and $\mathbf{U}'$ is unit upper triangular.

(d) $\mathbf{A} = (\mathbf{L}'')(\mathbf{L}'')^T$, where $\mathbf{L}''$ is lower triangular.

17. (Continuation) Evaluate the determinant of $\mathbf{A}$. *Hint:* $\det(\mathbf{A}) = \det(\mathbf{L}) \cdot \det(\mathbf{D}) \cdot \det(\mathbf{U}') = \det(\mathbf{D})$.

18. Given

$$\mathbf{A} = \begin{bmatrix} 3 & 2 & -1 \\ 5 & 3 & 2 \\ -1 & 1 & -3 \end{bmatrix}, \quad \mathbf{L}^{-1} = \begin{bmatrix} 1 & 0 & 0 \\ -\frac{5}{3} & 1 & 0 \\ -8 & 5 & 1 \end{bmatrix}, \quad \mathbf{U} = \begin{bmatrix} 3 & 2 & -1 \\ 0 & -\frac{1}{3} & \frac{11}{3} \\ 0 & 0 & 15 \end{bmatrix}$$

Evaluate the inverse of $\mathbf{A}$ by solving $\mathbf{U}\mathbf{x}^{(j)} = \mathbf{L}^{-1}\mathbf{e}_j$ for $j = 1, 2, 3$. Here $\mathbf{e}_j$ is the unit vector of all zero components except for one in the jth component. *Hint:* $\mathbf{A}^{-1} = (\mathbf{x}^{(1)}, \mathbf{x}^{(2)}, \mathbf{x}^{(3)})$.

19. Consider the 3×3 Hilbert matrix

$$\mathbf{A} = \begin{bmatrix} 1 & \frac{1}{2} & \frac{1}{3} \\ \frac{1}{2} & \frac{1}{3} & \frac{1}{4} \\ \frac{1}{3} & \frac{1}{4} & \frac{1}{5} \end{bmatrix}$$

Repeat Problems 16 and 17, using this matrix.

20. Solve the following system of equations, retaining only four significant figures in each step of the calculation, and compare to the solution obtained when eight significant figures are retained.

$$\begin{cases} 0.1036x + 0.2122y = 0.7381 \\ 0.2081x + 0.4247y = 0.9327 \end{cases}$$

Be consistent by either always rounding to the number of significant figures that are being carried or always chopping.

21. For what values of α does naive Gaussian elimination produce erroneous answers for this system?

$$\begin{cases} x_1 + x_2 = 2 \\ \alpha x_1 + x_2 = 2 + \alpha \end{cases}$$

(Explain what happens in the computer.)

22. How much time would be used on the computer to solve 2000 equations using Gaussian elimination with scaled partial pivoting? Give a rough estimate based on operation times in Sec. 6.2.

23. Consider the system of equations

$$\begin{cases} 6x_1 = 12 \\ 6x_2 + 3x_1 = -12 \\ 7x_3 - 2x_2 + 4x_1 = 14 \\ 21x_4 + 9x_3 - 3x_2 + 5x_1 = -2 \end{cases}$$

(a) Solve for x_1, x_2, x_3, x_4 (in order) by forward substitution.
(b) Write this system in matrix notation $\mathbf{A}\mathbf{x} = \mathbf{b}$, where $\mathbf{x} = (x_1, x_2, x_3, x_4)^T$. Determine the "*LU* factorization": $\mathbf{A} = \mathbf{L}\mathbf{U}$, where $\mathbf{L}$ is unit lower triangular and $\mathbf{U}$ is upper triangular. Why is this factorization associated with the Gaussian elimination method?

Supplementary Computer Problems (Chapter 6)

1. Define the $n \times n$ array A by $a_{ij} = -1 + 2\max\{i, j\}$. Set up array B in such a way that the solution of system (2) is $x_i = 1$ for $1 \leqq i \leqq n$. Test subroutines GAUSS and SOLVE on this system for a moderate value of n, say $n = 30$.
2. Select a modest value of n, say $5 \leqq n \leqq 20$, and let $a_{ij} = (i - 1)^{j-1}$ and $b_i = i - 1$. Solve the system $\mathbf{A}\mathbf{x} = \mathbf{b}$ on the computer. By looking at the output, guess what the correct solution is. Establish algebraically that your guess is correct. Account for the errors in the computed solution.
3. Consider the $n \times n$ lower-triangular system $\mathbf{A}\mathbf{x} = \mathbf{b}$, where $\mathbf{A} = (a_{ij})$, $\mathbf{x} = (x_i)$, $\mathbf{b} = (b_i)$, and $a_{ij} = 0$ for $i < j$.
(a) Write an algorithm (in mathematical terms) for solving for $\mathbf{x}$ by forward substitution.
(b) Write `SUBROUTINE FORSUB(NR, N, A, B, X)` that uses this algorithm. Here NR is the row dimension of the array A.

(c) Determine the number of divisions, multiplications, additions (or subtractions) in using this algorithm to solve for $\mathbf{x}$.

(d) Should Gaussian elimination with partial pivoting be used to solve such a system?

4. Write subroutine POLY(NR,N,KP1,A,C,WKSP,PA) for computing the $N \times N$ matrix $p_k(\mathbf{A})$ stored in array PA: $p_k(\mathbf{A}) = c_0\mathbf{I} + c_1\mathbf{A} + c_2\mathbf{A}^2 + \cdots + c_k\mathbf{A}^k$, where $\mathbf{A}$ is an $N \times N$ matrix and p_k is a kth-degree polynomial. Here KP1 is $k + 1$, C(I) are real constants c_{i-1}, for $i = 1, 2, \ldots, k + 1$, NR is the row dimension of A, and WKSP is a single-dimension workspace array. Use nested multiplication and write efficient code. Test POLY on the following data.

Case 1.

$$\mathbf{A} = \mathbf{I}_5, \qquad p_3(x) = 1 - 5x + 10x^3$$

Case 2.

$$\mathbf{A} = \begin{bmatrix} 1 & 2 \\ 3 & 4 \end{bmatrix}, \qquad p_2(x) = 1 - 2x + x^2$$

Case 3.

$$\mathbf{A} = \begin{bmatrix} 0 & 2 & 4 \\ 0 & 0 & 8 \\ 0 & 0 & 0 \end{bmatrix}, \qquad p_3(x) = 1 + 3x - 3x^2 + x^3$$

Case 4.

$$\mathbf{A} = \begin{bmatrix} 2 & -1 & 0 & 0 \\ -1 & 2 & -1 & 0 \\ 0 & -1 & 2 & -1 \\ 0 & 0 & -1 & 2 \end{bmatrix}, \qquad p_5(x) = 10 + x - 2x^2 + 3x^3 - 4x^4 + 5x^5$$

Case 5.

$$\mathbf{A} = \begin{bmatrix} -20 & -15 & -10 & -5 \\ 1 & 0 & 0 & 0 \\ 0 & 1 & 0 & 0 \\ 0 & 0 & 1 & 0 \end{bmatrix}, \qquad p_4(x) = 5 + 10x + 15x^2 + 20x^3 + x^4$$

Case 6.

$$\mathbf{A} = \begin{bmatrix} 5 & 7 & 6 & 5 \\ 7 & 10 & 8 & 7 \\ 6 & 8 & 10 & 9 \\ 5 & 7 & 9 & 10 \end{bmatrix}, \qquad p_4(x) = 1 - 100x + 146x^2 - 35x^3 + x^4$$

5. Consider the complex linear system

$$\mathbf{Az} = \mathbf{b}$$

where

$$\mathbf{A} = \begin{bmatrix} 5 + 9i & 5 + 5i & -6 - 6i & -7 - 7i \\ 3 + 3i & 6 + 10i & -5 - 5i & -6 - 6i \\ 2 + 2i & 3 + 3i & -1 + 3i & -5 - 5i \\ 1 + i & 2 + 2i & -3 - 3i & 4i \end{bmatrix}$$

By declaring certain variables complex and using the Fortran function CABS, establish a complex arithmetic version of GAUSS and SOLVE. Now solve this system four times with the following right-hand sides.

$$\begin{bmatrix} -10 + 2i \\ -5 + i \\ -5 + i \\ -5 + i \end{bmatrix}, \begin{bmatrix} 2 + 6i \\ 4 + 12i \\ 2 + 6i \\ 2 + 6i \end{bmatrix}, \begin{bmatrix} 7 - 3i \\ 7 - 3i \\ 0 \\ 7 - 3i \end{bmatrix}, \begin{bmatrix} -4 - 8i \\ -4 - 8i \\ -4 - 8i \\ 0 \end{bmatrix}$$

Verify that the solutions are $\mathbf{z} = \lambda^{-1}\mathbf{b}$ for scalars λ. The numbers λ are called *eigenvalues*, and the solutions $\mathbf{z}$ are *eigenvectors* of $\mathbf{A}$. Usually the $\mathbf{b}$ vector is not known and the solution of the problem $\mathbf{Az} = \lambda\mathbf{z}$ cannot be obtained by using a linear equation solver.

6. (Continuation) A common electrical engineering problem is calculating currents in an electric circuit. For example, the circuit shown in Fig. 6.5 with R_i (ohms), C_i (microfarads), L (millihenries), and ω (hertz), leads to the system

$$\begin{cases} (50 - 10i)I_1 + (50)I_2 + (50)I_3 = V_1 \\ (10i)I_1 + (10 - 10i)I_2 + (10 - 20i)I_3 = 0 \\ - (30i)I_2 + (20 - 50i)I_3 = -V_2 \end{cases}$$

Select V_1 to be 100 millivolts and solve two cases: (a) the two voltages are in phase—that is, $V_2 = V_1$; (b) the second voltage is a quarter of a cycle ahead of the first—that is, $V_2 = iV_1$.

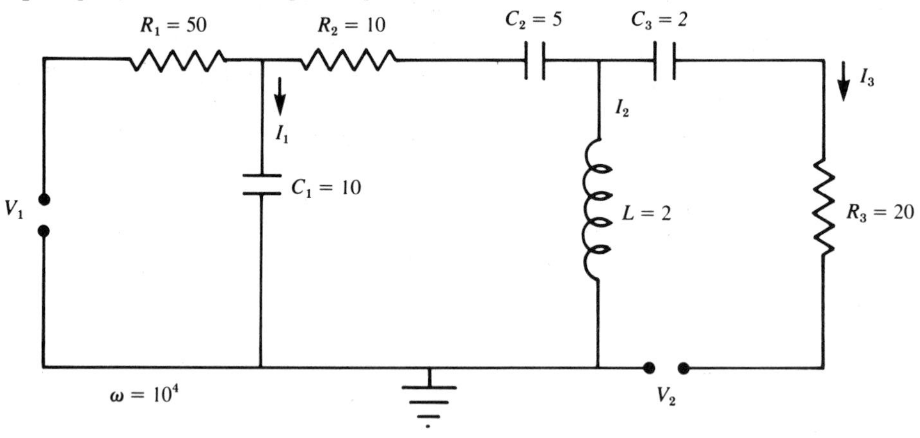

Figure 6.5

Using the complex arithmetic version of GAUSS and SOLVE, solve this system for the amplitude (milliamps) and the phase (degrees) for each current. The amplitude is CABS(X(J)) and the phase is 57.29577951*ATAN2(AIMAG(X(J)),REAL(X(J))) for J=1,2,3.

7. After subroutine GAUSS has been called, the determinant of the original matrix $\mathbf{A}$, denoted by det $(\mathbf{A})$, can be evaluated by the code

```
      DET = 1.0
      DO 2 I = 1,N
  2   DET = DET*A(L(I),I)
```

Numerically verify it by using the following test matrices with several values of n.

(a) $a_{ij} = |i - j|$

$\det(\mathbf{A}) = (-1)^{n-1}(n-1)2^{n-2}$

(b) $$a_{ij} = \begin{cases} 1, & j \geqq i \\ -j, & j < i \end{cases}$$

$\det(\mathbf{A}) = n!$

(c) $$\begin{cases} a_{1j} = a_{j1} = n^{-1}, & j \geqq 1 \\ a_{ij} = a_{i-1,j} + a_{i,j-1}, & i, j \geqq 2 \end{cases}$$

$\det(\mathbf{A}) = n^{-n}$

8. Modify subroutines GAUSS and SOLVE so that a system $\mathbf{AX} = \mathbf{B}$ can be solved in which $\mathbf{A}$, $\mathbf{X}$, and $\mathbf{B}$ are matrices of order $n \times n$, $n \times m$, and $n \times m$, respectively. Verify that the modified routines work on several test cases—one of which has $\mathbf{B} = \mathbf{I}$ so that the solution $\mathbf{X}$ is the inverse of $\mathbf{A}$.
9. Write and test a subroutine for computing the inverse of tridiagonal matrices.
10. (Continuation) Without modifying this routine invert symmetric tridiagonal matrices. Show that this subroutine works by multiplying the resulting matrix and the original matrix.
11. Consider the system of equations ($N = 7$)

$$\begin{bmatrix} d_1 & & & & & & a_7 \\ & d_2 & & 0 & & a_6 & \\ & & d_3 & & a_5 & & \\ & 0 & & d_4 & & 0 & \\ & & a_3 & & d_5 & & \\ & a_2 & & 0 & & d_6 & \\ a_1 & & & & & & d_7 \end{bmatrix} \begin{bmatrix} x_1 \\ x_2 \\ x_3 \\ x_4 \\ x_5 \\ x_6 \\ x_7 \end{bmatrix} = \begin{bmatrix} b_1 \\ b_2 \\ b_3 \\ b_4 \\ b_5 \\ b_6 \\ b_7 \end{bmatrix}$$

For N odd, write and test subroutine XGAUSS(N,A,D,B) that does the forward elimination phase of Gaussian elimination (without scaled partial pivoting) and subroutine XSOLVE(N,A,D,B,X) that does the back substitution for cross systems of this form.
12. Suppose that subroutine TRINOR(N,A,D,C,B) performs the normalized Gaussian elimination algorithm of Problem 3 on page 136 and subroutine TRI2N(N,A,D,C,B) does the algorithm outlined in Problem 7 on page 138. Using a timing routine on your computer, compare TRI, TRINOR, and TRI2N to determine which of them is faster for the tridiagonal system

$$a_i = i(n - i + 1)$$

$$d_i = (2i + 1)n - i - 2i$$

$$c_i = (i + 1)(n - i - 1)$$

$$b_i = i$$

with a large even value of n. *Cultural Note:* Due to computer architecture, more complicated algorithms may be faster than straightforward ones. This is true in particular for large pipeline or vector computers.

13. For a fixed value of n from 2 to 5, let

$$a_{ij} = (i + j)^2$$
$$b_i = in(i + n + 1) + n(1 + n(2n + 3))/6.$$

Show that the vector $\mathbf{x} = (1, 1, \ldots, 1)^T$ solves the system $\mathbf{Ax} = \mathbf{b}$. Test whether **GAUSS** and **SOLVE** can compute $\mathbf{x}$ correctly for $n = 2, 3, 4$. Explain what happens.

14. Using each value of n from 2 to 9, solve the $n \times n$ system $\mathbf{Ax} = \mathbf{b}$, where $\mathbf{A}$ and $\mathbf{b}$ are defined by

$$a_{ij} = (i + j - 1)^7$$
$$b_i = p(n + i - 1) - p(i - 1)$$

where

$$p(x) = \frac{x^2}{24}(2 + x^2(-7 + n^2(14 + n(12 + 3n))))$$

Explain what happens.

15. Recode **GAUSS** and **SOLVE** so that they are column oriented—that is, so that all inner loops vary the first index of $\mathbf{A}$. On some computer systems, this implementation may avoid paging or swapping between high-speed and secondary memory and be more efficient for large matrices.

16. Overflow and underflow often occur in evaluating determinants by the simple code of Prob. 7. To avoid this, one can compute $\log|\det(\mathbf{A})|$ as the sum of terms $\log|a_{\ell_i, i}|$ and use the exponential function at the end. Write a program incorporating this idea.

7

Approximation by Spline Functions

As a result of wind tunnel experiments, an airfoil is constructed by trial and error to have certain required characteristics. The cross section of the airfoil is then drawn as a curve on coordinate paper (See Fig. 7.1). In order to manufacture this airfoil, it is essential to have a formula for this curve. It can be obtained by reading the coordinates of a finite number of points on the curve and determining a cubic interpolating spline function for the data. This chapter describes methods for doing so.

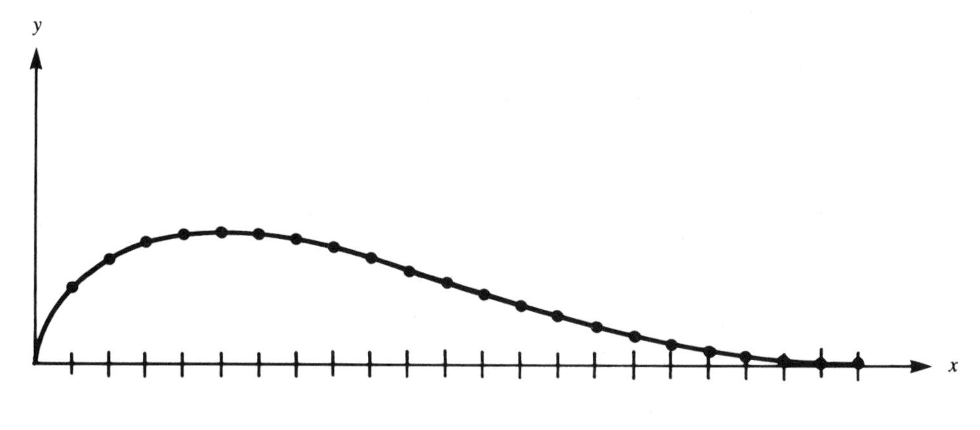

Figure 7.1 Airfoil cross section.

7.1 First-Degree Splines Jan 19

A *spline function* is a function consisting of polynomial pieces on subintervals, joined together with certain smoothness conditions. A simple example is the *polygonal* function, whose pieces are linear polynomials joined together to achieve continuity on an interval $[a, b]$, as in Fig. 7.2. The points $t_1, \ldots, t_8$ are termed *knots* in the theory of splines.

Such a function is somewhat complicated to define in explicit terms. We are forced to write

$$S(x) = \begin{cases} a_1 x + b_1, & x \in [t_1, t_2] \\ a_2 x + b_2, & x \in [t_2, t_3] \\ \vdots & \vdots \\ a_7 x + b_7, & x \in [t_7, t_8] \end{cases} \tag{1}$$

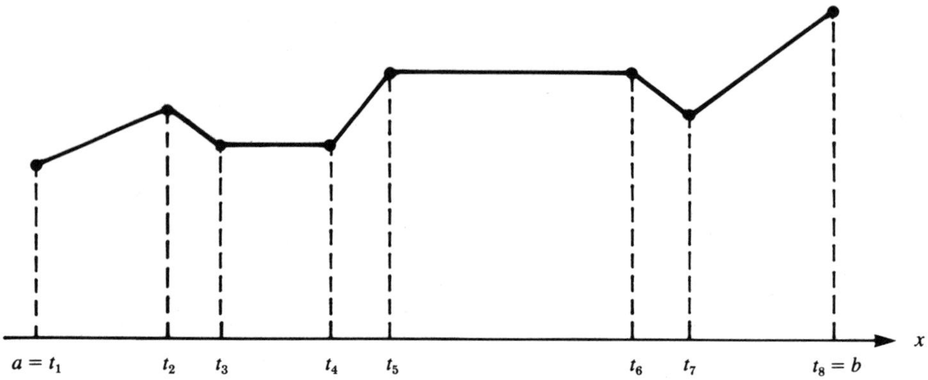

Figure 7.2 First-degree spline function.

If the knots $t_1, \ldots, t_8$ were given and if the coefficients $a_1, b_1, a_2, b_2, \ldots, a_7, b_7$ were all known, the evaluation of $S(x)$ at a specific x would proceed by first determining the interval containing x and then using the appropriate linear function for that interval.

If all a_i are zero, then $S(x)$ is a *spline of degree 0*. However, the function $S(x)$ just described is termed a *spline of degree 1*. It is characterized by the following properties.

1. The domain of S is an interval $[a, b]$.
2. S is continuous on $[a, b]$.
3. The interval $[a, b]$ can be partitioned by points called knots, $a = t_1 < t_2 < t_3 < \cdots < t_n = b$, in such a way that S is a linear polynomial on each subinterval $[t_i, t_{i+1}]$. (Such a function is "piecewise linear.")

The spline functions of degree 1 can be used for interpolation purposes. Suppose that a table of function values has been given.

x	t_1	t_2	$\cdots$	t_n
y	y_1	y_2	$\cdots$	y_n

There is no loss of generality in supposing that $t_1 < t_2 < t_3 < \cdots < t_n$, as this is only a matter of labelling the knots. The table can be represented by a set of n points in the plane, $(t_1, y_1), \ldots, (t_n, y_n)$, and these points have distinct abscissas. So we can draw a polygonal line through the points without ever drawing a *vertical* segment. This polygonal line is the graph of a function, and this function is obviously a spline of degree 1. What are the equations of the individual line segments that compose this graph? A glance at Fig. 7.3 shows that the answer is

$$y = y_i + m_i(x - t_i)$$

where

$$m_i = \frac{y_{i+1} - y_i}{t_{i+1} - t_i}$$

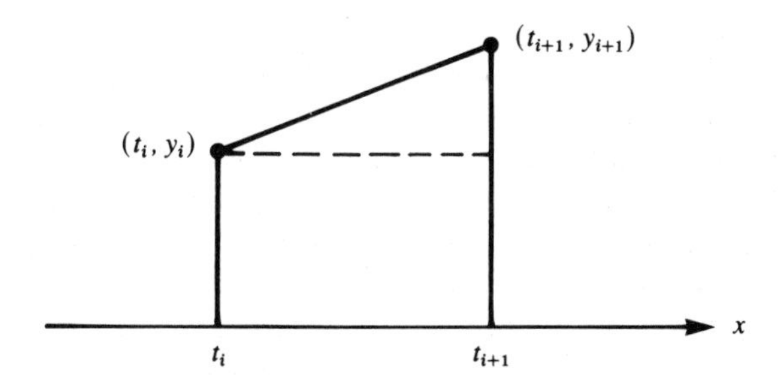

Figure 7.3

The resulting spline function S can be written in the form

$$S(x) = \begin{cases} y_1 + m_1(x - t_1), & x \in [t_1, t_2] \\ y_2 + m_2(x - t_2), & x \in [t_2, t_3] \\ \vdots & \vdots \\ y_{n-1} + m_{n-1}(x - t_{n-1}), & x \in [t_{n-1}, t_n] \end{cases} \tag{2}$$

The form of Eq. (2) is a better one than Eq. (1) for the explicit calculation of $S(x)$ because some of the quantities $x - t_i$ would need to be computed in any case simply to determine the interval containing x. The interval $[t_i, t_{i+1}]$ containing x is characterized by the fact that $x - t_{i+1}$ is the first of the quantities $x - t_1, x - t_2, \ldots$ that is *negative*.

Here is a function subprogram that accepts as input $n, t_1, \ldots, t_n, y_1, \ldots, y_n$, and x and returns $S(x)$ as given in Eq. (2). It is assumed that $t_1 < t_2 < \cdots < t_n$. If $x < t_1$, then $S(x) = y_1 + m_1(x - t_1)$; and if $x > t_n$, then $S(x) = y_{n-1} + m_{n-1}(x - t_{n-1})$.

```
      FUNCTION SPL1(N,T,Y,X)
      DIMENSION  T(N),Y(N)
      DO 2 I = 1,N-2
      TEMP = X - T(N-I)
      IF(TEMP .GE. 0.0)  GO TO 3
    2 CONTINUE
      I = N - 1
      TEMP = X - T(1)
    3 XM = (Y(N-I+1) - Y(N-I))/(T(N-I+1) - T(N-I))
      SPL1 = Y(N-I) + XM*TEMP
      RETURN
      END
```

Problems 7.1

1. Let p be a linear polynomial interpolating f at a and b. Show that $f(x) - p(x) = \frac{1}{2}f''(\xi) \times (x - a)(x - b)$ for some ξ in the interval (a, b). *Hint:* Use a result from Sec. 5.2.
2. (Continuation) Show that if $|f''(x)| \leqq M$ on the interval (a, b), then $|f(x) - p(x)| \leqq \frac{1}{8}M\ell^2$, where $\ell = b - a$.
3. (Continuation) Show that

$$f(x) - p(x) = \frac{(x - a)(x - b)}{b - a}\left[\frac{f(x) - f(b)}{x - b} - \frac{f(x) - f(a)}{x - a}\right]$$

4. (Continuation) If $|f'(x)| \leqq Q$ on (a, b), show that $|f(x) - p(x)| \leqq Q\ell/2$. *Hint:* Use the Mean Value Theorem on the result of Problem 3.
5. (Continuation) Let S be a spline function of degree 1 that interpolates f at $t_1, t_2, \ldots, t_n$. Let $t_1 < t_2 < \cdots < t_n$ and let $\delta = \max_{1 \leqq i \leqq n-1}(t_{i+1} - t_i)$. Then $|f(x) - S(x)| \leqq Q\,\delta/2$, where Q is an upper bound of $|f'(x)|$ on (t_1, t_n).
6. Let $t_1 < t_2 < \cdots < t_n$. Construct first-degree spline functions $S_1, S_2, \ldots, S_n$ by requiring that S_i vanish at $t_1, t_2, \ldots, t_{i-1}, t_{i+1}, \ldots, t_n$ but that $S_i(t_i) = 1$. Show that the first-degree spline function that interpolates f at $t_1, \ldots, t_n$ is $\sum_{i=1}^{n} f(t_i)S_i(x)$.
7. Let f be continuous on $[a, b]$. For a given $\varepsilon > 0$, let δ have the property that $|f(x) - f(y)| < \varepsilon$ whenever $|x - y| < \delta$ ("uniform continuity"). Let $n > 1 + (b - a)/\delta$. Show that there is a first-degree spline S having n knots such that $|f(x) - S(x)| < \varepsilon$ on $[a, b]$. *Hint:* Use Problem 3.
8. Show that the trapezoid rule for numerical integration results from approximating f by a first-degree spline S and then using

$$\int_a^b f(x)\,dx \approx \int_a^b S(x)\,dx$$

9. If the function $f(x) = \sin(100x)$ is to be approximated on the interval $[0, \pi]$ by an interpolating spline of degree 1, how many knots are needed to ensure that $|S(x) - f(x)| < 10^{-8}$?
10. Determine whether this function is a first-degree spline.

$$f(x) = \begin{cases} x, & -1 \leqq x \leqq 0.5 \\ 0.5 + 2(x - 0.5), & 0.5 \leqq x \leqq 2 \\ x + 1.5, & 2 \leqq x \leqq 4 \end{cases}$$

11. The simplest type of spline function is the piecewise constant function, which could be defined as

$$S(x) = \begin{cases} c_1, & t_1 \leqq x < t_2 \\ c_2, & t_2 \leqq x < t_3 \\ \vdots & \vdots \\ c_{n-1}, & t_{n-1} \leqq x \leqq t_n \end{cases}$$

Show that the indefinite integral of such a function is a polygonal function. What is the relationship between the piecewise constant functions and the rectangle rule of numerical integration? (See Problem 9, Sec. 4.2.)

Computer Problems 7.1

1. Recode subprogram SPL1 so that ascending subintervals are considered instead of descending ones. Test the code on a table of 15 unevenly spaced data points.
2. Recode subprogram SPL1 so that a *binary search* is used to find the desired interval. Test the revised code. What are the advantages and/or disadvantages of a binary search over the procedure in the text? A binary search is similar to the bisection method in that we choose t_k with $k = (i + j)/2$ or $k = (i + j + 1)/2$ and determine whether x is in $[t_i, t_k]$ or $[t_k, t_j]$.
3. Modify subprogram SPL1 so that a DO loop with a negative increment is used. Does this simplify the code?
4. Recode SPL1 in as few statements as possible.
5. A piecewise bilinear polynomial that interpolates points (x, y) specified in a rectangular grid is given by

$$p(x, y) = \frac{1}{(x_{i+1} - x_i)(y_{j+1} - y_j)}[(\ell_{ij} z_{i+1,j+1} + \ell_{i+1,j+1} z_{ij}) - (\ell_{i+1,j} z_{i,j+1} + \ell_{i,j+1} z_{i+1,j})]$$

where $\ell_{ij} = (x_i - x)(y_j - y)$. Here $x_i \leqq x \leqq x_{i+1}$ and $y_j \leqq y \leqq y_{j+1}$. The given grid (x_i, y_j) is specified by strictly increasing vectors (x_i) and (y_j) of length n and m, respectively. The given values z_{ij} at the grid points (x_i, y_j) are contained in the $n \times m$ matrix (z_{ij}), shown in Fig. 7.4. Write function subprogram BILIN(XI,N,YJ,M,Z,X,Y) to compute the value of $p(x, y)$. Test this routine on a set of 5×10 unequally spaced data points. Evaluate BILIN at four grid points and five nongrid points.

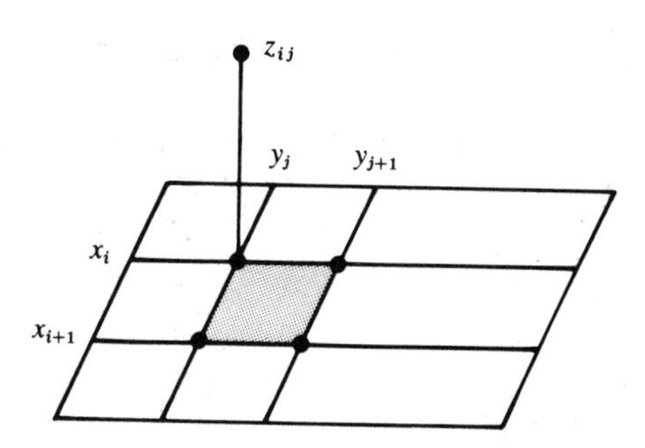

Figure 7.4

6. Write an adaptive spline interpolation subroutine. The input should be a function f, an interval $[a, b]$, and a tolerance ε. The output should be a set of knots $a = t_1 < t_2 < \cdots < t_n = b$ and a set of function values $y_i = f(t_i)$ such that the first-degree spline interpolating function S satisfies $|S(x) - f(x)| \leqq \varepsilon$ whenever x is any point $x_{ij} = t_i + j(t_{i+1} - t_i)/10$ $(1 \leqq i \leqq n - 1, 1 \leqq j \leqq 9)$.

7.2 Cubic and Quadratic Splines

The first-degree splines discussed in the preceding section, although useful in certain applications, suffer one obvious imperfection: they are not *smooth*. If the word "smooth" is interpreted intuitively, the lack of smoothness of the polygonal line (which is the graph of a first-degree spline) is immediately evident. Technically, the failure of smoothness is in the pronounced *discontinuity* of the first derivative. Thus at each knot the slope of the spline can change abruptly from one value to another.

Higher-degree splines are used whenever smoothness is needed in the approximating function. Explicitly, if we want the approximating spline to have a continuous mth derivative, a spline of degree at least $m + 1$ is selected. To see why, consider a situation in which knots $t_1 < \cdots < t_n$ have been prescribed. Suppose that a piecewise polynomial of degree m is to be defined, with its pieces joined at the knots in such a way that the resulting spline S has m continuous derivatives. At a typical interior knot t we have the following circumstances: to the left of t, $S(x) = p(x)$; and to the right of t, $S(x) = q(x)$, where p and q are mth-degree polynomials. The continuity of the mth derivative $S^{(m)}$ implies the continuity of the lower-order derivatives $S^{(m-1)}$, $S^{(m-2)}$, $\ldots$, S', S. So for any k

$$\lim_{x \to t^-} S^{(k)}(x) = \lim_{x \to t^+} S^{(k)}(x) \qquad (0 \leqq k \leqq m) \tag{1}$$

from which we conclude that

$$\lim_{x \to t^-} p^{(k)}(x) = \lim_{x \to t^+} q^{(k)}(x) \qquad (0 \leqq k \leqq m) \tag{2}$$

Here $\lim_{x \to t^+}$ means that the limit is taken over x values that converge to t from above t; that is, $(x - t)$ is positive for all x values. Similarly, $\lim_{x \to t^-}$ means that the x values converge to t from below. Since p and q are polynomials, their derivatives of all orders are continuous, and so Eq. (2) is the same as

$$p^{(k)}(t) = q^{(k)}(t) \qquad (0 \leqq k \leqq m) \tag{3}$$

This condition forces p and q to be the *same* polynomial, since by Taylor's theorem

$$p(x) = \sum_{k=0}^{m} \frac{1}{k!} p^{(k)}(t)(x - t)^k = \sum_{k=0}^{m} \frac{1}{k!} q^{(k)}(t)(x - t)^k = q(x)$$

This argument can be applied at each of the knots $t_2, \ldots, t_{n-1}$, and we see that S is simply one polynomial throughout the entire interval from t_1 to t_n. Thus we need a piecewise polynomial of degree $m + 1$ with m continuous derivatives in order to have a spline function that is not just a single polynomial throughout the entire interval. (We already know that ordinary polynomials do not serve well in curve fitting. See Sec. 5.2.)

The choice of degree most frequently made is 3. The resulting splines are, of course, termed *cubic splines*. In this case, we join cubic polynomials together in such a way that the resulting spline function has two continuous derivatives everywhere. At each knot three continuity conditions will be imposed. Since S, S', and S'' are continuous, the graph of the function will appear smooth to the eye. Discontinuities, of course, will occur in the third derivative but cannot be detected visually, which is one reason for choosing degree 3. Experience has shown, moreover, that seldom is any advantage gained by using splines of degree greater than 3. For technical reasons, odd-degree splines behave better than even-degree splines (when interpolating at the knots).

We turn next to algorithms for interpolating a given table of function values by a cubic spline whose knots coincide with the values of the independent variable in the table. As earlier, we start with the table

x	t_1	$\cdots$	t_n
y	y_1	$\cdots$	y_n

The t_i are the knots and are assumed to be arranged in ascending order.

The function S that we wish to construct consists of $n - 1$ cubic polynomial pieces, $S_1, \ldots, S_{n-1}$, with S_i being the polynomial used on interval $[t_i, t_{i+1}]$. The interpolation conditions are

$$S(t_i) = y_i \qquad (1 \leqq i \leqq n) \tag{4}$$

The continuity conditions are

$$\lim_{x \to t_i^-} S^{(k)}(x) = \lim_{x \to t_i^+} S^{(k)}(x) \qquad (0 \leqq k \leqq 2,\ 2 \leqq i \leqq n - 1) \tag{5}$$

It turns out that two further conditions must be imposed in order to use all the degrees of freedom. One choice is

$$S''(t_1) = S''(t_n) = 0 \tag{6}$$

The resulting spline function is then termed a *natural spline.*

We now verify that the number of conditions imposed equals the number of undetermined coefficients. There are n knots and hence $n - 1$ subintervals with a cubic polynomial over each. Since a cubic polynomial has four coefficients, a total of $4(n - 1)$ coefficients are available. As for conditions imposed, we have specified that within each interval the interpolating polynomial must go through two points, which gives $2(n - 1)$ conditions. The continuity adds no additional conditions. (Why?) The first and second derivatives must be continuous at the $n - 2$ interior points, for $2(n - 2)$ more conditions. The second derivatives must vanish at the two endpoints for a total of $2(n - 1) + 2(n - 2) + 2 = 4(n - 1)$ conditions.

Since S'' is continuous, the numbers $z_i = S''(t_i)$ are unambiguously defined. We do not know the values of $z_2, \ldots, z_{n-1}$, but, of course, $z_1 = z_n = 0$ by Eq. (6). If the z_i were known, we could construct S as now described.

On the interval $[t_i, t_{i+1}]$, S'' is a linear polynomial taking the values z_i and z_{i+1} at the endpoints. Thus

$$S_i''(x) = z_{i+1}\left(\frac{x - t_i}{t_{i+1} - t_i}\right) + z_i\left(\frac{t_{i+1} - x}{t_{i+1} - t_i}\right) \tag{7}$$

If this is integrated twice, we obtain S_i itself in the form

$$S_i(x) = \frac{z_{i+1}}{6(t_{i+1} - t_i)}(x - t_i)^3 + \frac{z_i}{6(t_{i+1} - t_i)}(t_{i+1} - x)^3 + C(x - t_i) + D(t_{i+1} - x) \tag{8}$$

Here C and D are constants of integration. The interpolation conditions $S_i(t_i) = y_i$ and $S_i(t_{i+1}) = y_{i+1}$ can be imposed now to determine the appropriate values of C and D. The reader should do so and verify that the result is as follows, with $h_i = t_{i+1} - t_i$.

$$S_i(x) = \frac{z_{i+1}}{6h_i}(x - t_i)^3 + \frac{z_i}{6h_i}(t_{i+1} - x)^3$$
$$+ \left(\frac{y_{i+1}}{h_i} - \frac{z_{i+1}h_i}{6}\right)(x - t_i) + \left(\frac{y_i}{h_i} - \frac{z_ih_i}{6}\right)(t_{i+1} - x) \tag{9}$$

When the values $z_1, \ldots, z_n$ have been determined, the spline function can be computed from Eq. (9).

One condition remains to be imposed, namely, the continuity of S'. At t_i $(2 \leqq i \leqq n - 1)$ we must have $S'_{i-1}(t_i) = S'_i(t_i)$, as can be seen in Fig. 7.5. We have from Eq. (9)

$$S'_i(x) = \frac{z_{i+1}}{2h_i}(x - t_i)^2 - \frac{z_i}{2h_i}(t_{i+1} - x)^2 + \frac{y_{i+1}}{h_i} - \frac{z_{i+1}h_i}{6} - \frac{y_i}{h_i} + \frac{z_ih_i}{6} \tag{10}$$

which gives

$$S'_i(t_i) = -\frac{h_i}{3}z_i - \frac{h_i}{6}z_{i+1} - \frac{y_i}{h_i} + \frac{y_{i+1}}{h_i} \tag{11}$$

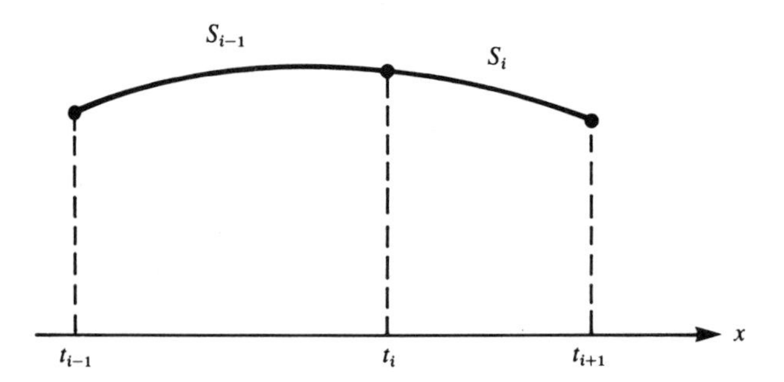

Figure 7.5

Analogously, we have

$$S'_{i-1}(t_i) = \frac{h_{i-1}}{6}z_{i-1} + \frac{h_{i-1}}{3}z_i - \frac{y_{i-1}}{h_{i-1}} + \frac{y_i}{h_{i-1}} \tag{12}$$

When these are set equal to each other, the resulting equation can be rearranged as

$$h_{i-1}z_{i-1} + 2(h_{i-1} + h_i)z_i + h_iz_{i+1} = \frac{6}{h_i}(y_{i+1} - y_i) - \frac{6}{h_{i-1}}(y_i - y_{i-1}) \tag{13}$$

for $2 \leqq i \leqq n - 1$. In this set of $n - 2$ equations everything is known except $z_2, \ldots, z_{n-1}$. The values $z_1 = z_n = 0$ have already been assigned. Observe that the system of equations is linear, symmetric, and tridiagonal. So in the ith equation only the variables z_{i-1}, z_i, and z_{i+1} appear. The subroutine TRI developed in Chapter 6 is applicable.

Alternatively, one can write a tridiagonal algorithm especially for system (13).

1. Compute $h_i = t_{i+1} - t_i$ for $i = 1, 2, \ldots, n - 1$.
2. Compute $b_i = 6(y_{i+1} - y_i)/h_i$ for $i = 1, 2, 3, \ldots, n - 1$.

3. Put $u_2 = 2(h_1 + h_2)$ and compute inductively $u_i = 2(h_i + h_{i-1}) - h_{i-1}^2/u_{i-1}$ for $i = 3, 4, \ldots, n-1$.
4. Put $v_2 = b_2 - b_1$ and compute inductively $v_i = b_i - b_{i-1} - h_{i-1}v_{i-1}/u_{i-1}$ for $i = 3, 4, \ldots, n-1$.
5. Put $z_n = 0$ and compute inductively $z_i = (v_i - h_i z_{i+1})/u_i$ for $i = n-1, n-2, \ldots, 2$.

This algorithm could conceivably fail because of zero divisions in steps 3, 4, and 5. Thus it is incumbent upon us to establish that $u_i \neq 0$ for all i. It is clear that $u_2 > h_2 > 0$. If $u_{i-1} > h_{i-1}$, then $u_i > h_i$ because

$$u_i = 2(h_i + h_{i-1}) - h_{i-1}\frac{h_{i-1}}{u_{i-1}} > 2(h_i + h_{i-1}) - h_{i-1} > h_i$$

Then by induction, $u_i > 0$ for $i = 2, \ldots, n-1$.

We use subroutine TRI from Chapter 6 to solve the $n \times n$ symmetric tridiagonal system for $z_1, z_2, \ldots, z_n$. Using $h_i = t_{i+1} - t_i$ and Eq. (13), we set the diagonal array D, the superdiagonal array C, and the right-hand side array Z as

$$d_1 = d_n = 1, \qquad c_1 = c_{n-1} = 0, \qquad z_1 = z_n = 0$$

$$d_i = 2(t_{i+1} - t_{i-1}) \qquad (2 \leqq i \leqq n-1)$$

$$c_i = t_{i+1} - t_i \qquad (2 \leqq i \leqq n-2)$$

$$z_i = 6\left(\frac{y_{i+1} - y_i}{t_{i+1} - t_i} - \frac{y_i - y_{i-1}}{t_i - t_{i-1}}\right) \qquad (2 \leqq i \leqq n-1)$$

Using the results of Problem 3 with nested multiplication to evaluate the spline function, we obtain the following routines for cubic spline interpolation of a table of values.

```
      SUBROUTINE ZSPL3(N,T,Y,D,C,Z)
      DIMENSION  T(N),Y(N),D(N),C(N),Z(N)
      D(1) = 1.0
      C(1) = 0.0
      Z(1) = 0.0
      DO 2 I = 2,N-1
      D(I) = 2.0*(T(I+1) - T(I-1))
      C(I) = T(I+1) - T(I)
      TEMP = (Y(I+1) - Y(I))/(T(I+1) - T(I))
    2 Z(I) = 6.0*(TEMP - (Y(I) - Y(I-1))/(T(I) - T(I-1)))
      D(N) = 1.0
      C(N-1) = 0.0
      Z(N) = 0.0
      CALL TRI(N,C,D,C,Z)
      RETURN
      END
```

```
      FUNCTION SPL3(N,T,Y,Z,X)
      DIMENSION  T(N),Y(N),Z(N)
      DO 2 J = 1,N-2
      I = N - J
      TEMP = X - T(I)
      IF(TEMP .GE. 0.0)  GO TO 3
    2 CONTINUE
      I = 1
      TEMP = X - T(1)
    3 H = T(I+1) - T(I)
      A = TEMP*(Z(I+1) - Z(I))/(6.0*H) + 0.5*Z(I)
      B = TEMP*A + (Y(I+1) - Y(I))/H - H*(2.0*Z(I) + Z(I+1))/6.0
      SPL3 = TEMP*B + Y(I)
      RETURN
      END
```

Notice that after the routine ZSPL3 is called, the subprogram SPL3 can be used to evaluate the resulting cubic spline function for particular X values. The parameters N, T, Y, and Z should not be changed between calls to SPL3.

To illustrate the use of these cubic spline routines, we rework an example from Sec. 5.1. Taking the same ten points over the interval $[0, \pi/2]$ for the sine function, we use ZSPL3/SPL3 to interpolate at $\pi/2$ and $\pi/6$. Here is the program.

```
      DIMENSION  T(10),Y(10),D(10),C(10),Z(10)
      PI2 = 2.0*ATAN(1.0)
      PI6 = PI2/3.0
      DO 2 I = 1,10
      T(I) = FLOAT(I-1)*0.1875
    2 Y(I) = SIN(T(I))
      CALL ZSPL3(10,T,Y,D,C,Z)
      R = SPL3(10,T,Y,Z,PI2)
      S = SPL3(10,T,Y,Z,PI6)
      PRINT 3,R,S
    3 FORMAT(5X,2F20.15)
      END
```

The results are

$$\sin\frac{\pi}{2} \approx 0.998703839768172 \qquad \sin\frac{\pi}{6} \approx 0.499998937158395$$

These values are *not* as accurate as those from the Newton interpolating polynomial! (Explain.)

Here is an example of curve fitting, using both the polynomial interpolation routine COEF from Chapter 5 and the cubic spline routine SPL3. In order to compare the results, we select 13 points on the well-known "serpentine curve" given by

$$y = \frac{x}{\frac{1}{4} + x^2}$$

So that the knots will not be equally spaced, we write the curve in parametric form

$$\begin{cases} x = \dfrac{1}{2}\cot\theta \\ y = \sin 2\theta \end{cases}$$

and take $\theta = i(\pi/14)$, where $i = -6, -5, \ldots, 5, 6, 7$.

Figure 7.6 shows the resulting cubic spline curve and the polynomial curve (dashed lines) from an automatic plotter. The polynomial becomes extremely erratic after the fourth knot from the origin and oscillates wildly, whereas the spline is a near-perfect fit.

A useful approximation process, first proposed by Subbotin, consists of interpolation with *quadratic* splines, where the nodes for interpolation are chosen to be the first and last knot and the midpoints between the knots. Remember that *knots* are, by definition, the points where the spline function is permitted to change in form from one polynomial to another. The *nodes* will be the points where values of the spline are specified. We will outline the theory here, leaving details to be filled in by the reader.

Suppose that knots $a = t_1 < t_2 < \cdots < t_n = b$ have been specified and let the nodes be the points

$$\begin{cases} \tau_0 = t_1, \qquad \tau_n = t_n \\ \tau_i = \dfrac{1}{2}(t_i + t_{i+1}) \qquad (1 \leqq i \leqq n - 1) \end{cases}$$

We seek a quadratic spline function S having the given knots and taking prescribed values at the nodes:

$$S(\tau_i) = y_i \qquad (0 \leqq i \leqq n)$$

The knots create $n - 1$ subintervals, and in each of them S can be a different quadratic polynomial. Let us say that on $[t_i, t_{i+1}]$ S is equal to the quadratic polynomial S_i. Since S is a quadratic spline, it and its first derivative should be continuous. Thus $z_i = S'(t_i)$ is well-defined, although as yet we do not know its value. It is easy to see that on $[t_i, t_{i+1}]$ our quadratic polynomial is representable in the form

$$S_i(x) = y_i + \frac{1}{2}(z_{i+1} + z_i)(x - \tau_i) + \frac{1}{2h_i}(z_{i+1} - z_i)(x - \tau_i)^2 \tag{14}$$

in which $h_i = t_{i+1} - t_i$. In order to verify the correctness of Eq. (14), we must check that $S_i(\tau_i) = y_i$, $S_i'(t_i) = z_i$, and $S_i'(t_{i+1}) = z_{i+1}$. When the polynomial pieces S_1, $S_2, \ldots, S_{n-1}$ are joined together to form S, the result may be discontinuous. Hence we impose continuity conditions at the interior knots:

$$\lim_{x \to t_i^-} S(x) = \lim_{x \to t_i^+} S(x) \qquad (2 \leqq i \leqq n - 1)$$

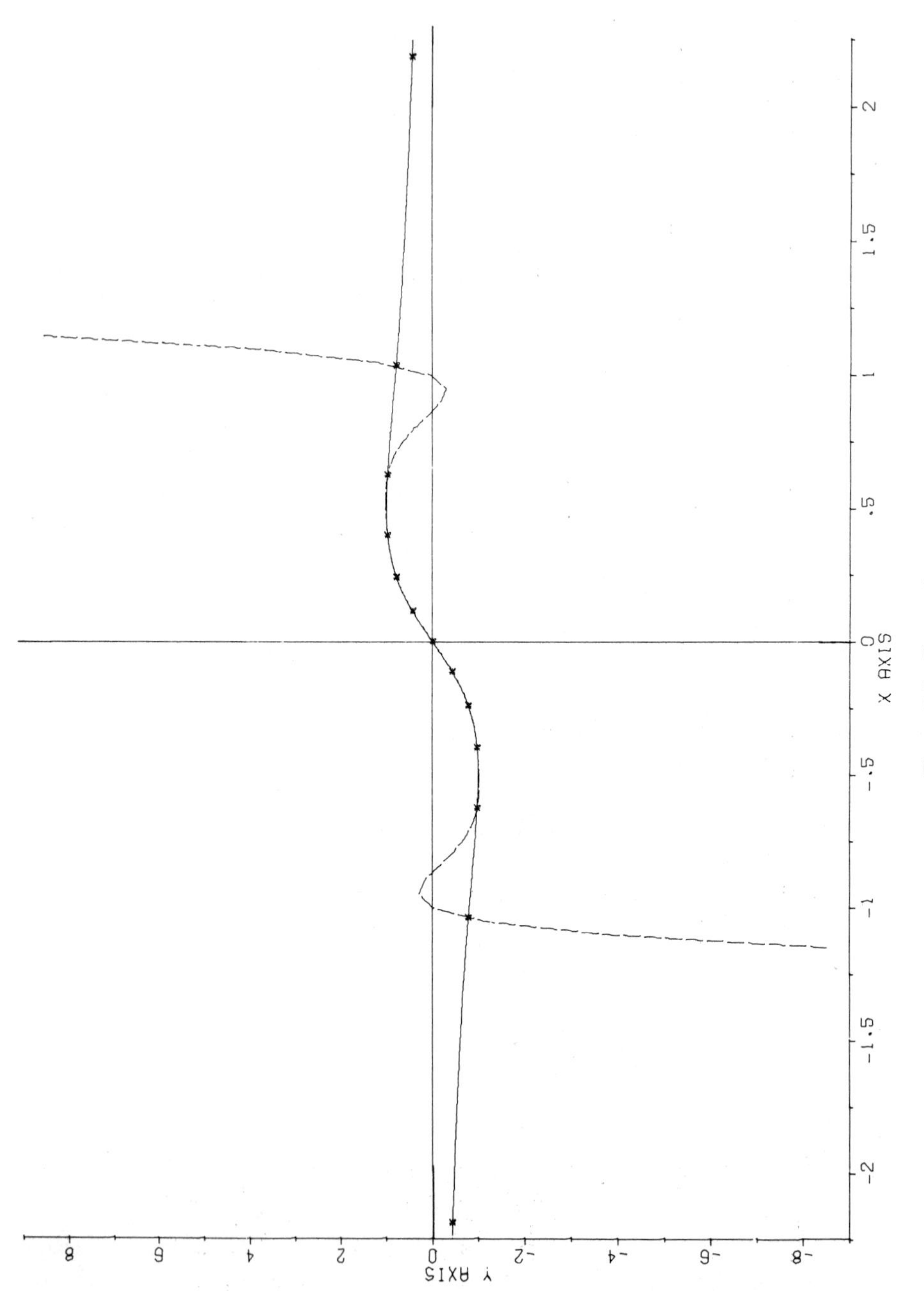

Figure 7.6 Serpentine curve

The reader will enjoy carrying out this analysis, which leads to

$$h_{i-1}z_{i-1} + 3(h_{i-1} + h_i)z_i + h_i z_{i+1} = 8(y_i - y_{i-1}) \qquad (2 \leqq i \leqq n-1) \tag{15}$$

The first and last interpolation condition must also be imposed:

$$S(\tau_0) = y_0 \qquad S(\tau_n) = y_n$$

These two equations lead to

$$\begin{aligned} 3h_1 z_1 + h_1 z_2 &= 8(y_1 - y_0) \\ 3h_{n-1}z_n + h_{n-1}z_{n-1} &= 8(y_n - y_{n-1}) \end{aligned}$$

The system of equations governing the **z** vector thus can be written in the matrix form

$$\begin{bmatrix} 3h_1 & h_1 & & & & & \\ h_1 & 3(h_1 + h_2) & h_2 & & & \mathbf{0} & \\ & h_2 & 3(h_2 + h_3) & h_3 & & & \\ & & \ddots & \ddots & \ddots & & \\ & \mathbf{0} & & & h_{n-2} & 3(h_{n-2} + h_{n-1}) & h_{n-1} \\ & & & & & h_{n-1} & 3h_{n-1} \end{bmatrix} \begin{bmatrix} z_1 \\ z_2 \\ z_3 \\ \vdots \\ z_{n-1} \\ z_n \end{bmatrix} = 8 \begin{bmatrix} y_1 - y_0 \\ y_2 - y_1 \\ y_3 - y_2 \\ \vdots \\ y_{n-1} - y_{n-2} \\ y_n - y_{n-1} \end{bmatrix}$$

This system of n equations in n unknowns can be conveniently solved by subroutine TRI in Chapter 6. After the **z** vector has been obtained, values of $S(x)$ can be computed from Eq. (14). The writing of suitable code to carry out this interpolation method is left as a programming project.

Why do spline functions serve the needs of data fitting better than ordinary polynomials? In order to answer this, one should understand that interpolation by polynomials of high degree is often unsatisfactory because they may exhibit wild *oscillations*. Polynomials are smooth in the technical sense of possessing continuous derivatives of all orders, whereas, in this same sense, spline functions are *not* smooth.

Wild oscillations in a function can be attributed to its derivatives being very *large*. Consider the function whose graph is shown in Fig. 7.7. The slope of the chord joining the points p and q is very large in magnitude. By the Mean Value Theorem, the slope

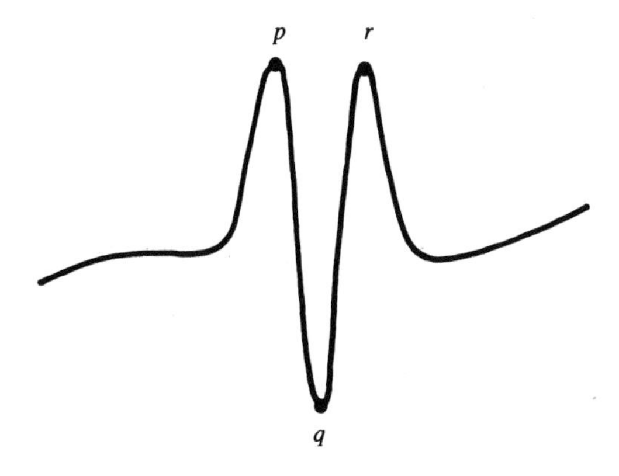

Figure 7.7

of that chord is the value of the derivative at some point. Thus the derivative must attain large values. Indeed, somewhere on the curve between p and q there is a point where $f'(x)$ is large and negative. Similarly, between q and r there is a point where $f'(x)$ is large and positive. Hence there is a point on the curve between p and r where $f''(x)$ is large. This reasoning can be continued if there are more oscillations. This is the behavior that spline functions do *not* exhibit. In fact, the following formal result shows that from a certain point of view natural cubic splines are the *best* functions to employ for curve fitting.

Let S be the natural cubic spline function that interpolates the function f at knots $t_1 < t_2 < \cdots < t_n$. Let f'' be continuous in an open interval (a, b) that contains the knots. Then

$$\int_a^b [S''(x)]^2\, dx \leqq \int_a^b [f''(x)]^2\, dx$$

We shall show why this is so presently. The interpretation of the integral inequality is that the average value of $(S'')^2$ on the interval $[a, b]$ is never larger than the corresponding quantity for any other function that takes the same values at the knots.

In order to verify the assertion about $(S'')^2$, we put $g = f - S$ so that $g(t_i) = 0$ $(1 \leqq i \leqq n)$ and $f'' = S'' + g''$. Therefore

$$\int_a^b (f'')^2\, dx = \int_a^b (S'')^2\, dx + \int_a^b (g'')^2\, dx + 2\int_a^b S''g''\, dx$$

If the last integral were zero, we would be finished, for then

$$\int_a^b (f'')^2\, dx = \int_a^b (S'')^2\, dx + \int_a^b (g'')^2\, dx \geqq \int_a^b (S'')^2\, dx$$

We apply the technique of integration by parts to the integral in question to show that it is zero.

$$\int_a^b S''g''\, dx = S''g' \Big|_a^b - \int_a^b S'''g'\, dx = -\int_a^b S'''g'\, dx$$

Here use has been made of the fact that S is a *natural* cubic spline. It is, therefore, a first-degree polynomial in the intervals $(-\infty, t_1)$ and (t_n, ∞). In particular, $S''(a) = S''(b) = 0$. Next, use the fact that S''' is zero in (a, t_1) and (t_n, b) so that

$$\int_a^b S'''g'\, dx = \int_{t_1}^{t_n} S'''g'\, dx = \int_{t_1}^{t_2} S'''g'\, dx + \cdots + \int_{t_{n-1}}^{t_n} S'''g'\, dx$$

Since S is a cubic polynomial in each interval (t_i, t_{i+1}), its third derivative there is a constant, c_i. So

$$\begin{aligned}\int_a^b S'''g'\, dx &= c_1 \int_{t_1}^{t_2} g'\, dx + \cdots + c_{n-1}\int_{t_{n-1}}^{t_n} g'\, dx \\ &= c_1[g(t_2) - g(t_1)] + \cdots + c_{n-1}[g(t_n) - g(t_{n-1})]\end{aligned}$$

This, of course, is zero, since $g(t_i) = 0$ for every i.

Problems 7.2

1. Let S be a cubic spline having knots $t_1 < t_2 < \cdots < t_n$. Suppose that on the two intervals $[t_1, t_2]$ and $[t_3, t_4]$ S reduces to linear polynomials. What can be said of S on $[t_2, t_3]$?
2. In the construction of the cubic interpolating spline, carry out the evaluation of constants C and D, and thus justify Eq. (9).
3. Show that S_i can also be written in the form

$$S_i(x) = y_i + A_i(x - t_i) + \frac{1}{2}z_i(x - t_i)^2 + \frac{z_{i+1} - z_i}{6h_i}(x - t_i)^3$$

with

$$A_i = -\frac{h_i}{3}z_i - \frac{h_i}{6}z_{i+1} - \frac{y_i}{h_i} + \frac{y_{i+1}}{h_i}$$

4. Carry out the details in deriving Eqs. (12) and (13).
5. Verify that the algorithm for computing the z vector is correct by showing that if z satisfies Eq. (13), then it satisfies the equation in step 5 of the algorithm.
6. Establish that $u_i > 2h_i + \frac{3}{2}h_{i-1}$ in the algorithm for determining the cubic spline interpolant.
7. By hand calculation, find the natural cubic spline interpolant for this table.

x	1	2	3	4	5
y	0	1	0	1	0

8. This problem and the next three concern quadratic interpolating splines. We want to interpolate a table (t_i, y_i), $1 \leqq i \leqq n$, where $t_1 < \cdots < t_n$, using a quadratic spline function. Show that the quadratic function

$$q_i(x) = \frac{z_{i+1}}{2h_i}(x - t_i)^2 - \frac{z_i}{2h_i}(t_{i+1} - x)^2 + y_i + \frac{z_i h_i}{2}$$

has these properties: $q_i(t_i) = y_i$, $q_i'(t_i) = z_i$, $q_i'(t_{i+1}) = z_{i+1}$. Here $h_i = t_{i+1} - t_i$.

9. (Continuation) Let $Q = q_i$ on $[t_i, t_{i+1}]$ with q_i as above. Impose the continuity condition on Q and get the system of equations $z_i + z_{i-1} = 2(y_i - y_{i-1})/h_{i-1}$ $(2 \leqq i \leqq n - 1)$.
10. (Continuation) Show that the quadratic function in Problem 8 can be written

$$q_i(x) = \frac{z_{i+1} - z_i}{2h_i}(x - t_i)^2 + z_i(x - t_i) + y_i$$

11. (Continuation) Show by induction that the following recursive definition of $z_1, \ldots, z_n$, together with the formula of Problem 10, produces an interpolating quadratic spline: z_1 arbitrary, $z_i = 2(y_i - y_{i-1})/h_{i-1} - z_{i-1}$ $(2 \leqq i \leqq n)$.
12. Using the three preceding problems, find a quadratic spline interpolant for these data.

x	-1	0	$\frac{1}{2}$	1	2	$\frac{5}{2}$
y	2	1	0	1	2	3

Computer Problems 7.2

1. Write and test a subroutine for the tridiagonal algorithm (given in this section) especially designed for system (13). Is there any advantage to using this algorithm instead of `TRI`?

2. Draw a free-form curve on graph paper, making certain that the curve is the graph of a function. Then read values of your function at a reasonable number of points, say 10 to 50, and compute the cubic spline function that takes those values. Compare the freely drawn curve to the graph of the cubic spline.
3. Let S be the cubic spline function that interpolates the function $f(x) = (x^2 + 1)^{-1}$ at 41 equally spaced knots in the interval $[-5, 5]$. Evaluate $S(x) - f(x)$ at 101 equally spaced points on the interval $[0, 5]$.
4. Write and test `ZSPL2(N,T,Y,Z)`, which computes the z vector of Problems 8 to 11 in the preceding problem set. (Seven Fortran statements will suffice.) Then write and test `FUNCTION SPL2(N,T,Y,Z,X)` to evaluate the quadratic interpolating spline at x. (Eight statements will suffice.) Problem 12 of the previous set provides a suitable test.
5. Repeat Computer Problem 2 of Sec. 7.1 for `ZSPL3/SPL3`.
6. Repeat Computer Problem 3 of Sec. 7.1 for `ZSPL3/SPL3`.
7. Draw a spiral (or other curve that is not a function) and reproduce it by spline functions as follows. Select points on the curve and label them $t = 1, 2, 3, \ldots$. For each value of t, read off the x and y coordinates of the point, thus producing a table.

t	1	2	$\cdots$	n
x	x_1	x_2	$\cdots$	x_n
y	y_1	y_2	$\cdots$	y_n

Then fit $x = S(t)$ and $y = \bar{S}(t)$, where S and $\bar{S}$ are cubic spline interpolants. S and $\bar{S}$ give a parametric representation of the curve (see Fig. 7.8).

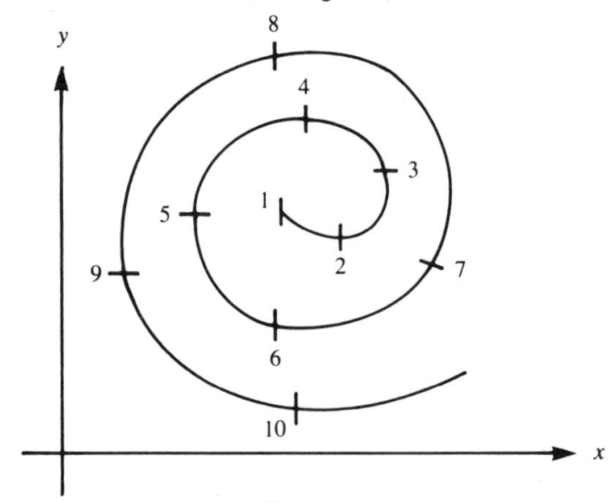

Figure 7.8

8. Write subroutines, *as simple as possible*, to perform natural cubic spline interpolation with equally spaced knots.
9. Verify the correctness of the equations in the text pertaining to Subbotin's spline interpolation process. Then program subroutines to carry out the procedure.
10. Write a program to estimate $\int_a^b f(x)\,dx$, assuming that we know only the values of f at certain prescribed knots $a = t_1 < t_2 < \cdots < t_n = b$. Approximate f first by an interpolating cubic spline and then compute the integral of it, using Eq. (9).
11. Write a subroutine to estimate $f'(x)$ for any x in $[a, b]$, assuming that we know only the values of f at knots $a = t_1 < \cdots < t_n = b$. Proceed as in Problem 10.
12. Construct a subroutine for evaluating $\int_a^b f(x)\,dx$ by interpolating f at n equally spaced knots with a natural cubic spline and taking the integral of the spline.

7.3 B Splines

In this section we give an introduction to the theory of *B splines*. These are special spline functions that are well-adapted to numerical tasks and are being used more and more frequently in production-type programs for approximating data. Thus the intelligent user of library codes should have some familiarity with them.

The B splines were so named because they formed a *basis* for the set of all splines. (We prefer the more romantic description of "Bell" splines because of their characteristic shape!)

Throughout this section we suppose that an infinite set of knots has been prescribed in such a way that

$$\begin{cases} \cdots < t_{-2} < t_{-1} < t_0 < t_1 < t_2 < \cdots \\ \lim_{i\to\infty} t_i = \infty = -\lim_{i\to\infty} t_{-i} \end{cases} \tag{1}$$

The B splines to be defined now depend on this set of knots, although the notation does not show that dependence. The B splines of degree 0 are defined by

$$B_i^0(x) = \begin{cases} 1 & \text{if } t_i \leqq x < t_{i+1} \\ 0 & \text{otherwise} \end{cases} \tag{2}$$

The graph of B_i^0 is shown in Fig. 7.9.

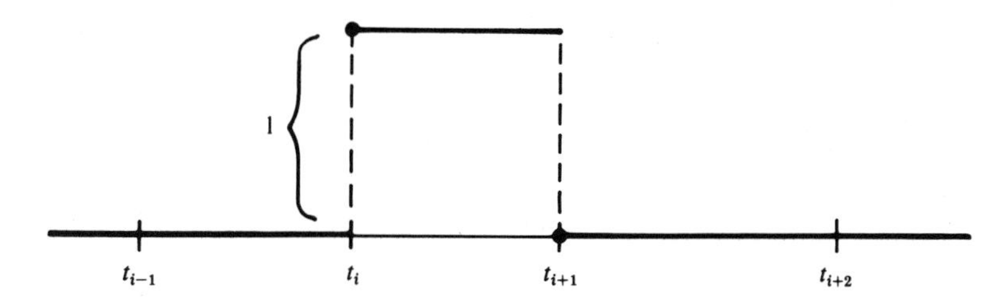

Figure 7.9

Obviously B_i^0 is discontinuous. However, it is continuous from the right at all points, even where the jumps occur. Thus

$$\lim_{x\to t_i^+} B_i^0(x) = 1 = B_i^0(t_i)$$

$$\lim_{x\to t_{i+1}^+} B_i^0(x) = 0 = B_i^0(t_{i+1})$$

If the "support" of a function f is defined as the set of points x where $f(x) \neq 0$, then we can say that the support of B_i^0 is the "half open" interval $[t_i, t_{i+1})$. Since B_i^0 is a piecewise constant function, it is a spline of degree 0 in the terminology of Sec. 7.1.

Two further observations can be made.

$$B_i^0(x) \geqq 0 \quad \text{for all } x \text{ and for all } i$$

$$\sum_{i=-\infty}^{\infty} B_i^0(x) = 1 \quad \text{for all } x$$

Although the second of these assertions contains an infinite series, there is no question of convergence, because for each x only one term in the series is different from zero. Indeed, for fixed x, there is a unique integer m such that $t_m \leqq x < t_{m+1}$, and then

$$\sum_{i=-\infty}^{\infty} B_i^0(x) = B_m^0(x) = 1$$

The reader should now see the reason for defining B_i^0 in the manner of Eq. (2).

A final remark concerning these B splines of degree 0 is this: any spline of degree 0 that is continuous from the right and is based on the knots (1) can be expressed as a linear combination of the B splines B_i^0. Indeed, if S is such a function, then it can be specified by a rule such as

$$S(x) = b_i \quad \text{if } t_i \leqq x < t_{i+1} \qquad (i = 0, \pm 1, \pm 2, \ldots)$$

Then S can be written

$$S = \sum_{i=-\infty}^{\infty} b_i B_i^0$$

With the functions B_i^0 as starting point, we now generate all the higher-degree B splines by a simple recursive definition.

$$B_i^k(x) = \left(\frac{x - t_i}{t_{i+k} - t_i}\right) B_i^{k-1}(x) + \left(\frac{t_{i+k+1} - x}{t_{i+k+1} - t_{i+1}}\right) B_{i+1}^{k-1}(x) \qquad (k \geqq 1) \tag{3}$$

Here $k = 1, 2, 3, \ldots$ and $i = 0, \pm 1, \pm 2, \ldots$.

To illustrate Eq. (3), let us determine B_i^1 in an alternative form.

$$B_i^1(x) = \left(\frac{x - t_i}{t_{i+1} - t_i}\right) B_i^0(x) + \left(\frac{t_{i+2} - x}{t_{i+2} - t_{i+1}}\right) B_{i+1}^0(x)$$

$$= \begin{cases} 0 & \text{if } x \geqq t_{i+2} \text{ or } x \leqq t_i \\ \dfrac{x - t_i}{t_{i+1} - t_i} & \text{if } t_i < x < t_{i+1} \\ \dfrac{t_{i+2} - x}{t_{i+2} - t_{i+1}} & \text{if } t_{i+1} \leqq x < t_{i+2} \end{cases}$$

The graph of B_i^1 is shown in Fig. 7.10. The support of B_i^1 is the open interval (t_i, t_{i+2}). It is true, but perhaps not so obvious, that $\sum_{i=-\infty}^{\infty} B_i^1(x) = 1$ and that every spline of degree 1 based on the knots (1) is a linear combination of B_i^1.

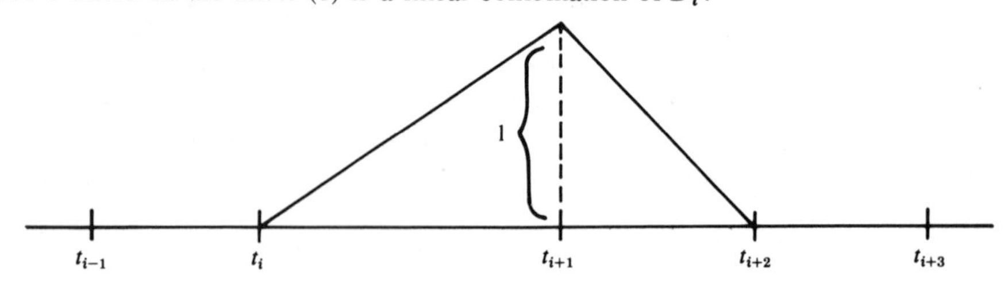

Figure 7.10

The functions B_i^k as defined by Eq. (3) are called *B splines of degree k*. Since each B_i^k is obtained by applying linear factors to B_i^{k-1} and B_{i-1}^{k-1}, we see that the degrees actually rise by one at each step. So B_i^1 is piecewise linear, B_i^2 is piecewise quadratic, and so on.

It is also easily shown by induction that $B_i^k(x) = 0$ outside the interval $[t_i, t_{i+k+1})$. To establish this, we start by observing that it is true when $k = 0$ because of definition (2). If it is true for index $k - 1$, then it is true for index k by the following reasoning. The inductive hypothesis tells us that $B_i^{k-1}(x) = 0$ if x is outside $[t_i, t_{i+k})$ and that $B_{i+1}^{k-1}(x) = 0$ if x is outside $[t_{i+1}, t_{i+k+1})$. If x is outside *both* intervals, it is outside their union, $[t_i, t_{i+k+1})$; then both terms on the right side of Eq. (3) are zero. So $B_i^k(x) = 0$ outside $[t_i, t_{i+k+1})$. That $B_i^k(t_i) = 0$ follows directly from Eq. (3), and so we know that $B_i^k(x) = 0$ for all x outside (t_i, t_{i+k+1}), if $k \geqq 1$.

Complementary to the property just established, we can show, again by induction, that $B_i^k(x) > 0$ on (t_i, t_{i+k+1}). By Eq. (2), this assertion is true when $k = 0$. If it is true for index $k - 1$, then $B_i^{k-1}(x) > 0$ on (t_i, t_{i+k}) and $B_{i+1}^{k-1}(x) > 0$ on (t_{i+1}, t_{i+k+1}). In Eq. (3) the factors that multiply $B_i^{k-1}(x)$ and $B_{i+1}^{k-1}(x)$ are positive when $t_i < x < t_{i+k+1}$. Thus $B_i^k(x) > 0$ on this interval.

The principal use of the B splines B_i^k $(i = 0, \pm 1, \pm 2, \ldots)$ is as a basis for the set of all kth-degree splines having the same knot sequence. Thus linear combinations $\sum_{i=-\infty}^{\infty} c_i B_i^k$ are important objects of study. (We use c_i for fixed k and C_i^k to emphasize the degree k of the corresponding B splines.) Our first task is to develop an efficient method to evaluate a function of the form

$$f(x) = \sum_{i=-\infty}^{\infty} C_i^k B_i^k(x) \tag{4}$$

under the supposition that the coefficients C_i^k are given (as well as the knot sequence t_i). Using definition (3) and some simple series manipulations, we have

$$\begin{aligned} f(x) &= \sum_{i=-\infty}^{\infty} C_i^k \left[\left(\frac{x - t_i}{t_{i+k} - t_i} \right) B_i^{k-1}(x) + \left(\frac{t_{i+k+1} - x}{t_{i+k+1} - t_{i+1}} \right) B_{i+1}^{k-1}(x) \right] \\ &= \sum_{i=-\infty}^{\infty} \left[C_i^k \left(\frac{x - t_i}{t_{i+k} - t_i} \right) + C_{i-1}^k \left(\frac{t_{i+k} - x}{t_{i+k} - t_i} \right) \right] B_i^{k-1}(x) \\ &= \sum_{i=-\infty}^{\infty} C_i^{k-1} B_i^{k-1}(x) \end{aligned} \tag{5}$$

where C_i^{k-1} is defined to be the appropriate coefficient from the preceding line.

This algebraic manipulation shows how a linear combination of $B_i^k(x)$ can be expressed as a linear combination of $B_i^{k-1}(x)$. Repeating this process $k - 1$ times, we eventually express $f(x)$ in the form

$$f(x) = \sum_{i=-\infty}^{\infty} C_i^0 B_i^0(x) \tag{6}$$

If $t_m \leqq x < t_{m+1}$, then $f(x) = C_m^0$. The formula by which the coefficients C_i^{j-1} are obtained is

$$C_i^{j-1} = \frac{C_i^j(x - t_i) + C_{i-1}^j(t_{i+j} - x)}{t_{i+j} - t_i} \tag{7}$$

A nice feature of Eq. (4) is that only the $k + 1$ coefficients $C_m^k, C_{m-1}^k, \ldots, C_{m-k}^k$ are needed to compute $f(x)$ if $t_m \leqq x < t_{m+1}$. (See Problem 6.) Thus if f is defined by Eq. (4) and we want to compute $f(x)$, we use Eq. (7) to calculate the entries in the following triangular array.

$$\begin{array}{llll} C_m^k & C_m^{k-1} & \cdots & C_m^0 \\ C_{m-1}^k & C_{m-1}^{k-1} & \cdots & \\ \vdots & \vdots & & \\ C_{m-k}^k & & & \end{array}$$

Although our notation does not show it, the coefficients in Eq. (4) are independent of x, whereas the C_i^{j-1} calculated subsequently by Eq. (7) do depend on x.

It is now a simple matter to establish that $\sum_{i=-\infty}^{\infty} B_i^k(x) = 1$ for all x and all k. If $k = 0$, we already know this. If $k > 0$, we use Eq. (4) with $C_i^k = 1$ for all i. By Eq. (7), all subsequent coefficients $C_i^k, C_i^{k-1}, C_i^{k-2}, \ldots, C_i^0$ are also equal to 1 (induction is needed here!). Thus at the end Eq. (6) is true with $C_i^0 = 1$ and so $f(x) = 1$. Therefore from (4) the sum of all B splines of degree k is unity.

The smoothness of the B splines B_i^k increases with the index k. In fact, we can show by induction that B_i^k has a continuous $(k - 1)$st derivative.

The B splines can be used as substitutes for complicated functions in many mathematical situations. Differentiation and integration are important examples. A basic result about the derivatives of B splines is

$$\frac{d}{dx} B_i^k(x) = \left(\frac{k}{t_{i+k} - t_i}\right) B_i^{k-1}(x) - \left(\frac{k}{t_{i+k+1} - t_{i+1}}\right) B_{i+1}^{k-1}(x) \tag{8}$$

This equation can be proved by induction, using the recursive formula (3). Once (8) is established, we get the useful formula

$$\begin{cases} \dfrac{d}{dx} \displaystyle\sum_{i=-\infty}^{\infty} c_i B_i^k(x) = \sum_{i=-\infty}^{\infty} d_i B_i^{k-1}(x) \\ \\ d_i = k \dfrac{c_i - c_{i-1}}{t_{i+k} - t_i} \end{cases} \tag{9}$$

The verification is as follows. By Eq. (8),

$$\begin{aligned} \frac{d}{dx} \sum_{i=-\infty}^{\infty} c_i B_i^k(x) &= \sum_{i=-\infty}^{\infty} c_i \frac{d}{dx} B_i^k(x) \\ &= \sum_{i=-\infty}^{\infty} c_i \left[\left(\frac{k}{t_{i+k} - t_i}\right) B_i^{k-1}(x) - \left(\frac{k}{t_{i+k+1} - t_{i+1}}\right) B_{i+1}^{k-1}(x)\right] \\ &= \sum_{i=-\infty}^{\infty} \left[\left(\frac{c_i k}{t_{i+k} - t_i}\right) - \left(\frac{c_{i-1} k}{t_{i+k} - t_i}\right)\right] B_i^{k-1}(x) \\ &= \sum_{i=-\infty}^{\infty} d_i B_i^{k-1}(x) \end{aligned}$$

For numerical integration, the B splines are also recommended, especially for indefinite integration. Here is the basic result needed for integration.

$$\int_{-\infty}^{x} B_i^k(s)\,ds = \left(\frac{t_{i+k+1} - t_i}{k+1}\right) \sum_{j=i}^{\infty} B_j^{k+1}(x) \tag{10}$$

It can be verified by differentiating both sides with respect to x and simplifying by use of Eq. (9). In order to be sure that the two sides of (10) do not differ by a constant, we note that for any $x < t_i$ both sides reduce to zero.

The basic result (10) produces this useful formula

$$\begin{cases} \displaystyle\int_{-\infty}^{x} \sum_{i=-\infty}^{\infty} c_i B_i^k(s)\,ds = \sum_{i=-\infty}^{\infty} e_i B_i^{k+1}(x) \\ \displaystyle e_i = \frac{1}{k+1} \sum_{j=-\infty}^{i} c_j(t_{j+k+1} - t_j) \end{cases} \tag{11}$$

It should be emphasized that this formula gives an indefinite integral (antiderivative) of any function expressed as a linear combination of B splines. Any definite integral can be obtained by selecting a specific value of x. For example, if x is a knot, say $x = t_m$, then

$$\int_{-\infty}^{t_m} \sum_{i=-\infty}^{\infty} c_i B_i^k(s)\,ds = \sum_{i=-\infty}^{\infty} e_i B_i^{k+1}(t_m) = \sum_{i=m-k-1}^{m} e_i B_i^{k+1}(t_m)$$

Problems 7.3

1. Show that the functions $f_n(x) = \cos nx$ are generated by this recursive definition.

$$\begin{cases} f_0(x) = 1, \qquad f_1(x) = \cos x \\ f_{n+1}(x) = 2f_1(x)f_n(x) - f_{n-1}(x) \qquad (n \geqq 1) \end{cases}$$

2. What functions are generated by the following recursive definition?

$$\begin{cases} f_0(x) = 1, \qquad f_1(x) = x \\ f_{n+1}(x) = 2xf_n(x) - f_{n-1}(x) \qquad (n \geqq 1) \end{cases}$$

3. Find an expression for $B_i^2(x)$ and verify that it is piecewise quadratic. Show that $B_i^2(x)$ is zero at every knot except

$$B_i^2(t_{i+1}) = \frac{t_{i+1} - t_i}{t_{i+2} - t_i} \quad \text{and} \quad B_i^2(t_{i+2}) = \frac{t_{i+3} - t_{i+2}}{t_{i+3} - t_{i+1}}$$

4. Verify Eq. (5).
5. Establish that $\sum_{i=-\infty}^{\infty} f(t_i)B_{i-1}^1(x)$ is a first-degree spline that interpolates f at every knot. What is the zero-degree spline that does so?
6. Show that if $t_m \leqq x < t_{m+1}$, then

$$\sum_{i=-\infty}^{\infty} c_i B_i^k(x) = \sum_{i=m-k}^{m} c_i B_i^k(x)$$

7. Put $h_i = t_{i+1} - t_i$. Show that if

$$S(x) = \sum_{i=-\infty}^{\infty} c_i B_i^2(x) \quad \text{and if} \quad c_{i-1}h_{i-1} + c_{i-2}h_i = y_i(h_i + h_{i-1})$$

for all i, then $S(t_m) = y_m$ for all m. *Hint:* Use Problem 3.

8. Show that the coefficients C_i^{j-1} generated by Eq. (7) satisfy $\min_i C_i^{j-1} \leqq f(x) \leqq \max_i C_i^{j-1}$.
9. For equally spaced knots, show that $k(k+1)^{-1}B_i^k(x)$ lies in the interval whose end points are $B_i^{k-1}(x)$ and $B_{i+1}^{k-1}(x)$.
10. Letting the knots be the integers on the real line, $(t_i = i)$, show that $B_i^k(x) = B_0^k(x - t_i)$.
11. Show that

$$\int_{-\infty}^{\infty} B_i^k(x)\,dx = \frac{t_{i+k+1} - t_i}{k+1}$$

12. Show that the class of all spline functions of degree m having knots $x_1, \ldots, x_n$ *includes* the class of polynomials of degree m.

Computer Problems 7.3

1. Using an automatic plotter, graph B_0^k for $k = 0, 1, 2, 3, 4$. Use integer knots $t_i = i$ over the interval $[0, 5]$.
2. Let $t_i = i$ (so the knots are the integer points on the real line). Print a table of 100 values of the function $3B_7^1 + 6B_8^1 - 4B_9^1 + 2B_{10}^1$ on the interval $[6, 14]$. Using a plotter, construct the graph of this function on the given interval.
3. (Continuation) Repeat Problem 2 for the function

$$3B_7^2 + 6B_8^2 - 4B_9^2 + 2B_{10}^2$$

4. Write a subroutine to evaluate $S'(x)$ at a specified x, assuming that $S(x) = \sum_{i=1}^n c_i B_i^k(x)$. Input will be $n, k, x, t_1, \ldots, t_{n+k+1}$, and $c_1, \ldots, c_n$.
5. Write a subroutine to evaluate $\int_a^b S(x)\,dx$, assuming that $S(x) = \sum_{i=1}^n c_i B_i^k(x)$. Input will be $n, k, a, b, c_1, \ldots, c_n, t_1, \ldots, t_{n+k+1}$.

7.4 Interpolation and Approximation by B Splines

In the preceding section, we developed a number of properties of B splines and showed how B splines are used in various numerical tasks. The problem of obtaining a B spline representation of a given function was not discussed. Here we consider the problem of interpolating a table of data; later a noninterpolatory method of approximation is described.

A basic question is how to determine the coefficients in the expression

$$S(x) = \sum_{i=-\infty}^{\infty} A_i B_{i-k}^k(x) \tag{1}$$

so that the resulting spline function interpolates a prescribed table.

x	t_1	t_2	$\cdots$	t_n
y	y_1	y_2	$\cdots$	y_n

We mean by "interpolate" that

$$S(t_i) = y_i \qquad (1 \leqq i \leqq n) \tag{2}$$

The natural starting point is with the simplest splines, corresponding to $k = 0$. Since $B_i^0(t_j) = \delta_{ij}$ (1 if $i = j$ and 0 if $i \neq j$), the solution to the problem is immediate: just set $A_i = y_i$ for $1 \leqq i \leqq n$. All other coefficients in (1) are arbitrary. In particular, they can be zero. We arrive then at this result.

> The zero-degree spline
>
> $$S(x) = \sum_{i=1}^{n} y_i B_i^0(x)$$
>
> has the interpolation property
>
> $$S(t_i) = y_i \qquad (1 \leqq i \leqq n).$$

The next case, $k = 1$, also has a simple solution. We use the fact that $B_{i-1}^1(t_j) = \delta_{ij}$. Hence the following is true.

> The first-degree spline
>
> $$S(x) = \sum_{i=1}^{n} y_i B_{i-1}^1(x)$$
>
> has the interpolation property
>
> $$S(t_i) = y_i \qquad (1 \leqq i \leqq n).$$

If the table has four entries, for instance, we use B_0^1, B_1^1, B_2^1, and B_3^1. They, in turn, require for their definition knots $t_0, t_1, \ldots, t_5$. Knots t_0 and t_5 can be arbitrary. Fig. 7.11 shows the graphs of the four splines. In such a problem, if t_0 and t_5 are not prescribed, it is natural to define them in such a way that t_1 is the midpoint of the interval $[t_0, t_2]$ and t_4 is the midpoint of $[t_3, t_5]$.

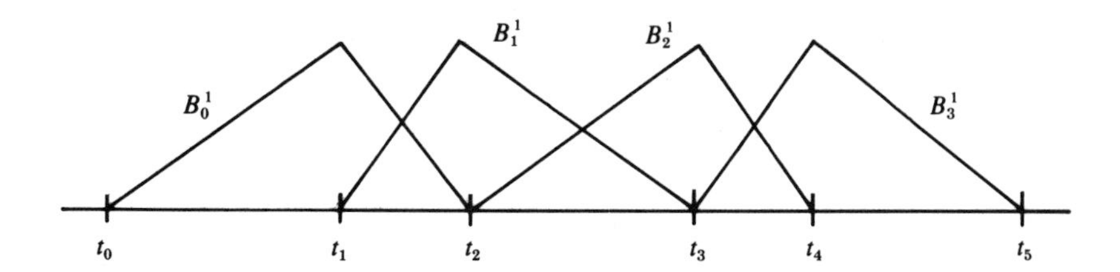

Figure 7.11

In both elementary cases considered, the unknown coefficients $A_1, \ldots, A_n$ in Eq. (1) were uniquely determined by the interpolation conditions (2). If terms were present in Eq. (1) corresponding to values of i *outside* the range $\{1, \ldots, n\}$, they had no influence on the values of $S(x)$ at $t_1, t_2, \ldots, t_n$.

For higher-degree splines, we shall see that some arbitrariness exists in choosing coefficients. In fact, *none* of the coefficients is uniquely determined by the interpolation conditions. This fact can be advantageous if other properties are desired of the solution.

In the quadratic case, we begin with the equation

$$\sum_{i=-\infty}^{\infty} A_i B_{i-2}^2(t_j) = \frac{1}{t_{j+1} - t_{j-1}} [A_j(t_{j+1} - t_j) + A_{j+1}(t_j - t_{j-1})] \tag{3}$$

Its justification is left to Problem 3. If the interpolation conditions (2) are now imposed, we obtain the following system of equations. It gives the necessary and sufficient conditions on the coefficients.

$$A_j(t_{j+1} - t_j) + A_{j+1}(t_j - t_{j-1}) = y_j(t_{j+1} - t_{j-1}) \qquad (1 \leqq j \leqq n) \tag{4}$$

This is a system of n linear equations in $n + 1$ unknowns $A_1, \ldots, A_{n+1}$.

One way to solve (4) is to assign any value to A_1 and then use (4) to solve for A_2, $A_3, \ldots,$ recursively. For this purpose, the equations could be rewritten as

$$A_{j+1} = \alpha_j + \beta_j A_j \qquad (1 \leqq j \leqq n) \tag{5}$$

where these abbreviations have been used:

$$\begin{cases} \alpha_j = y_j \dfrac{t_{j+1} - t_{j-1}}{t_j - t_{j-1}} & (1 \leqq j \leqq n) \\[2ex] \beta_j = \dfrac{t_j - t_{j+1}}{t_j - t_{j-1}} & (1 \leqq j \leqq n) \end{cases}$$

In order to keep the coefficients small in magnitude, we recommend selecting A_1 so that the expression $\Phi = \sum_{i=1}^{n+1} A_i^2$ will be a minimum. To determine this value of A_1, we proceed as follows. By successive substitution using Eq. (5), we can show that

$$A_{j+1} = \gamma_j + \delta_j A_1 \qquad (1 \leqq j \leqq n) \tag{6}$$

where the coefficients γ_j and δ_j are obtained recursively by this algorithm:

$$\gamma_1 = \alpha_1, \ \delta_1 = \beta_1, \ \gamma_j = \alpha_j + \beta_j \gamma_{j-1}, \ \delta_j = \beta_j \delta_{j-1} \qquad (2 \leqq j \leqq n) \tag{7}$$

Then Φ is a quadratic function of A_1 as follows.

$$\begin{aligned} \Phi &= A_1^2 + A_2^2 + \cdots + A_{n+1}^2 \\ &= A_1^2 + (\gamma_1 + \delta_1 A_1)^2 + (\gamma_2 + \delta_2 A_1)^2 + \cdots + (\gamma_n + \delta_n A_1)^2 \end{aligned}$$

To find the minimum of Φ, we take its derivative with respect to A_1 and set it equal to zero.

$$\frac{d\Phi}{dA_1} = 2A_1 + 2(\gamma_1 + \delta_1 A_1)\,\delta_1 + \cdots + 2(\gamma_n + \delta_n A_1)\,\delta_n = 0$$

This is equivalent to $qA_1 + p = 0$, where

$$\begin{cases} q = 1 + \delta_1^2 + \delta_2^2 + \cdots + \delta_n^2 \\ p = \gamma_1 \delta_1 + \gamma_2 \delta_2 + \cdots + \gamma_n \delta_n \end{cases}$$

A Fortran subroutine that computes coefficients $A_1, \ldots, A_{n+1}$ in the manner outlined is given next. Its calling sequence is `ASPL2(NP1,T,Y,A,H)`. Here `NP1` $= n + 1$, `T` is the knot array $(t_1, \ldots, t_n)$, `Y` is the array $(y_1, \ldots, y_n)$, `A` is the array of coefficients

$(A_1, \ldots, A_{n+1})$, and H is an array containing $h_i = t_i - t_{i-1}$ $(1 \leqq i \leqq n)$. Only NP1, T, and Y are input. They are available unchanged as output. Arrays A and H are computed and available as output.

```
      SUBROUTINE ASPL2(NP1,T,Y,A,H)
      DIMENSION  T(NP1),Y(NP1),A(NP1),H(NP1)
      DO 2 J = 2,NP1-1
    2 H(J) = T(J) - T(J-1)
      H(1) = H(2)
      H(NP1) = H(2)
      D = -1.0
      G = 2.0*Y(1)
      P = D*G
      Q = 2.0
      DO 3 I = 2,NP1-1
      R = H(I+1)/H(I)
      D = -R*D
      G = -R*G + (R + 1.0)*Y(I)
      P = P + D*G
    3 Q = Q + D*D
      A(1) = -P/Q
      DO 4 J = 2,NP1
    4 A(J) = ((H(J-1) + H(J))*Y(J-1) - H(J)*A(J-1))/H(J-1)
      RETURN
      END
```

Next is a Fortran function subroutine for computing values of the quadratic spline given by $S(x) = \sum_{i=1}^{n+1} A_i B_{i-2}^2(x)$. Its calling sequence is BSPL2(NP1,T,A,H,X), with the same variables as in the preceding code. The input variable X is a single real number that should lie between t_1 and t_n.

```
      FUNCTION BSPL2(NP1,T,A,H,X)
      DIMENSION  T(NP1),A(NP1),H(NP1)
      DO 2 J = 2,NP1-1
      IF(X .LE. T(J))  GO TO 3
    2 CONTINUE
    3 TEMP = A(J+1)*(X - T(J-1)) + A(J)*(T(J) - X + H(J+1))
      C2 = TEMP/(H(J) + H(J+1))
      TEMP = A(J)*(X - T(J-1) + H(J-1)) + A(J-1)*(T(J-1) - X + H(J))
      C1 = TEMP/(H(J-1) + H(J))
      BSPL2 = (C2*(X - T(J-1)) + C1*(T(J) - X))/H(J)
      RETURN
      END
```

An efficient process due to Schoenberg can also be used to obtain B spline approximations to a given function. Its quadratic version is defined by

$$S(x) = \sum_{i=-\infty}^{\infty} f(\tau_i) B_i^2(x) \qquad \tau_i = \frac{1}{2}(t_{i+1} + t_{i+2}) \tag{8}$$

Here, of course, the knots are $\{t_i\}_{-\infty}^{\infty}$, and the points where f must be evaluated are midpoints between the knots.

Equation (8) is useful in producing a quadratic spline function that approximates f. The salient properties of this process are as follows.

1. If $f(x) = ax + b$, then $S(x) = f(x)$.
2. If $f(x) \geqq 0$ everywhere, then $S(x) \geqq 0$ everywhere.
3. $\max_x |S(x)| \leqq \max_x |f(x)|$
4. If f is continuous on $[a, b]$ and if $\delta = \max_i |t_{i+1} - t_i|$, then for x in $[a, b]$,
$$|S(x) - f(x)| \leqq \frac{3}{2} \max_{a \leqq u \leqq v \leqq u + \delta \leqq b} |f(u) - f(v)|$$
5. The graph of S does not cross any line in the plane a greater number of times than does the graph of f.

Some of these properties are elementary; others are more abstruse. Property 1 is outlined in Problem 6. Property 2 is obvious because $B_i^2(x) \geqq 0$ for all x. Property 3 follows easily from (8) because if $|f(x)| \leqq M$, then

$$\begin{aligned}|S(x)| &\leqq \left| \sum_{i=-\infty}^{\infty} f(\tau_i) B_i^2(x) \right| \\ &\leqq \sum_{i=-\infty}^{\infty} |f(\tau_i)| B_i^2(x) \\ &\leqq M \sum_{i=-\infty}^{\infty} B_i^2(x) = M\end{aligned}$$

Properties 4 and 5 will be accepted without proof. Their significance, however, should not be overlooked. By 4, we can make the function S close to a continuous function f simply by making the "mesh size" δ small. This is because $f(u) - f(v)$ can be made as small as we wish simply by imposing the inequality $|u - v| \leqq \delta$ (uniform continuity property).

Finally, Property 5 can be interpreted as a shape-preserving attribute of the approximation process. In a crude interpretation, S should not exhibit more undulations than f. This property is exhibited in Computer Problem 5.

A program to obtain a spline approximation of the form $\sum_{i=1}^{n+2} D_i B_{i-1}^2(x)$; using Schoenberg's process, is given next. It is assumed that f is defined on $[a, b]$ and that f can be evaluated at equally spaced points $\tau_i = a + ih$ $(1 \leqq i \leqq n)$, with $h = (b - a)/(n - 1)$. In the calling sequence for SCH, F is an external function and NP2 is $n + 2$. After execution, the $n + 2$ desired coefficients are in the D array.

```
      SUBROUTINE SCH(F, A, B, NP2, D)
      DIMENSION  D(NP2)
      H = (B - A)/FLOAT(NP2-3)
      DO 2 J = 2, NP2-1
    2 D(J) = F(A + H*FLOAT(J-2))
      D(1) = 2.0*D(2) - D(3)
      D(NP2) = 2.0*D(NP2-1) - D(NP2-2)
      RETURN
      END
```

After the coefficients D_i have been obtained by the preceding subroutine, we recover values of the spline $S(x) = \sum_{i=1}^{n+2} D_i B_{i-1}^2(x)$ by means of the following function subroutine. In the calling sequence for ESCH, A and B are the ends of the interval, NP2 $= n + 2$, and X is a point where the value of $S(x)$ is desired. The subprogram determines knots t_i in such a way that the equally spaced points τ_i in the preceding subroutine satisfy $\tau_i = \frac{1}{2}(t_{i+1} + t_{i+2})$.

```
FUNCTION ESCH(A,B,NP2,D,X)
DIMENSION  D(NP2)
H = (B - A)/FLOAT(NP2-3)
C = A - 2.5*H
Y = (X - C)/H
J = INT(Y)
Y = Y - AINT(Y)
P = D(J+1)*Y + D(J)*(2.0 - Y)
Q = D(J)*(Y + 1.0) + D(J-1)*(1.0 - Y)
ESCH = (P*Y + Q*(1.0 - Y))*0.5
RETURN
END
```

We now illustrate the use of spline functions in fitting a curve to a set of data. Consider the following table.

T	0.0	0.6	1.5	1.7	1.9	2.1	2.3	2.6	2.8	3.0
Y	−0.8	−0.34	−0.59	0.59	0.23	0.1	0.28	1.03	1.5	1.44

3.6	4.7	5.2	5.7	5.8	6.0	6.4	6.9	7.6	8.0
0.74	−0.82	−1.27	−0.92	−0.92	−1.04	−0.79	−0.06	1.0	0.0

These 20 points were selected from a free-hand curve drawn on graph paper. We seek to reproduce the curve, using an automatic plotter. An excellent fit is provided by using the cubic spline routines ZSPL3/SPL3, as Fig. 7.12 shows. The quadratic B spline routines ASPL2/BSPL2 produce a reasonable curve (see Fig. 7.13), yet not as smooth as the cubic spline graph. These curves show once again that cubic splines are simple and elegant functions for curve fitting.

Problems 7.4

1. Establish formulas

$$B_{i-2}^2(t_i) = \frac{t_{i+1} - t_i}{t_{i+1} - t_{i-1}} = \frac{h_i}{h_i + h_{i-1}}$$

$$B_{i-1}^2(t_i) = \frac{t_i - t_{i-1}}{t_{i+1} - t_{i-1}} = \frac{h_{i-1}}{h_i + h_{i-1}}$$

where $h_i = t_{i+1} - t_i$.

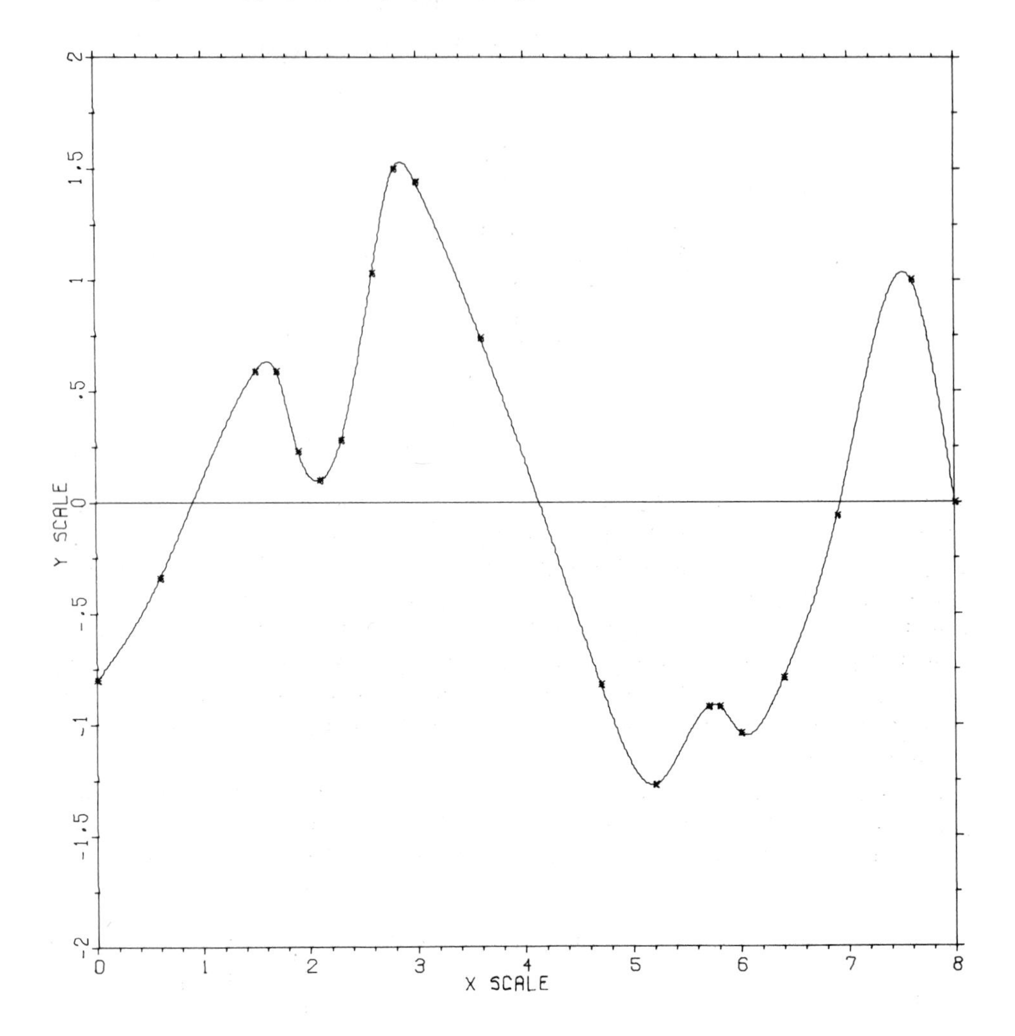

Figure 7.12 Cubic spline.

2. Show by induction that if

$$A_j = \frac{1}{t_{j-1} - t_{j-2}} [y_{j-1}(t_j - t_{j-2}) - A_{j-1}(t_j - t_{j-1})]$$

for $j = 2, \ldots, n + 1$, then

$$\sum_{i=1}^{n+1} A_i B_{i-2}^2(t_j) = y_j \qquad (1 \leqq j \leqq n)$$

3. If $S(x) = \sum_{i=-\infty}^{\infty} A_i B_{i-2}^2(x)$ and $t_{k-1} \leqq x \leqq t_k$, then

$$S(x) = \frac{1}{t_k - t_{k-1}} [d(x - t_{k-1}) + e(t_k - x)]$$

with

$$d = \frac{1}{t_{k+1} - t_{k-1}} [A_{k+1}(x - t_{k-1}) + A_k(t_{k+1} - x)]$$

and

$$e = \frac{1}{t_k - t_{k-2}} [A_k(x - t_{k-2}) + A_{k-1}(t_k - x)]$$

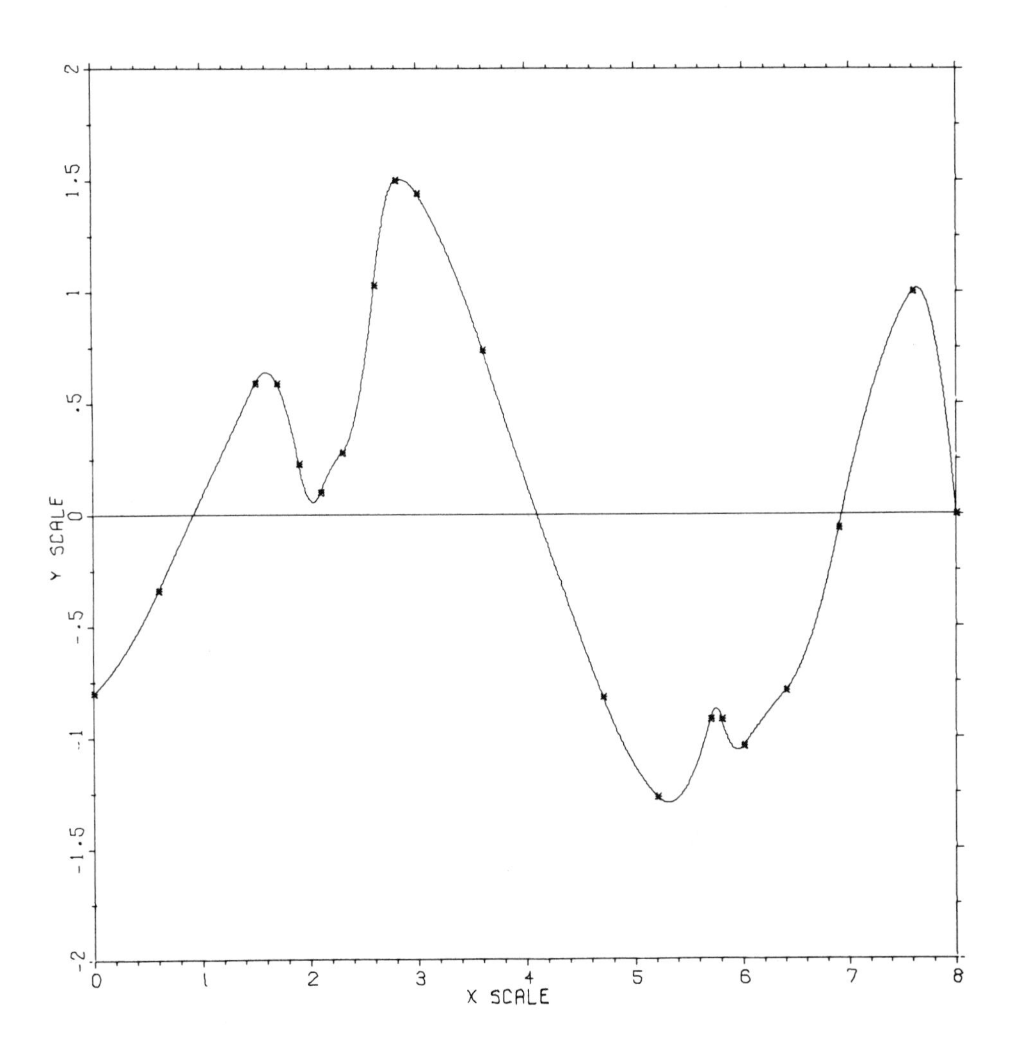

Figure 7.13 Quadratic spline.

4. Verify Eqs. (6) and (7) by induction, using Eq. (5).
5. If points $\tau_1 < \tau_2 < \cdots < \tau_n$ are given, can we always determine points t_i such that $t_i < t_{i+1}$ and $\tau_i = \frac{1}{2}(t_{i+1} + t_{i+2})$?

6. Show that if $f(x) = x$, then Schoenberg's process produces $S(x) = x$.
7. Show that $x^2 = \sum_{i=-\infty}^{\infty} t_{i+1}t_{i+2}B_i^2(x)$.
8. Let $f(x) = x^2$. Assume that $t_{i+1} - t_i \leqq \delta$ for all i. Show that the quadratic spline approximation to f given by Eq. (8) differs from f no more than $\delta^2/4$. *Hint:* Use the preceding problem and the fact that $\sum_{i=-\infty}^{\infty} B_i^2 \equiv 1$.

Computer Problems 7.4

1. Recode ASPL2/BSPL2 so that the array H is not used.
2. Recode ASPL2/BSPL2 for the special case of equally spaced knots, simplifying the code where possible.
3. Write a subroutine to produce a spline approximation to the function $F(x) = \int_a^x f(t)\,dt$. Assume that $a \leqq x \leqq b$. Begin by finding a quadratic spline interpolant to f at the n points $t_i = a + (i - 1)(b - a)/(n - 1)$. Test your program on
   (a) $f(x) = \sin x, \quad 0 \leqq x \leqq \pi$
   (b) $f(x) = e^x, \quad 0 \leqq x \leqq 4$
   (c) $f(x) = (x^2 + 1)^{-1}, \quad 0 \leqq x \leqq 2.$
4. Write a subroutine to produce a spline function that approximates $(d/dx)f(x)$ for a given f on a given interval $[a, b]$. Begin by finding a quadratic spline interpolant to f at n points evenly spaced in $[a, b]$, including endpoints. Test your subroutine on the functions suggested in Problem 3.
5. Define f on $[0, 6]$ to be a polygonal line joining points (0, 0), (1, 2), (3, 3), (5, 3), (6, 0). Determine spline approximations to f, using Schoenberg's process and taking 7, 13, 19, 25, 31 knots.
6. Write suitable codes to calculate $\sum_{i=-\infty}^{\infty} f(s_i)B_i^2(x)$, with $s_i = \frac{1}{2}(t_{i+1} + t_{i+2})$. Assume that f is defined on $[a, b]$ and that x will lie in $[a, b]$. Assume also that $t_2 < a < t_3$ and $t_{n+1} < b < t_{n+2}$. (Make no assumption about the spacing of knots.)
7. Write a subroutine to carry out this approximation scheme.

$$S(x) = \sum_{i=-\infty}^{\infty} f(\tau_i)B_i^3(x) \qquad \tau_i = \frac{1}{3}(t_{i+1} + t_{i+2} + t_{i+3})$$

   Assume that f is defined on $[a, b]$ and that $\tau_i = (i - 1)(b - a)/(n - 1)$.
8. The size of the Army of Flanders has been estimated by historians as follows.

Date	Sept. 1572	Dec. 1573	Mar. 1574	Jan. 1575	May 1576	Feb. 1578
Number	67,259	62,280	62,350	59,250	51,457	27,603

Sept. 1580	Oct. 1582	Apr. 1588	Nov. 1591	Mar. 1607
45,435	61,162	63,455	62,164	41,471

   Fit the table with a quadratic B spline and use it to find the average size of the army during the period given.

Supplementary Problems (Chapter 7)

1. Describe explicitly the ***natural cubic spline*** that interpolates a table with only two entries:

x	t_1	t_2
y	y_1	y_2

(t_1 and t_2 are the knots)

   Give a formula for it.

2. Suppose that $f(0) = 0$, $f(1) = 1.1752$, $f'(0) = 1$, $f'(1) = 1.5431$. Determine the cubic interpolating polynomial $p_3(x)$ for these data. Is it a natural cubic spline?
3. A *periodic* cubic spline having knots $t_1, \ldots, t_n$ is defined as a cubic spline function $S(x)$ such that $S(t_1) = S(t_n)$, $S'(t_1) = S'(t_n)$, and $S''(t_1) = S''(t_n)$. It would be employed to fit data known to be periodic. Carry out the analysis necessary to obtain a periodic cubic spline interpolant for a table

x	t_1	$\cdots$	t_n
y	y_1	$\cdots$	y_n

assuming that $y_n = y_1$.
4. Verify (for $k > 0$) that $B_i^k(t_j) = 0$ if and only if $j \leqq i$ or $j \geqq i + k + 1$.
5. Find a cubic spline over knots $-1, 0, 1$ such that $S''(-1) = S''(1) = 0$, $S(-1) = S(1) = 0$, and $S(0) = 1$.
6. Show that no quadratic spline S interpolates the table of Problem 12 of Sec. 7.2 and, in addition, satisfies $S'(t_1) = S'(t_5)$.
7. What equations must be solved if a quadratic spline S having knots $t_1, \ldots, t_n$ is required to take prescribed values at points $\frac{1}{2}(t_i + t_{i+1})$, $1 \leqq i \leqq n - 1$?
8. Let knots $t_1 < \cdots < t_n$ and numbers y_i and z_i be given. Determine formulas for a piecewise cubic function f having the given knots such that $f(t_i) = y_i$ $(1 \leqq i \leqq n)$, $\lim_{x \to t_i^+} f''(x) = z_i$ $(1 \leqq i \leqq n - 1)$, and $\lim_{x \to t_i^-} f''(x) = z_i$ $(2 \leqq i \leqq n)$. Why is f not generally a cubic spline?
9. Define a function f by

$$f(x) = \begin{cases} x^3 + x - 1, & -1 \leqq x \leqq 0 \\ x^3 - x - 1, & 0 < x \leqq 1 \end{cases}$$

Show that $\lim_{x \to 0^+} f(x) = \lim_{x \to 0^-} f(x)$ and that $\lim_{x \to 0^+} f''(x) = \lim_{x \to 0^-} f''(x)$. Are f and f'' continuous? Does it follow that f is a cubic spline? Explain.
10. Show that there is a unique cubic spline S having knots $t_1 < \cdots < t_n$, interpolating data $S(t_i) = y_i$ $(1 \leqq i \leqq n)$, and satisfying end conditions $S'(t_1) = S'(t_n) = 0$.
11. Establish Eq. (8) in Sec. 7.3 by induction.
12. Which B splines B_i^k have a nonzero value on the interval (t_n, t_m)?
13. Show that on $[t_i, t_{i+1}]$ we have

$$B_i^k(x) = \frac{(x - t_i)^k}{(t_{i+1} - t_i)(t_{i+2} - t_i) \cdots (t_{i+k} - t_i)}$$

14. Is a spline of the form $S(x) = \sum_{i=-\infty}^{\infty} c_i B_i^k(x)$ *uniquely* determined by a finite set of interpolation conditions $S(t_i) = y_i$ $(1 \leqq i \leqq n)$?
15. If the spline function $S(x) = \sum_{i=-\infty}^{\infty} c_i B_i^k(x)$ vanishes at each knot, must it be identically zero?
16. What is the necessary and sufficient condition on the coefficients in order that $\sum_{i=-\infty}^{\infty} c_i B_i^k = 0$?
17. Expand the function $f(x) = x$ in an infinite series $\sum_{i=-\infty}^{\infty} c_i B_i^1$.
18. Establish that $\sum_{i=-\infty}^{\infty} B_i^k$ is a constant function by means of Eq. (9) in Sec. 7.3.
19. Show that if $k \geqq 2$,

$$\frac{d^2}{dx^2} \sum_{i=-\infty}^{\infty} c_i B_i^k =$$

$$k(k-1) \sum_{i=-\infty}^{\infty} \left[\frac{c_i - c_{i-1}}{(t_{i+k} - t_i)(t_{i+k-1} - t_i)} - \frac{c_{i-1} - c_{i-2}}{(t_{i+k-1} - t_{i-1})(t_{i+k-1} - t_i)} \right] B_i^{k-2}$$

20. The derivatives and integrals of polynomials are polynomials. State and prove a similar result about spline functions.
21. What is the maximum value of B_i^2 and where does it occur?
22. Given a differentiable function f and knots $t_1 < \cdots < t_n$, show how to obtain a cubic spline S that interpolates f at the knots and satisfies the end condition $S'(t_1) = f'(t_1)$, $S'(t_n) = f'(t_n)$. *Note:* This procedure produces a better fit to f when applicable. If f' is not known, finite-difference approximations to $f'(t_1)$ and $f'(t_n)$ can be used.
23. Analyze the Subbotin interpolation scheme in this alternative manner. Let $v_i = S(t_i)$. Show that

$$S_i(x) = A_i(x - t_i)^2 + B_i(x - t_{i+1})^2 + C_i$$

where

$$C_i = 2y_i - \frac{1}{2}v_i - \frac{1}{2}v_{i+1}, \quad B_i = \frac{v_i - C_i}{h_i^2}, \quad A_i = \frac{v_{i+1} - C_i}{h_i^2}, \quad h_i = t_{i+1} - t_i$$

Hint: Show that $S_i(t_i) = v_i$, $S_i(t_{i+1}) = v_{i+1}$, and $S_i(\tau_i) = y_i$.
24. (Continuation) When continuity conditions on S' are imposed in Problem 23, the result is the following equation, in which $i = 2, \ldots, n - 1$.

$$h_i v_{i-1} + 3(h_i + h_{i+1})v_i + h_{i-1}v_{i+1} = 4h_{i-1}y_i + 4h_i y_{i-1}$$

25. Determine a, b, and c so that this is a cubic spline function.

$$S(x) = \begin{cases} x^3, & 0 \leqq x \leqq 1 \\ \frac{1}{2}(x-1)^3 + a(x-1)^2 + b(x-1) + c, & 1 \leqq x \leqq 3 \end{cases}$$

26. Is this a quadratic spline?

$$S(x) = \begin{cases} x, & -\infty < x \leqq 1 \\ x^2, & 1 \leqq x \leqq 2 \\ 4, & 2 \leqq x < \infty \end{cases}$$

27. Is there a choice of coefficients for which the following function is a natural cubic spline?

$$f(x) = \begin{cases} x + 1, & -2 \leqq x \leqq -1, \\ ax^3 + bx^2 + cx + d, & -1 \leqq x \leqq 1 \\ x - 1, & 1 \leqq x \leqq 2 \end{cases}$$

28. Determine the coefficients in the function

$$S(x) = \begin{cases} x^3 - 1, & -9 \leqq x \leqq 0 \\ ax^3 + bx^2 + cx + d, & 0 \leqq x \leqq 5 \end{cases}$$

so that it is a cubic spline taking the value 2 when $x = 1$.
29. Determine the coefficients so that the function

$$S(x) = \begin{cases} x^2 + x^3, & 0 \leqq x \leqq 1 \\ a + bx + cx^2 + dx^3, & 1 \leqq x \leqq 2 \end{cases}$$

is a cubic spline and has the property $S'''(x) = 12$.
30. Describe the function f that interpolates a table of values (x_i, y_i), $1 \leqq i \leqq m$, and minimizes the expression $\int_a^b |f'(x)|\,dx$. Assume that $a = x_1 < x_2 < \cdots < x_m = b$.
31. How many auxiliary conditions are needed to specify uniquely a spline of degree four over n knots?

8

Ordinary Differential Equations

In a simple electrical circuit the current in amperes is a function of time: $I(t)$. The function $I(t)$ will satisfy a differential equation of the form

$$\frac{dI}{dt} = f(t, I)$$

Here the right side is a function of t and I that depends on the circuit and on the nature of the electromotive force supplied to the circuit. Using methods developed in this chapter, the differential equation can be solved numerically to produce a table of I as a function of t.

A *differential equation* is an equation in which there appears an unknown function, together with one or more of its derivatives. A *solution* of a differential equation is a specific function which satisfies that equation. Here are some examples of differential equations with their solutions. In each case, t is the independent variable and x is the dependent variable. Thus x is the name of the unknown function of the independent variable t.

Equation	*Solution*
$x' - x = e^t$	$x(t) = te^t + ce^t$
$x'' + 9x = 0$	$x(t) = c_1 \sin 3t + c_2 \cos 3t$
$x' + \dfrac{1}{2x} = 0$	$x(t) = \sqrt{c - t}$

In these three examples, the letter c denotes an arbitrary constant. The fact that such constants appear in the solutions is an indication that a differential equation does not, in general, determine a unique solution function. When occurring in a scientific problem, a differential equation is usually accompanied by auxiliary conditions that (together with the differential equation) serve to specify the unknown function precisely.

In this chapter we focus on one type of differential equation and one type of auxiliary condition—the *initial value problem* for a first-order differential equation. The standard form adopted is

$$x' = f(t, x), \qquad x(a) \text{ prescribed} \tag{1}$$

It is understood that x is a function of t so that the differential equation written in more detail looks like

$$\frac{dx(t)}{dt} = f(t, x(t))$$

Problem (1) is termed an *initial value* problem, since t can be interpreted as time and $t = a$ can be thought of as the initial instant. We want to be able to determine the value of x at any time t before or after a.

Here are some examples of initial value problems, together with their solutions.

Equation	*Initial Value*	*Solution*
$x' = x + 1$	$x(0) = 0$	$x = e^t - 1$
$x' = 6t - 1$	$x(1) = 6$	$x = 3t^2 - t + 4$
$x' = \dfrac{t}{x + 1}$	$x(0) = 0$	$x = \sqrt{t^2 + 1} - 1$

Although many methods exist for obtaining analytical solutions of differential equations they are primarily limited to special differential equations. When applicable, they produce a solution in the form of a formula, such as shown in the preceding table. In practical problems, however, frequently a differential equation is *not* amenable to solution by special methods and a *numerical solution* must be sought. Even when a formal solution can be obtained, a numerical solution may be preferable, especially if the formal solution is very complicated. A numerical solution of a differential equation is usually obtained in the form of a table; the function remains unknown insofar as a specific formula is concerned.

The form of the differential equation adopted here permits the function f to depend on t and x. If f does not involve x, as in the second example above, then the differential equation can be solved by a direct process of indefinite integration. To illustrate, consider the initial value problem

$$x' = 3t^2 - 4t^{-1} + (1 + t^2)^{-1}, \qquad x(5) = 17$$

The differential equation can be integrated to produce

$$x(t) = t^3 - 4 \ln t + \text{Arctan } t + C$$

The constant C can then be chosen so that $x(5) = 17$. In such examples a numerical solution may still be preferable, for the function of t on the right side of differential equation (1) may not be integrable in terms of elementary functions. Consider, for instance, the differential equation

$$x' = e^{-\sqrt{t^3 - \sin t}} + \ln|\sin t + \tanh t^3| \tag{2}$$

The solution is obtained by taking the integral or antiderivative of the right-hand side. It can be done in principle but not in practice. In other words, a function x exists for which dx/dt is the right member of Eq. (2), but it is not possible to write $x(t)$ in terms of familiar functions.

8.1 Taylor Series Method

Fob 2

The method described first is not of utmost generality, but it is natural and capable of high precision. Its principle is to represent the solution of a differential equation locally by a few terms of its Taylor series.

Recall that if $x(t)$ has enough smoothness (continuous derivatives), then we can write

$$x(t + h) = x(t) + hx'(t) + \frac{1}{2}h^2x''(t) + \frac{1}{3!}h^3x'''(t) + \frac{1}{4!}h^4x^{iv}(t) + \cdots \quad (1)$$

For numerical purposes, the Taylor series enables us to compute $x(t + h)$ rather accurately if h is small and if $x(t)$, $x'(t)$, $x''(t)$, . . . are known.

This idea is exploited in the *Taylor series method*. An example is used to explain the procedure. Consider the initial value problem

$$x' = 1 + x^2 + t^3, \qquad x(0) = 0 \quad (2)$$

If the differential equation is differentiated several times with respect to t, the results are as follows. (Remember that a function of x must be differentiated with respect to t by using the chain rule.)

$$\begin{aligned} x' &= 1 + x^2 + t^3 \\ x'' &= 2xx' + 3t^2 \\ x''' &= 2xx'' + 2x'x' + 6t \\ x^{iv} &= 2xx''' + 6x'x'' + 6 \end{aligned} \quad (3)$$

If t and $x(t)$ are known, these four formulas, applied in order, will yield $x'(t)$, $x''(t)$, $x'''(t)$, and $x^{iv}(t)$. Thus, it is possible from this work to use the first five terms in the Taylor series, Eq. (1). Since $x(0) = 0$, we have a suitable starting point. Using a small value of h (for example, $h = 2^{-7} = 1/128 = 0.0078125$) we can compute an approximation to $x(0 + h)$ from formulas (1) and (3). The same step can be repeated to compute $x(2h)$, using $x(h)$, $x'(h)$, ... Here is the program.

```
      DATA  T,X/2*0.0/, H/7.8125E-3/
      PRINT 3,T,X
      DO 2 K = 1,128
      X1 = 1.0 + X*X + T**3
      X2 = 2.0*X*X1 + 3.0*T*T
      X3 = 2.0*X*X2 + 2.0*X1*X1 + 6.0*T
      X4 = 2.0*X*X3 + 6.0*X1*X2 + 6.0
      X = X + H*(X1 + H*(X2/2.0 + H*(X3/6.0 + H*X4/24.0)))
      T = FLOAT(K)*H
    2 PRINT 3,T,X
    3 FORMAT(5X,F10.5,5X,E20.13)
      END
```

A few words of explanation may be helpful. First, it is good practice to choose the step size h to be a machine number in order to avoid roundoff due to conversion. So for an h

approximately equal to 0.01, the value 1/128 is a good choice on a binary machine. Before writing the program, determine the interval in which you want to compute the solution of the differential equation. In the example this interval is chosen as $0 \leqq t \leqq 1$, and 128 steps are needed. In each step the current value of t is an integer multiple of the step size h. The statements defining `X1`, `X2`, `X3`, and `X4` are simply carrying out calculations of the derivatives according to Eq. (3). The final calculation carries out the evaluation of the Taylor series in Eq. (1), using five terms. Since this equation is a polynomial in h, it is evaluated most efficiently by using nested multiplication, which explains the formula for `X` in the code. The computation `T=T+H` may cause a small amount of roundoff error to accumulate in the value of `T`. It is avoided by coding `T=FLOAT(K)*H`.

When terms through $h^n x^{(n)}(t)/n!$ are included in the Taylor series, we say the Taylor series method is of *order n*.

What sort of accuracy can we expect in such a procedure? Are all the digits printed by the machine for the variable `X` accurate? Of course not! On the other hand, it is not easy to say how many digits *are* reliable. Here is a coarse assessment. Since terms up to $(1/24)h^4x^{(4)}(t)$ are included, the first term *not* included in the Taylor series is $(1/120) \times h^5x^{(5)}(t)$. The error may be larger than this. But the factor $h^5 = (2^{-7})^5 = 2^{-35} \approx (1/3)10^{-10}$ is affecting only the tenth decimal place. The printed solution is perhaps accurate to eight decimals. Bridges or airplanes should not be built on such shoddy analysis, but for now our attention is focused on the general form of the procedure.

Actually, there are two types of errors to consider. At each step, if $x(t)$ is known and $x(t + h)$ is computed from the first few terms of the Taylor series, an error occurs because we have truncated the Taylor series. This error, then, is called the *trucation error* or, to be more precise, the *local truncation error*. In the preceding example it is roughly $(1/120)h^5x^{(5)}(\xi)$. In this situation we say that the local truncation error is "of the order of h^5," abbreviated by $\mathcal{O}(h^5)$.

The second type of error obviously present is due to the *accumulated* effects of *all* local truncation errors. Indeed, the calculated value of $x(t + h)$ is in error because $x(t)$ is *already* wrong (because of previous truncation errors) and because another local truncation error occurs in the computation of $x(t + h)$ by means of the Taylor series.

Additional sources of errors must be considered in a complete theory. One is *roundoff error*. Although not serious in any *one* step of the solution procedure, after hundreds or thousands of steps it may accumulate and contaminate the calculated solution seriously. Remember that an error made at a certain step is carried forward into all succeeding steps. Depending on the differential equation and the method used to solve it, such errors may be magnified by succeeding steps.

Problems 8.1

1. Give the solutions of these differential equations.
   (a) $x' = t^3 + 7t^2 - t^{1/2}$
   (b) $x' = x$
   (c) $x' = -x$
   (d) $x'' = -x$
   (e) $x'' = x$
   (f) $x'' + x' - 2x = 0$ *Hint:* Try $x = e^{at}$.

2. Give the solutions of these initial value problems.
   (a) $x' = t^2 + t^{1/3}$, $x(0) = 7$
   (b) $x' = 2x$, $x(0) = 15$
   (c) $x'' = -x$, $x(\pi) = 0$, $x'(\pi) = 3$
3. Solve the following differential equations.
   (a) $x' = 1 + x^2$ *Hint:* $1 + \tan^2 t = \sec^2 t$
   (b) $x' = \sqrt{1 - x^2}$ *Hint:* $\sin^2 t + \cos^2 t = 1$
   (c) $x' = t^{-1} \sin t$ *Hint:* Problem 1 in Sec. 4.1
   (d) $x' + tx = t^2$ *Hint:* Multiply the equation by $f(t) = \exp(t^2/2)$. The left side becomes $(xf)'$.
4. The general first-order linear equation is $x' + px + q = 0$, where p and q are functions of t. Show that the solution is $x = -y^{-1}(z + c)$, where y and z are functions obtained as follows. Let u be an antiderivative of p, $y = e^u$, and z be an antiderivative of yq.
5. Here is an example of an initial value problem having two solutions: $x' = x^{1/3}$, $x(0) = 0$. Verify that two solutions are $x_1(t) = 0$ and $x_2(t) = (\frac{2}{3}t)^{3/2}$ for $t \geqq 0$. If the Taylor series method is applied, what happens?
6. The Taylor series method of order 1 is known as *Euler's method*. If applied to the initial value problem $x' = f(t, x)$, $x(a) = s$, what formulas are needed to determine $x(b)$ when $h = (b - a)/(n - 1)$?
7. Consider the problem $x' = x$. If the initial condition is $x(0) = c$, then the solution is $x(t) = ce^t$. If a roundoff error of ε occurs in reading the value of c into the computer, what effect is there on the solution at the point $t = 10$? At $t = 20$? Do the same for $x' = -x$.
8. If the Taylor series method is used on the initial value problem $x' = t^2 + x^3$, $x(0) = 0$, and if we intend to use the derivatives of x up to and including $x^{(iv)}$, what are the five main equations that must be programmed?
9. Solve part (d) of Problem 3 by substituting a Taylor series $x(t) = \sum_{n=0}^{\infty} a_n t^n$ and then determining the appropriate values of the coefficients.
10. Find a polynomial p with the property $p - p' = t^3 + t^2 - 2t$.
11. Show that the integral $\int_a^b f(s)ds$ can be computed by solving the initial value problem $x' = f(t)$, $x(a) = 0$.

Computer Problems 8.1

1. Write and test a program for applying the Taylor series method to the initial value problem

$$x' = x + x^2 \qquad x(1) = \frac{e}{16 - e} = 0.2046\ 6341\ 7289\ 1552\ 6943$$

Generate the solution in the interval $[1, 2.77]$. Use derivatives up to $x^{(5)}$ in the Taylor series. Use $h = 1/128$. Print out for comparison the values of the exact solution $x(t) = e^t/(16 - e^t)$. Verify that it is the exact solution.

2. Write a program to solve each problem on the indicated intervals. Use the Taylor series method, $h = 1/128$, and include terms to h^3. Account for any difficulties.

$$\text{(a)} \begin{cases} x' = t + x^2; \text{ interval } [0, 0.9] \\ x(0) = 1 \end{cases}$$

$$\text{(b)} \begin{cases} x' = x - t; \text{ interval } [1, 1.75] \\ x(1) = 1 \end{cases}$$

(c) $\begin{cases} x' = tx + t^2x^2; \text{ interval } [2, 5] \\ x(2) = -0.63966\ 25333 \end{cases}$

3. Solve the differential equation $x' = x$ with initial value $x(0) = 1$ by the Taylor series method on the interval $[0, 10]$. Compare the result with the exact solution $x(t) = e^t$. Use derivatives up to and including the tenth. Use step size $h = 1/128$.
4. Solve for $x(1)$, using the Taylor series method of order 5 with $h = 1/128$ and compare with the exact solution given.
   (a) $x' = 1 + x^2$, $x(0) = 0$; $x(t) = \tan t$
   (b) $x' = (1 + t)^{-1}x$, $x(0) = 1$; $x(t) = 1 + t$
5. Solve the initial value problem $x' = t + x + x^2$ on the interval $[0, 1]$ with initial condition $x(1) = 1$. Use the Taylor series method of order 5.
6. Solve the initial value problem $x' = (x + t)^2$, $x(0) = -1$ on the interval $[0, 1]$, using the Taylor series method with derivatives up to and including the fourth. Compare this to Taylor series methods of order 1, 2, and 3.
7. Write a program to solve on the interval $[0, 1]$ the initial value problem

$$x' = tx \qquad x(0) = 1$$

   using the Taylor series method of order 20—that is, include terms in the Taylor series up to and including h^{20}. Observe that a simple recursive formula can be used to obtain $x^{(n)}$, $n = 1, \ldots, 20$.
8. Write a program to solve the initial value problem $x' = \sin x + \cos t$, using the Taylor series method. Continue the solution from $t = 2$ to $t = 5$, starting with $x(2) = 0.32$. Include terms up to and including h^3.
9. Write a short program to solve the initial value problem $x' = e^t x$, $x(2) = 1$ on the interval $0 \leqq t \leqq 2$, using the Taylor series method. Include terms up to h^4.
10. Write a program to solve $x' = tx + t^4$ on interval $0 \leqq t \leqq 5$ with $x(5) = 3$. Use the Taylor series method with terms to h^4.
11. Write and run a program to print an accurate table of the function

$$\text{Si}(t) = \int_0^t \frac{\sin r}{r}\, dr$$

   The table should cover the interval $0 \leqq t \leqq 1$ in steps of 0.01. Use the idea of Problem 11 in the preceding problem set to obtain a suitable initial value problem. Solve the latter with the Taylor series method of order 8, using $h = 0.01$. If you start with $x' = t^{-1} \sin t$, successive derivatives are easily related to each other. For example, $x'' = t^{-1}(\cos t - x')$. To avoid $\sin 0/0$, you can find the value of $x(0.01)$ from the Taylor series given in Computer Problem 1, Sec. 4.1.

8.2 Runge–Kutta Methods

The methods named after Carl Runge and Wilhelm Kutta are designed to imitate the Taylor series method without requiring analytic differentiation of the original differential equation. Recall that in using the Taylor series method on the initial value problem

$$x' = f(t, x) \qquad x(a) = s \tag{1}$$

we need to obtain x'', x''', . . . by differentiating the function f. This requirement can be a serious obstacle to using the method. The user of this method must do some preliminary work of an analytical nature before writing a computer program. Ideally, a method for solving Eq. (1) should involve nothing more than writing a subprogram to evaluate the function f. The Runge–Kutta methods accomplish this.

For purposes of exposition, the Runge–Kutta method of order 2 is presented, although its low precision usually precludes it from actual scientific calculations. Later the Runge–Kutta method of order 4 is given *without* a derivation. It is in common use. The order two Runge–Kutta procedure does find application in real-time calculations on small computers. For example, it is used in some aircraft by the on-board minicomputer.

At the heart of any method for solving an initial value problem is a procedure for advancing the solution function one step at a time. That is, a formula must be given for $x(t + h)$ in terms of known quantities. As examples of known quantities we can cite $x(t)$, $x(t - h)$, $x(t - 2h)$, . . . if the solution process has gone through a number of steps. At the beginning only $x(a)$ is known. Of course, we assume that $f(t, x)$ can be computed for any point (t, x).

Before explaining the Runge–Kutta method of order 2, let us review the Taylor series for two variables. The infinite series is

$$f(x + h, y + k) = \sum_{i=0}^{\infty} \frac{1}{i!}\left(h\frac{\partial}{\partial x} + k\frac{\partial}{\partial y}\right)^i f(x, y) \tag{2}$$

The mysterious looking terms are interpreted as follows.

$$\left(h\frac{\partial}{\partial x} + k\frac{\partial}{\partial y}\right)^0 f(x, y) = f$$

$$\left(h\frac{\partial}{\partial x} + k\frac{\partial}{\partial y}\right)^1 f(x, y) = h\frac{\partial f}{\partial x} + k\frac{\partial f}{\partial y}$$

$$\left(h\frac{\partial}{\partial x} + k\frac{\partial}{\partial y}\right)^2 f(x, y) = h^2\frac{\partial^2 f}{\partial x^2} + 2hk\frac{\partial^2 f}{\partial x\,\partial y} + k^2\frac{\partial^2 f}{\partial y^2}$$

$$\vdots$$

where f and all partial derivatives are evaluated at (x, y). As in the one-variable case, if the Taylor series is truncated, an error term or remainder term is needed to restore the equality.

$$\begin{aligned} &f(x + h, y + k) \\ &= \sum_{i=0}^{n-1} \frac{1}{i!}\left(h\frac{\partial}{\partial x} + k\frac{\partial}{\partial y}\right)^i f(x,y) + \frac{1}{n!}\left(h\frac{\partial}{\partial x} + k\frac{\partial}{\partial y}\right)^n f(\bar{x}, \bar{y}) \end{aligned} \tag{3}$$

Here the point $(\bar{x}, \bar{y})$ lies on the line segment joining (x, y) to $(x + h, y + k)$ in the plane.

In applying Taylor series, subscripts are used to denote partial derivatives. So, for instance,

$$f_x = \frac{\partial f}{\partial x}, \quad f_t = \frac{\partial f}{\partial t}, \quad f_{xx} = \frac{\partial^2 f}{\partial x^2}, \quad f_{xt} = \frac{\partial^2 f}{\partial t\,\partial x} \tag{4}$$

We are dealing with functions for which the order of these subscripts is immaterial—e.g., $f_{xt} = f_{tx}$. Thus we have

$$f(x+h, y+k) = f + hf_x + kf_y$$
$$+ \frac{1}{2!}(h^2 f_{xx} + 2hkf_{xy} + k^2 f_{yy})$$
$$+ \frac{1}{3!}(h^3 f_{xxx} + 3h^2 k f_{xxy} + 3hk^2 f_{xyy} + k^3 f_{yyy})$$
$$+ \cdots$$

As special cases, we notice that

$$f(x+h, y) = f + hf_x + \frac{h^2}{2!} f_{xx} + \frac{h^3}{3!} f_{xxx} + \cdots$$

$$f(x, y+k) = f + kf_y + \frac{k^2}{2!} f_{yy} + \frac{k^3}{3!} f_{yyy} + \cdots$$

In the Runge–Kutta method of order 2 a formula is adopted having two function evaluations of the form

$$\begin{cases} F_1 = hf(t, x) \\ F_2 = hf(t + \alpha h, x + \beta F_1) \end{cases}$$

and a linear combination is added to the value of x at t to obtain the value at $t + h$

$$x(t+h) = x(t) + w_1 F_1 + w_2 F_2$$

or, equivalently,

$$x(t+h) = x(t) + w_1 hf(t, x) + w_2 hf(t + \alpha h, x + \beta hf(t, x)) \tag{5}$$

The objective is to determine constants w_1, w_2, α, β so that (5) is as accurate as possible. Explicitly, we want to reproduce as many terms as possible in the Taylor series

$$x(t+h) = x(t) + hx'(t) + \frac{1}{2} h^2 x''(t) + \frac{1}{3!} h^3 x'''(t) + \cdots \tag{6}$$

Now compare Eq. (5) with (6). One way to force them to agree up through the term in h is to set $w_1 = 1$ and $w_2 = 0$, since $x' = f$. However, agreement up through the h^2 term is possible by a more adroit choice of parameters. To see how, apply the two-variable form of the Taylor series to the final term in Eq. (5). With t, αh, x, βhf playing the role of x, h, y, k, respectively, we use the first three terms ($n = 2$) of the two-variable Taylor series given by formula (3)

$$f(t + \alpha h, x + \beta hf) = f + \alpha hf_t + \beta hff_x + \frac{1}{2}\left(\alpha h \frac{\partial}{\partial t} + \beta hf \frac{\partial}{\partial x}\right)^2 f(\bar{x}, \bar{y})$$

The result is a new form for Eq. (5).

$$x(t+h) = x(t) + (w_1 + w_2)hf + \alpha w_2 h^2 f_t + \beta w_2 h^2 ff_x + \mathcal{O}(h^3) \tag{7}$$

Equation (6) is also given a new form by using differential equation (1). Since $x' = f$, we have

$$x'' = \frac{dx'}{dt} = \frac{df(t, x)}{dt} = \left(\frac{\partial f}{\partial t}\right)\left(\frac{dt}{dt}\right) + \left(\frac{\partial f}{\partial x}\right)\left(\frac{dx}{dt}\right) = f_t + f_x f$$

So Eq. (6) implies

$$x(t + h) = x + hf + \frac{1}{2} h^2 f_t + \frac{1}{2} h^2 f f_x + \mathcal{O}(h^3) \tag{8}$$

Agreement between (7) and (8) is achieved by stipulating that

$$w_1 + w_2 = 1, \quad \alpha w_2 = \frac{1}{2}, \quad \beta w_2 = \frac{1}{2} \tag{9}$$

A convenient solution of these equations is

$$\alpha = 1, \quad \beta = 1, \quad w_1 = \frac{1}{2}, \quad w_2 = \frac{1}{2}$$

The resulting Runge–Kutta formula is then, from Eq. (5),

$$x(t + h) = x(t) + \frac{h}{2} f(t, x) + \frac{h}{2} f(t + h, x + hf(t, x)) \tag{10}$$

or, equivalently,

$$x(t + h) = x(t) + \frac{1}{2}(F_1 + F_2)$$

$$\begin{cases} F_1 = hf(t, x) \\ F_2 = hf(t + h, x + F_1) \end{cases}$$

As can be seen, the solution function at $t + h$ is computed at the expense of two evaluations of the function f.

Notice that other solutions for the nonlinear equations (9) are possible. For example, α can be arbitrary and then

$$\beta = \alpha, \quad w_1 = 1 - \frac{1}{2\alpha}, \quad w_2 = \frac{1}{2\alpha}$$

One can show (Problem 4) that the error term for Runge–Kutta methods of order 2 is

$$h^3\left(\frac{1}{6} - \frac{\alpha}{4}\right)\left(\frac{\partial}{\partial t} + f\frac{\partial}{\partial x}\right)^2 f + \frac{h^3}{6} f_x \left(\frac{\partial}{\partial t} + f\frac{\partial}{\partial x}\right) f \tag{11}$$

So the method with $\alpha = \frac{2}{3}$ is especially interesting. However, none of the second-order Runge–Kutta methods are widely used on large computers, since the error is only $\mathcal{O}(h^3)$.

One algorithm in common use for the initial value problem (1) is the Runge–Kutta method of order 4. Its formulas are as follows.

$$x(t + h) = x(t) + \frac{1}{6}(F_1 + 2F_2 + 2F_3 + F_4) \tag{12}$$

$$\begin{cases} F_1 = hf(t, x) \\ F_2 = hf\left(t + \frac{h}{2}, x + \frac{1}{2}F_1\right) \\ F_3 = hf\left(t + \frac{h}{2}, x + \frac{1}{2}F_2\right) \\ F_4 = hf(t + h, x + F_3) \end{cases}$$

As can be seen, the solution at $x(t + h)$ is obtained at the expense of evaluating the function f four times. The final formula agrees with the Taylor expansion up to and including the term in h^4. The error therefore contains h^5 but no lower powers of h. Without knowing the coefficient of h^5 in the error, it is difficult to be precise about the local truncation error. In treatises devoted to this subject, these matters are further explored.

Here is a subroutine to implement the Runge–Kutta method of order 4.

```
      SUBROUTINE RK4(F, T, X, H, NSTEP)
      PRINT 3, T, X
      H2 = 0.5*H
      START = T
      DO 2 K = 1, NSTEP
      F1 = H*F(T, X)
      F2 = H*F(T + H2, X + 0.5*F1)
      F3 = H*F(T + H2, X + 0.5*F2)
      F4 = H*F(T + H, X + F3)
      T = START + H*FLOAT(K)
      X = X + (F1 + F2 + F2 + F3 + F3 + F4)/6.0
    2 PRINT 3, T, X
    3 FORMAT(5X, 2E20.13)
      RETURN
      END
```

To illustrate the use of the preceding code, consider the initial value problem

$$x' = 2 + (x - t - 1)^2 \qquad x(1) = 2 \tag{13}$$

whose exact solution is $x(t) = 1 + t + \tan(t - 1)$. A driver program and function subprogram to solve this problem on the interval $[1 - \pi/2, 1 + \pi/2]$ by the Runge–Kutta subroutine is given below. The number of steps needed is calculated by dividing the length of the intervals $\pi/2 \approx 1.57$ by step size $h = 1/128$.

```
      EXTERNAL F
      DATA  T/1.0/, X/2.0/, H/7.8125E-3/, NSTEP/201/
      CALL RK4(F, T, X, H, NSTEP)
      CALL RK4(F, 1.0, 2.0, -H, NSTEP)
      END
      FUNCTION F(T, X)
      F = 2.0 + (X - T - 1.0)**2
      RETURN
      END
```

General-purpose routines incorporating the Runge–Kutta algorithm usually include additional programming to monitor the truncation error and make necessary changes in step size as the solution progresses. In general terms, the step size can be large when the solution is slowly varying but should be small when rapidly varying. Such a program is presented in Sec. 8.3.

Problems 8.2

1. Derive the equations needed in applying the fourth-order Taylor series method to the differential equation $x' = tx^2 + x - 2t$. Compare them in complexity to the equations required for the fourth-order Runge–Kutta method.
2. An important theorem of calculus states that the equation $f_{tx} = f_{xt}$ is true, provided that at least one of these two partial derivatives exists and is continuous. Test this equation on some functions, such as $f(t, x) = xt^2 + x^2t + x^3t^4$, $e^x \sinh(t + x) + \cos(2x - 3t)$, and $\log(x - t^{-1})$.
3. If $x' = f(t, x)$, then

$$x''' = f_{tt} + 2ff_{tx} + f^2 f_{xx} + f_x f_t + ff_x^2 = \left(\frac{\partial}{\partial t} + f\frac{\partial}{\partial x}\right)^2 f + f_x\left(\frac{\partial}{\partial t} + f\frac{\partial}{\partial x}\right) f$$

   Verify and determine $x^{(iv)}$ in a similar form.
4. Establish the error term (11) for Runge–Kutta methods of order 2.
5. Derive the two-variable form of the Taylor series from the one-variable form by considering the function of one variable $\phi(t) = f(x + th, y + tk)$ and expanding it by Taylor's theorem.
6. The Taylor series expansion about point (a, b) in terms of two variables x and y is given by

$$f(x, y) = \sum_{i=0}^{\infty} \frac{1}{i!}\left((x - a)\frac{\partial}{\partial x} + (y - b)\frac{\partial}{\partial y}\right)^i f(a, b)$$

   Show that formula (2) can be obtained from this form by a change of variables.
7. Using the form given in Problem 6, determine the first four nonzero terms in the Taylor series for $f(x, y) = \sin x + \cos y$ about the point $(0, 0)$. Compare the result to the known series for $\sin x$ and $\cos y$. Make a conjecture about the Taylor series for functions having the special form $f(x, y) = g(x) + h(y)$.
8. On a certain computer it was found that in using the fourth-order Runge–Kutta method over an interval $[a, b]$ with $h = (b - a)/(n - 1)$, the total error due to roundoff was about $36n2^{-50}$ and the total truncation error was $9nh^5$, where n is the number of steps and h is the step size. What is an optimum value of h? *Hint:* Minimize the total error, roundoff plus truncation.
9. Put these differential equations into a form suitable for numerical solution by the Runge–Kutta method.
   (a) $x + 2xx' - x' = 0$
   (b) $\log x' = t^2 - x^2$
   (c) $(x')^2(1 - t^2) = x$
10. Solve the differential equation

$$\begin{cases} \dfrac{dx}{dt} = -tx^2 \\ x(0) = 2 \end{cases}$$

   at $t = -0.2$, correct to two decimal places with one step by using (a) the Taylor series method and (b) the Runge–Kutta method of order 2.

11. Describe how the fourth-order Runge–Kutta method can be used to produce a table of values for the function

$$f(x) = \int_0^x e^{-t^2}\, dt$$

at 100 equally spaced points in the unit interval. *Hint:* Find an appropriate initial value problem whose solution is f.

Computer Problems 8.2

1. Run the sample program given in the text to illustrate the Runge–Kutta method.
2. Solve the initial value problem $x' = x/t + t \sec(x/t)$ with $x(0) = 0$ by the fourth-order Runge–Kutta algorithm. Continue the solution to $t = 1$, using $h = 2^{-7}$. Compare the numerical solution to the exact solution, which is $x(t) = t \arcsin t$. Define $f(0, 0) = 0$, where $f(t, x) = x/t + t \sec(x/t)$.
3. Select one of the following initial value problems and compare the numerical solutions obtained with fourth-order Runge–Kutta formulas and fourth-order Taylor series. Use different values of $h = 2^{-n}$ $(n = 2, 3, \ldots, 7)$ to compute the solution on the interval $[1, 2]$.
   (a) $x' = 1 + x/t$, $x(1) = 1$
   (b) $x' = 1/x^2 - xt$, $x(1) = 1$
   (c) $x' = 1/t^2 - x/t - x^2$, $x(1) = -1$
4. Select a Runge–Kutta routine from your computer center's program library and test it on the initial value problem $x' = (2 - t)x$, $x(2) = 1$. Compare to the exact solution, $x = \exp[-(\frac{1}{2})(t - 2)^2]$.
5. (A "Stiff" Differential Equation) Solve the differential equation $x' = 10x + 11t - 5t^2 - 1$ with initial value $x(0) = 0$. Continue the solution from $x = 0$ to $x = 3$, using the fourth-order Runge–Kutta method with $h = 2^{-8}$. Print the numerical solution and the exact solution $(t^2/2 - t)$ at every tenth step and draw a graph of the two solutions. Verify that the solution of the same problem with initial value $x(0) = \varepsilon$ is $\varepsilon e^{10t} + t^2/2 - t$ and thus account for the discrepancy between the numerical and exact solutions of the original problem.
6. Another fourth-order Runge–Kutta method is given by

$$x(t + h) = x(t) + w_1F_1 + w_2F_2 + w_3F_3 + w_4F_4$$

$$\begin{cases} F_1 = hf(t, x) \\ F_2 = hf\left(t + \dfrac{2}{5}h,\ x + \dfrac{2}{5}F_1\right) \\ F_3 = hf\left(t + \dfrac{14 - 3\sqrt{5}}{16}h,\ x + c_{31}F_1 + c_{32}F_2\right) \\ F_4 = hf(t + h,\ x + c_{41}F_1 + c_{42}F_2 + c_{43}F_3) \end{cases}$$

Here the appropriate constants are

$$c_{31} = 3(-963 + 476\sqrt{5})/1024 \qquad c_{32} = 5(757 - 324\sqrt{5})/1024$$

$$c_{41} = (-3365 + 2094\sqrt{5})/6040 \qquad c_{42} = (-975 - 3046\sqrt{5})/2552$$

$$c_{43} = 32(14595 + 6374\sqrt{5})/240845$$

$$w_1 = (263 + 24\sqrt{5})/1812 \qquad w_2 = 125(1 - 8\sqrt{5})/3828$$

$$w_3 = 1024(3346 + 1623\sqrt{5})/5924787 \qquad w_4 = 2(15 - 2\sqrt{5})/123$$

Select a differential equation with a known solution and compare the two fourth-order Runge–Kutta methods. Print the errors at each step. Is the ratio of the two errors a constant at each step? What are the advantages and/or disadvantages of each method? *Cultural note:* There are any number of Runge–Kutta methods of any order. The higher the order, the more complicated the formulas. Since the one given by Eq. (12) has error $\mathcal{O}(h^5)$ and is rather simple, it is the most popular fourth-order Runge–Kutta method. The error term for the method of this problem is also $\mathcal{O}(h^5)$, but it is optimum in a certain sense (see Ralston [1965] for details).

7. Solve the initial value problem $x' = x\sqrt{x^2 - 1}$, $x(0) = 1$, by the Runge–Kutta method on the interval $0 \leqq t \leqq 1.6$ and explain any difficulties. Then using negative h, solve the same differential equation on the same interval with initial value $x(1.6) = 1.0$.
8. A fifth-order Runge–Kutta method is given by

$$x(t + h) = x(t) + \frac{1}{24}F_1 + \frac{5}{48}F_4 + \frac{27}{56}F_5 + \frac{125}{336}F_6$$

$$\begin{cases} F_1 = hf(t, x) \\ F_2 = hf\left(t + \frac{1}{2}h,\ x + \frac{1}{2}F_1\right) \\ F_3 = hf\left(t + \frac{1}{2}h,\ x + \frac{1}{4}F_1 + \frac{1}{4}F_2\right) \\ F_4 = hf(t + h,\ x - F_2 + 2F_3) \\ F_5 = hf\left(t + \frac{2}{3}h,\ x + \frac{7}{27}F_1 + \frac{10}{27}F_2 + \frac{1}{27}F_4\right) \\ F_6 = hf\left(t + \frac{1}{5}h,\ x + \frac{28}{625}F_1 - \frac{1}{5}F_2 + \frac{546}{625}F_3 + \frac{54}{625}F_4 - \frac{378}{625}F_5\right) \end{cases}$$

Write and test a subroutine to use these equations.

9. The following pathological example has been given by Dahlquist and Bjorck [1974]. Consider the differential equation $x' = 100(\sin t - x)$ with initial value $x(0) = 0$. Integrate it with the fourth-order Runge–Kutta method on the interval $[0, 3]$, using $h = 0.015, 0.020, 0.025, 0.030$. Observe the numerical instability!
10. Consider the differential equation

$$x' = \begin{cases} x + t, & -1 \leqq t \leqq 0 \\ x - t, & 0 \leqq t \leqq 1 \end{cases}$$

$$x(-1) = 1$$

Using RK4 Runge–Kutta subroutine with step size $h = 0.1$, solve this problem over interval $[-1, 1]$. Now solve by using $h = 0.09$. Which numerical solution is more accurate and why? *Hint:* The true solution is given by $x = e^{(t+1)} - (t + 1)$ if $t \leqq 0$ and $x = e^{(t+1)} - 2e^t + (t + 1)$ if $t \geqq 0$.

11. Solve $t - x' + 2xt = 0$, $x(0) = 0$ on interval $[0, 10]$, using Runge–Kutta formulas with $h = 1/10$. Compare to the true solution, which is $\frac{1}{2}(e^{t^2} - 1)$. Draw a graph or have one created by an automatic plotter. Then graph the logarithm of the solution.
12. Write a program to solve $x' = \sin(xt) + \text{Arctan}\, t$ on $1 \leqq t \leqq 7$ with $x(2) = 4$, using Runge–Kutta subroutine RK4.

8.3 Stability and Adaptive Runge–Kutta Methods

Let us now resume the discussion of errors that inevitably occur in the numerical solution of an initial value problem

$$x' = f(t, x) \qquad x(a) = s \tag{1}$$

The exact solution is a function $x(t)$. It depends on the initial value s, and, in order to show this, we write $x(t, s)$. The differential equation gives rise, therefore, to a family of solution curves, each corresponding to one value of parameter s. For example, the differential equation $x' = x$ gives rise to the family of solution curves $x = se^{(t-a)}$ that differ in their initial values $x(a) = s$. A few such curves are shown in Fig. 8.1. The fact that the curves there diverge from each other as t increases has important numerical significance. Suppose, for instance, that initial value s is read into the computer with some roundoff error. Then even if all subsequent calculations are precise and *no truncation errors* occur, the computed solution will be wrong. An error made at the beginning has the effect of selecting the wrong *curve* from the family of all solution curves. Since these curves diverge from each other, the minute error made at the beginning is responsible for an eventual complete loss of accuracy. This phenomenon is not restricted to errors made in the first step, for each point in the numerical solution can be interpreted as the initial value for succeeding points.

For an example in which this difficulty does not arise, consider $x' = -x$ with $x(a) = s$. Its solutions are $x = se^{-(t-a)}$. As t increases, these curves come closer together, as in Fig 8.2. Thus errors made in the numerical solution still result in selecting the wrong curve, but the effect is not as serious, since the curves converge to one another.

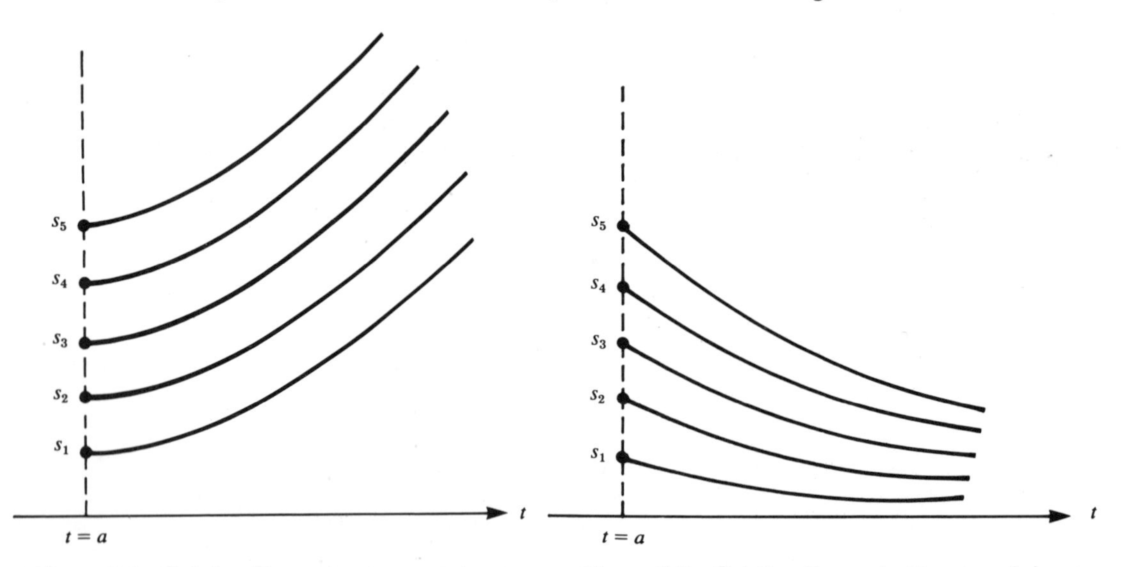

Figure 8.1 Solution Curves to $x' = x$, $x(a) = s$

Figure 8.2 Solution Curves to $x' = -x$, $x(a) = s$

For the general differential equation (1), how can the two modes of behavior just discussed be distinguished? It is simple. If $f_x > \delta > 0$, the curves diverge, whereas if $f_x < \delta < 0$, they converge. In order to see why, consider two nearby solution curves corresponding to initial values s and $s + h$. By Taylor series,

$$x(t, s + h) = x(t, s) + h\frac{\partial}{\partial s}x(t, s) + \frac{1}{2}h^2\frac{\partial^2}{\partial s^2}x(t, s) + \cdots$$

whence

$$x(t, s + h) - x(t, s) \approx h\frac{\partial}{\partial s}x(t, s)$$

Thus the condition of divergence of the curves, which means that

$$\lim_{t\to\infty} |x(t, s + h) - x(t, s)| = \infty$$

can be written as

$$\lim_{t\to\infty} \left| \frac{\partial}{\partial s}x(t, s) \right| = \infty$$

To calculate this partial derivative, start with the differential equation satisfied by $x(t, s)$:

$$\frac{\partial}{\partial t}x(t, s) = f(t, x(t, s))$$

and differentiate partially with respect to s:

$$\frac{\partial}{\partial s}\frac{\partial}{\partial t}x(t, s) = \frac{\partial}{\partial s}f(t, x(t, s))$$

Hence

$$\frac{\partial}{\partial t}\frac{\partial}{\partial s}x(t, s) = f_x(t, x(t, s))\frac{\partial}{\partial s}x(t, s) + f_t(t, x(t, s))\frac{\partial t}{\partial s} \tag{2}$$

But s and t are independent variables (a change in s produces no change in t) so $\partial t/\partial s = 0$. If s is now fixed and if we put $u(t) = (\partial/\partial s)x(t, s)$ and $q(t) = f_x(t, x(t, s))$, then Eq. (2) becomes

$$u' = qu \tag{3}$$

This is a linear differential equation with solution $u(t) = ce^{Q(t)}$, where Q is the indefinite integral (antiderivative) of q. Obviously the condition $\lim_{t\to\infty}|u(t)| = \infty$ is met if $\lim_{t\to\infty}Q(t) = \infty$. This situation, in turn, occurs if $q(t)$ is positive and bounded away from zero, for then

$$Q(t) = \int_a^t q(\theta)\,d\theta > \int_a^t \delta d\theta = \delta(t - a) \to \infty$$

as $t \to \infty$ if $f_x = q > \delta > 0$.

To illustrate, consider the differential equation $x' = t + \tan x$. Since $f_x(t, x) = \sec^2 x > 1$, the solution curves diverge from one another as $t \to \infty$.

In realistic situations involving the numerical solution of initial value problems, there is always a need to estimate the precision attained in the computation. Usually a tolerance is prescribed, and the numerical solution must not deviate from the true solution beyond this tolerance. Once a method is selected, the error tolerance dictates the allowable step size. Even if we consider only the local truncation error, determining an appropriate step size may be difficult. Moreover, often a small step size is needed on one portion of the solution curve, whereas a larger one may suffice elsewhere.

For the reasons given, various methods have been developed for *automatically* adjusting the step size in algorithms for the initial value problem. One is easily described. Consider the fourth-order Runge–Kutta method discussed in Sec. 8.2. To advance the solution curve from t to $t + h$, we can take one step of size h, using Runge–Kutta formulas. But we can also take *two* steps of size $h/2$ to arrive at $t + h$. If there were no truncation error, the value of the numerical solution $x(t + h)$ would be the same for both procedures. The difference in the numerical results can be taken as an estimate of the local truncation error. So, in practice, if this difference is within the prescribed tolerance, the current step size h is satisfactory. If this difference exceeds the tolerance, the step size is halved. If the difference is very much less than the tolerance, the step size is doubled.

The procedure just outlined is easily programmed but rather wasteful of computing time. A more sophisticated method is due to E. Fehlberg [1969]. The Fehlberg method of order 4 is of Runge–Kutta type and uses these formulas.

$$x(t + h) = x(t) + \frac{25}{216}F_1 + \frac{1408}{2565}F_3 + \frac{2197}{4104}F_4 - \frac{1}{5}F_5$$

$$\begin{cases} F_1 = hf(t, x) \\ F_2 = hf\left(t + \frac{1}{4}h,\ x + \frac{1}{4}F_1\right) \\ F_3 = hf\left(t + \frac{3}{8}h,\ x + \frac{3}{32}F_1 + \frac{9}{32}F_2\right) \\ F_4 = hf\left(t + \frac{12}{13}h,\ x + \frac{1932}{2197}F_1 - \frac{7200}{2197}F_2 + \frac{7296}{2197}F_3\right) \\ F_5 = hf\left(t + h,\ x + \frac{439}{216}F_1 - 8F_2 + \frac{3680}{513}F_3 - \frac{845}{4104}F_4\right) \end{cases}$$

Since this scheme requires one more function evaluation than the standard Runge–Kutta method of order 4, it is of questionable value alone. However, with an additional function evaluation

$$F_6 = hf\left(t + \frac{1}{2}h,\ x - \frac{8}{27}F_1 + 2F_2 - \frac{3544}{2565}F_3 + \frac{1859}{4104}F_4 - \frac{11}{40}F_5\right)$$

we can obtain a formula of order 5, namely,

$$x(t + h) = x(t) + \frac{16}{135}F_1 + \frac{6656}{12825}F_3 + \frac{28561}{56430}F_4 - \frac{9}{50}F_5 + \frac{2}{55}F_6$$

The difference between $x(t + h)$ from the fourth and fifth procedure is an estimate of the local truncation error in the fourth-order procedure. So six function evaluations give a fourth-order approximation, together with an error estimate. A subroutine for the Runge–Kutta–Fehlberg method follows.

```
      SUBROUTINE RK45(F,T,X,H,EST)
      DATA  C21,C31,C32, C41,C42,C43, C51,C52,C53,C54,
     A       C61,C62,C63,C64,C65, A1,A3,A4,A5, B1,B3,B4,B5, B6, C40
     B  /0.25,0.09375,0.28125,
     C  0.87938097405553,-3.27719617660446,3.32089212562585,
     D  2.0324074074074,-8.0,7.17348927875244,-0.20589668615984,
     E  -0.2962962962963,2.0,-1.38167641325536,0.45297270955166,-0.275,
     F  0.11574074074074,0.54892787524366,0.5353313840156,-0.2,
     G  0.11851851851852,0.51898635477583,0.50613149034201,-0.18,
     H  0.0363636363636364, 0.92307692307692/
C
     F1 = H*F(T,X)
     F2 = H*F(T+ 0.25*H,X + C21*F1)
     F3 = H*F(T+0.375*H,X + C31*F1 + C32*F2)
     F4 = H*F(T+C40*H,X + C41*F1 + C42*F2 + C43*F3)
     F5 = H*F(T+H ,X + C51*F1 + C52*F2 + C53*F3 + C54*F4)
     F6 = H*F(T+0.5*H,X + C61*F1 + C62*F2 + C63*F3 + C64*F4 + C65*F5)
C
     X5 = X + B1*F1 + B3*F3 + B4*F4 + B5*F5 + B6*F6
     X  = X + A1*F1 + A3*F3 + A4*F4 + A5*F5
     T = T + H
     EST = ABS(X - X5)
     RETURN
     END
```

The coefficients were computed and the numerical values set in a DATA statement so that they need not be recomputed each time the subroutine is called.

The error estimate can tell us when to adjust the step size and thus control the single-step error.

We now describe a simple "adaptive" procedure. First, compute the six function evaluations from the preceding formulas and then the fourth- and fifth-order approximations for $x(t+h)$, say x_4 and x_5. The user specifies bounds for $\varepsilon = |x_4 - x_5|$ and the range for the allowable step size: ε_{min}, ε_{max}, h_{min}, and h_{max}. The step size is doubled or halved so as to keep ε within the bounds set. Clearly we must set large enough bounds so that the adaptive procedure does not get caught in a loop, trying repeatedly and double the step size from the same point in order to meet an error bound that is too restrictive for the given function. The error analysis being used is given in the flowchart shown in Fig. 8.3 (see p. 197). The output variable IFLAG has the four values shown.

IFLAG	Meaning
0	successful procedure
1	h out of range
2	h out of range at endpoint
3	maximum number of iterations reached

The code for this adaptive procedure is the RK45AD subroutine.

```
      SUBROUTINE RK45AD(F,T,X,H,TB,ITMAX,EMIN,EMAX,HMIN,HMAX,IFLAG)
      EXTERNAL  F
      DATA  EPSI/0.5E-10/
      IFLAG = 3
      NSTEP = 0
      PRINT 7
      PRINT 6,H,T,X
C
    2 DT = ABS(TB - T)
      IF(DT .GT. ABS(H))  GO TO 3
      IFLAG = 0
      IF(DT .LE. EPSI*AMAX1(ABS(TB),ABS(T)))  RETURN
      H = SIGN(DT,H)
      IF(HMIN .LE. ABS(H) .AND. ABS(H) .LE. HMAX)  GO TO 3
      IFLAG = 2
      RETURN
C
    3 XSAVE = X
      TSAVE = T
      CALL RK45(F,T,X,H,EST)
      NSTEP = NSTEP+1
      PRINT 6,H,T,X,EST
      IF(NSTEP .GT. ITMAX .OR. IFLAG .EQ. 0)  RETURN
      IF(EST .LT. EMIN)  GO TO 4
      IF(EST .LE. EMAX)  GO TO 2
      H = 0.5*H
      GO TO 5
    4 H = 2.0*H
    5 X = XSAVE
      T = TSAVE
      IF(HMIN .LE. ABS(H) .AND. ABS(H) .LE. HMAX)  GO TO 2
      IFLAG = 1
      RETURN
C
    6 FORMAT(5X,E10.3,2(5X,E20.13),5X,E10.3)
    7 FORMAT(9X,1HH,20X,1HT,24X,1HX,17X,3HEST)
      END
```

Here F is the external right-hand side function, T and X the initial values for t and x, H the initial step size, TB the final value for t, and ITMAX the maximum number of steps to be taken in going from a to b. On return, T and X are the exit values for t and x, and H is the final value of h considered or used.

Problems 8.3

1. Establish Eq. (3).
2. The initial value problem $x' = (1 + t^2)x$ with $x(0) = 1$ is to be solved on the interval $[0, 9]$. How sensitive is $x(9)$ to perturbations in the initial value $x(0)$?

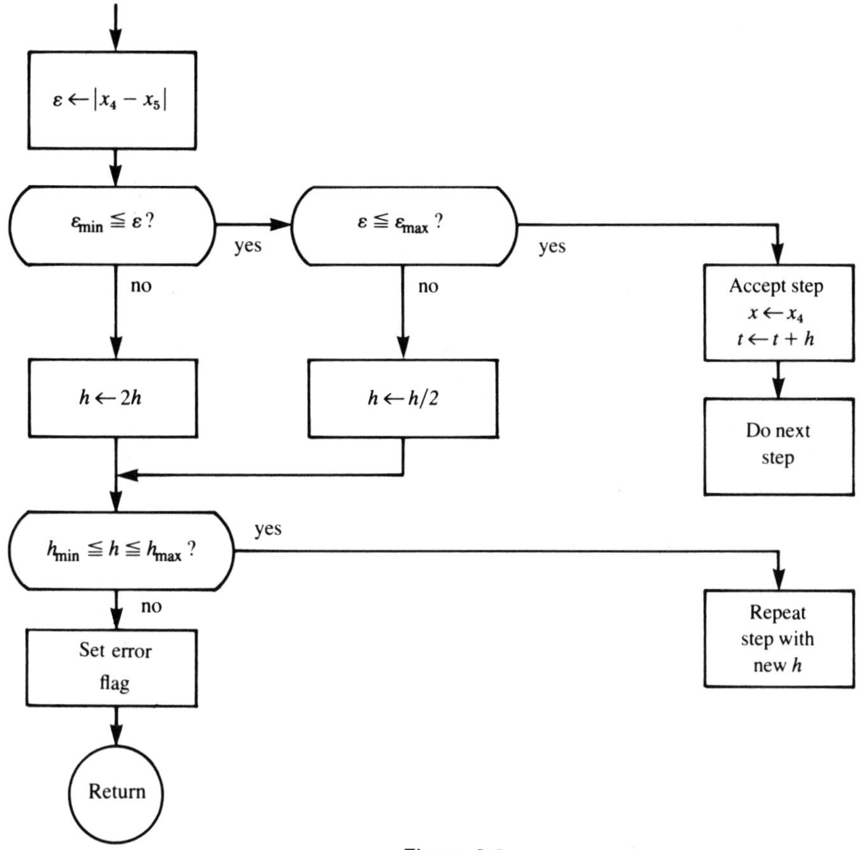

Figure 8.3

3. For each differential equation, determine regions in which solution curves tend to diverge from each other as t increases.
   (a) $x' = \sin t + e^x$
   (b) $x' = x + te^{-t}$
   (c) $x' = xt$
   (d) $x' = x^3(t^2 + 1)$
   (e) $x' = \cos t - e^x$
   (f) $x' = (1 - x^3)(1 + t^2)$
4. Consider the numerical solution of

$$x' = -x, \qquad x(0) = 1$$

by the midpoint rule

$$x(t + h) = x(t - h) + 2hx'(t)$$

with $x(h) = -h + \sqrt{1 + h^2}$. Are there any difficulties in using this method here? *Hint:* Consider fixed h and assume $x_n = [x(h)]^n$.

Computer Problems 8.3

1. Design and carry out a numerical experiment to verify that a slight perturbation in the initial value problem can cause catastrophic errors in a numerical solution.

2. Solve

$$x' = \frac{3x}{t} + \frac{9}{2}t - 13, \quad x(3) = 6$$

at $x(\frac{1}{2})$, using the RK45AD routine to obtain the desired solution to nine decimals. Compare to the true solution.

$$x = t^3 - \frac{9}{2}t^2 + \frac{13}{2}t$$

3. (Continuation) Repeat Problem 2 for $x(-\frac{1}{2})$.
4. It is known that the fourth-order Runge–Kutta method described earlier has a local truncation error that is $\mathcal{O}(h^5)$. Devise and carry out a numerical experiment to test this. *Suggestions:* Take just one step in the numerical solution of a nontrivial differential equation whose solution is known beforehand. However, use a variety of values for h, such as 4^{-n}, $1 \leqq n \leqq 24$. Test whether the ratio of errors to h^5 remains bounded as $h \to 0$. A double-precision calculation may be needed. Print the indicated ratios.
5. Compute and print a table of the function

$$f(\phi) = \int_0^{\phi} \sqrt{1 - \tfrac{1}{4}\sin^2\theta}\, d\theta$$

by solving an appropriate initial value problem. Cover the interval $[0, 90°]$ with steps of 1° and use the Runge–Kutta method of order 4. Check values: $f(30°) = 0.51788\ 193$, $f(90°) = 1.46746\ 221$. *Cultural note:* This is an example of an "elliptic integral of the second kind." It arises in finding arclength on an ellipse and in many engineering problems.
6. Compute a table, at 101 equally spaced points in the interval $[0, 2]$, of the "Dawson integral"

$$f(x) = \exp(-x^2) \int_0^x \exp(t^2)\, dt$$

by numerically solving, with the Taylor series method of suitable order, an initial value problem of which f is the solution. Make the table accurate to eight decimals and print only eight decimal places. *Hint:* Find the relationship between $f'(x)$ and $xf(x)$. The Fundamental Theorem of Calculus is useful. Check points are $f(1) = 0.53807\ 95069$ and $f(2) = 0.30134\ 03889$.

Supplementary Problems (Chapter 8)

1. Determine whether the solution curves of the differential equation $x' = t(x^3 - 6x^2 + 15x)$ diverge from one another as $t \to \infty$.
2. Determine whether the solution curves of the differential equation $x' = (1 + t^2)^{-1}x$ diverge from each other as $t \to \infty$.
3. Suppose that a differential equation is solved numerically on an interval $[a, b]$ and that the local truncation error is ch^p. Show that if all truncation errors have the same sign (the worst possible case), then the total truncation error is $(b - a)ch^{p-1}$ where $h = (b - a)/n$.
4. Consider the function e^{x^2+y}. Determine its Taylor series about the point (0, 1) through second partial derivative terms. Use this result to obtain an approximate value for $f(0.001, 0.998)$.
5. Consider the ordinary differential equation $x' = t^3x^2 - 2x^3/t^2$ with $x(1) = 0$. Determine the equations that would be used in applying (a) the Taylor series method of order 3 and (b) the Runge–Kutta method of order 4.
6. Consider the third-order Runge–Kutta method.

$$x(t+h) = x(t) + \frac{1}{9}(2F_1 + 3F_2 + 4F_3)$$

$$\begin{cases} F_1 = hf(t, x) \\ F_2 = hf\left(t + \frac{1}{2}h,\ x + \frac{1}{2}F_1\right) \\ F_3 = hf\left(t + \frac{3}{4}h,\ x + \frac{3}{4}F_2\right) \end{cases}$$

Show that it agrees with the Taylor series method of the same order for the differential equation $x' = x + t$.

7. Derive a nontrivial formula of the form

$$x(t+h) = ax(t) + bx(t-h) + h[cx'(t+h) + dx''(t) + ex'''(t-h)]$$

that is accurate for polynomials of as high a degree as possible. *Hint:* Use polynomials $1, t, t^2$, and so on.

8. How would you solve the initial value problem

$$x' = \sin x + \sin t \qquad x(0) = 0$$

on the interval $[0, 1]$ if ten decimals of accuracy are required? Assume that you have a computer in which unit roundoff error is $\frac{1}{2} \times 10^{-15}$ and assume that the fourth-order Runge–Kutta method will involve local truncation errors of magnitude $100h^5$.

9. Consider the differential equation $dx/ds = f(s, x)$. Integrating from t to $t + h$, we have

$$x(t+h) - x(t) = \int_t^{t+h} \left(\frac{dx}{ds}\right) ds = \int_t^{t+h} f(s, x(s))\, ds$$

Replace the integral with one of the numerical integration rules from Chapter 4 and obtain a formula for solving the differential equation.

10. Determine the coefficients of an "implicit" one-step ordinary differential equation method of the form

$$x(t+h) = ax(t) + bx'(t) + cx'(t+h)$$

so that it is exact for polynomials of as high a degree as possible. What is the order of the error term?

Supplementary Computer Problems (Chapter 8)

1. (A Predictor–Corrector Scheme) Using Supplementary Problem 9 and the Adams–Bashforth–Moulton formulas of Supplementary Problem 26 of Chapter 4, derive the predictor–corrector scheme given by the following equations.

$$\begin{aligned} \tilde{x}(t+h) = x(t) + \frac{h}{24}\{&55f(t, x(t)) - 59f(t-h, x(t-h)) \\ &+ 37f(t-2h, x(t-2h)) - 9f(t-3h, x(t-3h))\} \end{aligned}$$

$$\begin{aligned} x(t+h) = x(t) + \frac{h}{24}\{&9f(t+h, \tilde{x}(t+h)) + 19f(t, x(t)) \\ &- 5f(t-h, x(t-h)) + f(t-2h, x(t-2h))\} \end{aligned}$$

Write and test a subroutine for carrying it out. *Cultural note:* This is a multistep process since values of x at t, $t-h$, $t-2h$, $t-3h$ are used to determine the "predicted" value $\tilde{x}(t+h)$, which, in turn, is used with values of x at t, $t-h$, $t-2h$ to obtain the "corrected" value $x(t+h)$. The error terms for these formulas are $(251/720)h^5 f^{(iv)}(\xi)$ and $-(19/720)h^5 f^{(iv)}(\eta)$, respectively.

2. The Adams–Moulton method of order 2 is given by

$$\tilde{x}(t+h) = x(t) + \frac{h}{2}\{3f(t, x(t)) - f(t-h, x(t-h))\}$$

$$x(t+h) = x(t) + \frac{h}{2}\{f(t+h, \tilde{x}(t+h)) + f(t, x(t))\}$$

The absolute value of the step-by-step error is $\varepsilon \equiv K|x(t+h) - \tilde{x}(t+h)|$, where $K = \frac{1}{6}$. Using ε to monitor the convergence, write and test an adaptive subroutine for this procedure.

3. (Continuation) Repeat for the Adams–Moulton method of order 3

$$\tilde{x}(t+h) = x(t) + \frac{h}{12}\{23f(t, x(t)) - 16f(t-h, x(t-h)) + 5f(t-2h, x(t-2h))\}$$

$$x(t+h) = x(t) + \frac{h}{12}\{5f(t+h, \tilde{x}(t+h)) + 8f(t, x(t)) - f(t-h, x(t-h))\}$$

where $K = 1/10$.

4. If we plan to use the Taylor series method with terms up to h^{20}, how should the computation $\sum_{n=0}^{20} x^{(n)}(t)h^n/n!$ be carried out in Fortran? Assume that $x(t)$ is available in X and $x^{(1)}(t)$, $\ldots$, $x^{(20)}(t)$ are available in an array Y of dimension 20. *Hint:* Five Fortran statements suffice.

5. Use the Runge–Kutta method and the ideas of Problem 11 in Sec. 8.1 to compute $\int_0^1 \sqrt{1+s^3}\,ds$.

6. Tabulate and graph the function $(1 - \ln v(x))v(x)$ on $[0, e]$, where $v(x)$ is the solution of the initial value problem $(dv/dx)\ln v(x) = 2x$, $v(0) = 1$. *Check point:* $v(1) = e$.

7. By solving an appropriate initial value problem, make a table of the function

$$f(x) = \int_{1/x}^{\infty} \frac{dt}{te^t}$$

on the interval $[0, 1]$. Determine how well f is approximated by $xe^{-1/x}$. *Hint:* Let $t = -\ln s$.

8. By solving an appropriate initial value problem, make a table of the function

$$f(x) = \frac{2}{\sqrt{\pi}}\int_0^x e^{-t^2}\,dt$$

on the interval $0 \leqq x \leqq 2$. Determine how accurately $f(x)$ is approximated on this interval by the function

$$g(x) = 1 - (ay + by^2 + cy^3)\frac{2}{\sqrt{\pi}}e^{-x^2}$$

where $a = 0.3084284$, $b = -0.0849713$, $c = 0.6627698$, and $y = (1 + 0.47047x)^{-1}$.

9. Compute a table of the function

$$\mathrm{Shi}(x) = \int_0^x \frac{\sinh t}{t}\,dt$$

by finding an initial value problem that it satisfies and then solving the initial value problem by the Taylor series method. Your table should be accurate to nearly machine precision.

9

Monte Carlo Methods and Simulation

A highway engineer wishes to simulate the flow of traffic in a proposed design for a major freeway intersection. The information obtained will then be used to determine the capacity of "storage lanes" (in which cars must slow down to yield the right of way). The intersection has the form shown in Fig. 9.1, and various flows (cars per minute) are postulated at the points where arrows are drawn. By running the simulation program, the engineer can study the effect of different speed limits, determine which flows lead to saturation (bottle necks), and so on. Some techniques for constructing such programs are developed in this chapter.

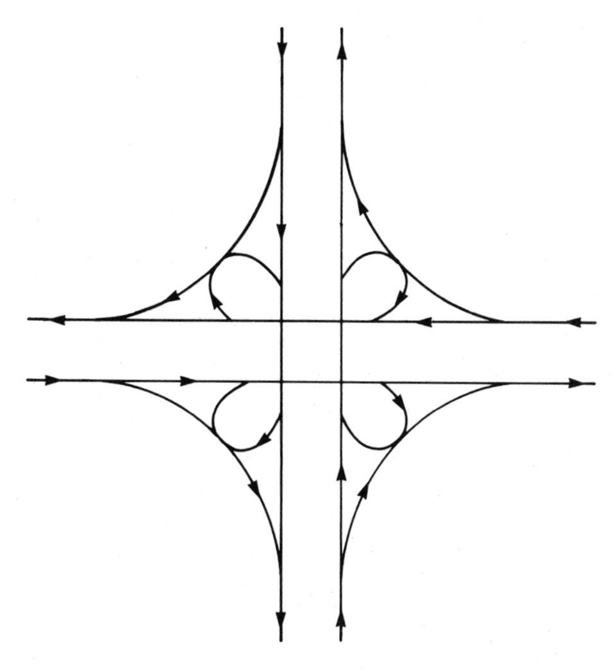

Figure 9.1 Traffic flow.

9.1 Random Numbers

This chapter differs from most of the others in its point of view. Instead of addressing clear-cut mathematical problems, it attempts to develop methods for simulating complicated processes or phenomena. If the computer can be made to imitate an experiment or a process, then by repeating the computer simulation with different data, we can draw statistical conclusions. In such an approach the conclusions may lack a high degree of mathematical precision but may still be sufficiently accurate to enable us to understand the process being simulated.

Particular emphasis is given to problems in which the computer simulation involves an element of chance. The whimsical name of ''Monte Carlo methods'' was applied some years ago by S. Ulam to this way of imitating reality by a computer. Since chance or randomness is part of the method, we begin with the elusive concept of *random numbers*.

Consider a sequence of real numbers $x_1, x_2, x_3, \ldots$ all lying in the unit interval (0, 1). Expressed informally, the sequence is *random* if the numbers seem to be distributed haphazardly throughout the interval and if there seems to be no pattern in the progression $x_1, x_2, x_3, \ldots$. For example, if all the numbers in decimal form began with the digit 3, then the numbers would be clustered in the subinterval $0.3 \leqq x < 0.4$ and would not be randomly distributed in (0, 1). If the numbers are monotonically increasing, they are not random. If each x_i is obtained from its predecessor by a simple continuous function, say $x_i = f(x_{i-1})$, then the sequence is not random (although it might appear so). A precise definition of ''randomness'' is quite difficult to formulate, and the interested reader may wish to consult an article by Chaitlin [1975] in which randomness is related to the complexity of computer algorithms! It thus seems best, at least in introductory material, to accept intuitively the notion of a random sequence of numbers in an interval, and to accept certain algorithms for generating sequences that are more or less random.

Most computer systems have *random number generators*, which are subprograms that provide a new random number with each call. In this chapter we call such a function subprogram `RANDOM`. The reader can use a random number generator available at his/her own computing installation or one of the generators described below. Random number generators are often available in mathematical program libraries. For the problems in this chapter, one should select a routine to provide random numbers uniformly distributed in the interval (0, 1). A sequence of numbers is *uniformly distributed* in the interval (0, 1) if no subset of the interval contains more than its share of the numbers. In particular, the probability that an element x drawn from the sequence falls within the subinterval $[a, a + h]$ should be h and hence independent of the number a. Similarly, if $p_i = (x_i, y_i)$ are random points in the plane uniformly distributed in some rectangle, then the number of these points falling inside a small square of area h should depend only on h and not on where the square is situated inside the rectangle.

Random numbers produced by a computer code cannot be truly random, for the manner in which they are produced is completely *deterministic*; that is, no element of chance is actually present. But the sequences produced by these routines appear to be random, and they do pass certain tests for randomness. Some authors prefer to emphasize this point by calling such sequences *pseudorandom*.

If the reader wishes to program a random number generator, the following one should be satisfactory. This algorithm generates random numbers $x_1, x_2, \ldots$ uniformly distributed in the open interval (0, 1) by means of the recursive formula

$$\ell_i \equiv 7^5 \ell_{i-1} \text{ modulo } (2^{31} - 1)$$

$$x_i = \frac{\ell_i}{2^{31} - 1} \qquad (i \geqq 1)$$

The initial integer ℓ_0 is called the *seed* for the sequence and is selected as any integer between 1 and the Mersenne prime number $2^{31} - 1 = 2{,}147{,}483{,}647$. A function subprogram to carry out this algorithm is as follows.

```
FUNCTION RANDOM(L)
L = MOD(16807*L, 2147483647)
RANDOM = FLOAT(L)*4.6566128752458E-10
RETURN
END
```

The intrinsic function MOD(N, M), for two positive integers N and M, is defined as the remainder when N is divided by M. Thus, for example, MOD(44, 7) = 2 and MOD(3, 11) = 3.

All ℓ_i are integers in the range $1 < \ell_i < 2^{31} - 1$. Most large computers can represent all integers in this range using the INTEGER format. However for computers with a short word length, the following version of RANDOM may be needed.

```
FUNCTION RANDOM(L)
DOUBLE PRECISION  DL
DL = DMOD(16807.0D0*DFLOAT(L),2147483647.0D0)
L = IDINT(DL)
RANDOM = SNGL(DL*4.6566128752458D-10)
RETURN
END
```

In this text we will use the first version of RANDOM given above.

Many minicomputers cannot represent integers in the desired range with their INTEGER format. If they have another format such as DOUBLE PRECISION, then a random number generator such as

```
DOUBLE PRECISION FUNCTION DRANDM(DL)
DOUBLE PRECISION  DL
DL = DMOD(16807.0D0*DL,2147483647.0D0)
DRANDM = DL*4.6566128752458D-10
RETURN
END
```

can be implemented. In many situations a machine independent random number generator is needed. The interested reader should consult an article by Schrage [1979] in which he discusses a more portable random number generator.

An example of a program to compute and print 100 random numbers using the function RANDOM follows. The seed has been arbitrarily selected as 256.

```
      L = 256
      DO 2 I = 1,100
      X = RANDOM(L)
    2 PRINT 3,X
    3 FORMAT(5X,F20.13)
      END
```

Observe that each call of the function RANDOM(L) modifies the value of L. The sequence of X's is completely determined by the initial choice of seed. This is a desirable attribute in many situations—for example, in debugging, when it is necessary to repeat a computer run and produce the same sequence of random numbers. On the other hand, a new set of random numbers is easily generated simply by changing the seed. Of course, the usage X = RANDOM(256) is ill-advised. (Why?)

As a coarse check on the random number generator, let us compute a long sequence of random numbers and determine what proportion of them lies in the interval $(0, \frac{1}{2}]$. The computed answer should be approximately 50%. The results with different seeds and different sequence lengths are tabulated below. Here is the program to carry out this experiment.

```
      DIMENSION  A(5)
      NPTS = 16000
      DO 3 I = 1,5
      L = 256*I
      N = 1000
      J = 1
      M = 0
      DO 2 K = 1,NPTS
      IF(RANDOM(L) .LE. 0.5)  M = M + 1
      IF(K .LT. N)  GO TO 2
      A(J) = 100.0*FLOAT(M)/FLOAT(N)
      J = J + 1
      N = 2*N
    2 CONTINUE
    3 PRINT 4,(A(J),J = 1,5)
    4 FORMAT(5X,5F6.1)
      END
```

In the program the seed is taken successively to be L = 256, 512, 768, 1024, and 1280. For each seed, a sequence of random numbers of length 16,000 is generated. Along the way the current proportion of numbers less than $\frac{1}{2}$ is computed at the 1000th step and then at steps 2000, 4000, 8000, 16,000. The results of the experiment are given in Table 9.1.

The experiment described can also be interpreted as a computer simulation of the tossing of a coin. A single toss corresponds to the selection of a random number x in the interval $(0,1)$. We arbitrarily associate "heads" with event $0 < x \leq \frac{1}{2}$ and "tails" with event $\frac{1}{2} < x < 1$. One thousand tosses of the coin correspond to 1000 choices of random numbers. The entries in the table show the proportion of "heads" resulting in repeated tossing of the coin. Random integers can be used to simulate coin tossing. (See Prob. 12.)

Table 9.1. Coin toss percentages

n / Seed	1000	2000	4000	8000	16,000
256	45.1	47.8	48.1	48.9	49.5
512	51.0	50.7	49.3	49.1	49.8
768	50.8	49.9	50.0	50.0	50.2
1024	49.8	49.6	50.3	50.6	49.9
1280	48.8	48.1	49.1	49.5	49.9

Observe from the table that (at least in this experiment) reasonable precision is attained for each seed with only a moderate number of trials (4000). Repeating the experiment 16,000 times has only a marginal influence on the precision. Of course, theoretically, if the random numbers were truly random, the limiting value in the table, as the number of trials increase without bound, would be exactly 50%.

Two random number routines that can be easily programmed are as follows.

1. Let the seed x_0 be any number in the interval (0, 1). For $i = 1, 2, 3, \ldots$, let x_i be the fractional part of $(\pi + x_{i-1})^5$.
2. Let $u_0 = 1$. For $i = 1, 2, 3, \ldots$, let $u_i \equiv (8t - 3)u_{i-1}$ (modulo 28) and $x_i = u_i/2^i$. Here t can be any large integer. In computing u_i, we retain only q binary bits in the indicated product. This algorithm is suitable for a binary machine with word length q.

A few words of caution about random number generators in computing systems are needed. The fact that the sequences produced by these programs are not truly random has already been noted. In some simulations the failure of randomness can lead to erroneous conclusions. Here are three specific points to remember.

First, the algorithms of the type illustrated here by **RANDOM** and those above produce *periodic* sequences. That is, the sequences eventually repeat themselves. The period is of the order 2^{30} for **RANDOM**.

Secondly, if the random number generator is used to produce random points in n-dimensional space, these points lie on a relatively small number of planes or hyperplanes. As Marsaglia [1968] reports, points obtained in this way in three-space lie on a set of 119,086 planes for a 48-bit word length computer. In ten-space they lie on a set of 126 planes.

Thirdly, the individual digits that compose random numbers generated by routines such as **RANDOM** are not, in general, independent random digits. For example, it might happen that the digit 3 follows the digit 5 more (or less) often than would be expected.

Now we consider some basic questions about generating random points in various geometric configurations. First, if uniformly distributed random points are needed on some interval (a, b), the statement

```
X = (B - A)*RANDOM(L) + A
```

accomplishes this. Secondly, the code

```
      I = INT((N + 1)*RANDOM(L))
```

produces random integers in the set $\{0, 1, 2, \ldots, n\}$. Finally, the Fortran statements

```
      X = RANDOM(L)
      DO 2 I = 1,4
      X = 10.0*X
      N(I) = INT(X)
    2 X = X - AINT(X)
```

can be used to obtain the first four digits in a random number. We arbitrarily select the seed L to be 256 in the examples in this chapter.

Consider now the problem of generating 1000 random points, uniformly distributed inside the ellipse $x^2 + 4y^2 = 4$. One way to do so is to generate random points in the rectangle $-2 \leqq x \leqq 2$, $-1 \leqq y \leqq 1$, and discard those that do not lie in the ellipse (Fig. 9.3).

```
      DIMENSION  X(1000),Y(1000)
      L = 256
      DO 3 K = 1,1000
    2 U = 4.0*RANDOM(L) - 2.0
      V = 2.0*RANDOM(L) - 1.0
      IF(U*U+4.0*V*V .GT. 4.0)  GO TO 2
      X(K) = U
    3 Y(K) = V
      . . .
```

In an attempt to be less wasteful, we might be tempted to *force* the $|y|$ value to be less than $\frac{1}{2}\sqrt{4 - x^2}$ as in the following program, *which produces erroneous results* (Fig. 9.4).

```
      DIMENSION  X(1000),Y(1000)
      L = 256
      DO 2 K = 1,1000
      X(K) = 4.0*RANDOM(L) - 2.0
      T = 0.5*SQRT(4.0 - X(K)**2)
    2 Y(K) = 2.0*T*RANDOM(L) - T
      . . .
```

This program does *not* produce uniformly distributed points inside the ellipse. To be convinced of this, consider two vertical strips taken inside the ellipse (see Fig. 9.2). If each strip is of width h, then approximately $1000(h/4)$ of the random points lie in each strip because the random variable x is uniformly distributed in $(-2, 2)$ and with each x a corresponding y is generated by the program so that (x, y) is inside the ellipse. But the two strips shown should *not* contain approximately the same number of points, for they do not have the same area. The points generated by the second program tend to be clustered at the left and right extremities of the ellipse.

For the same reasons, the following code does *not* produce uniformly distributed random points in the circle $x^2 + y^2 = 1$ (Fig. 9.5).

```
      DIMENSION  X(1000),Y(1000)
      DATA  L/256/,PI2/6.28318530717959/
      DO 2 K = 1,1000
      THETA = PI2*RANDOM(L)
      R = RANDOM(L)
      X(K) = R*COS(THETA)
    2 Y(K) = R*SIN(THETA)
      . . .
```

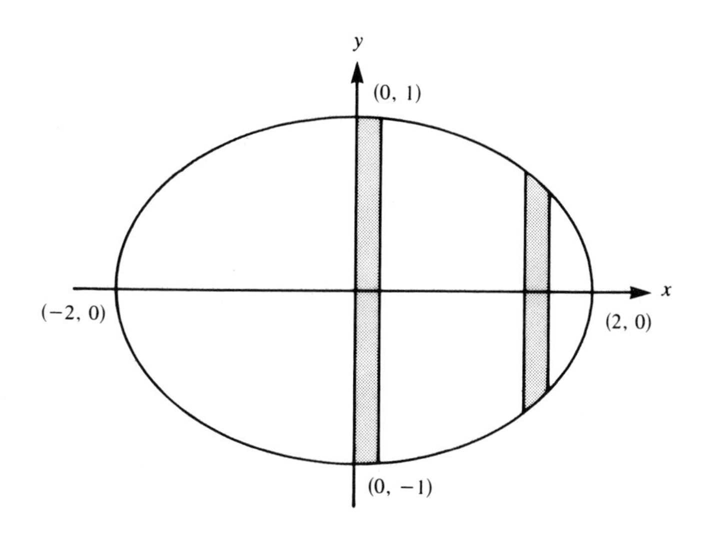

Figure 9.2

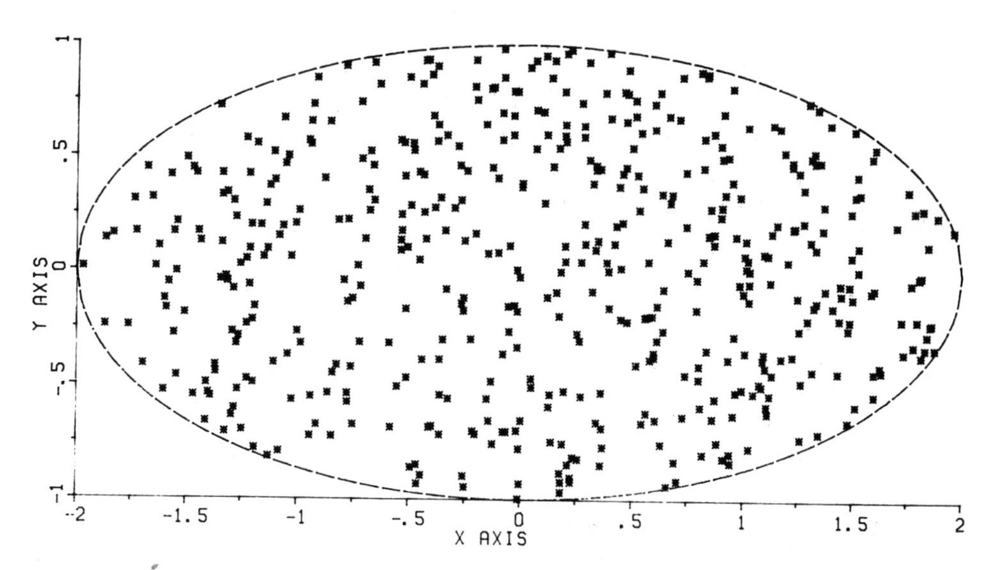

Figure 9.3 Uniformly distributed random points in ellipse $x^2 + 4y^2 = 4$.

In this program θ is uniformly distributed in $(0, 2\pi)$ and r is uniformly distributed in $(0, 1)$. However, in transferring from polar to rectangular coordinates by the equations $x = r \cos \theta$ and $y = r \sin \theta$, the uniformity is lost. The random points are strongly clustered near the origin.

The plots in Figs. 9.3, 9.4, and 9.5 illustrate the uniform distribution of random points using **RANDOM** correctly and the nonuniform distribution from incorrect usage.

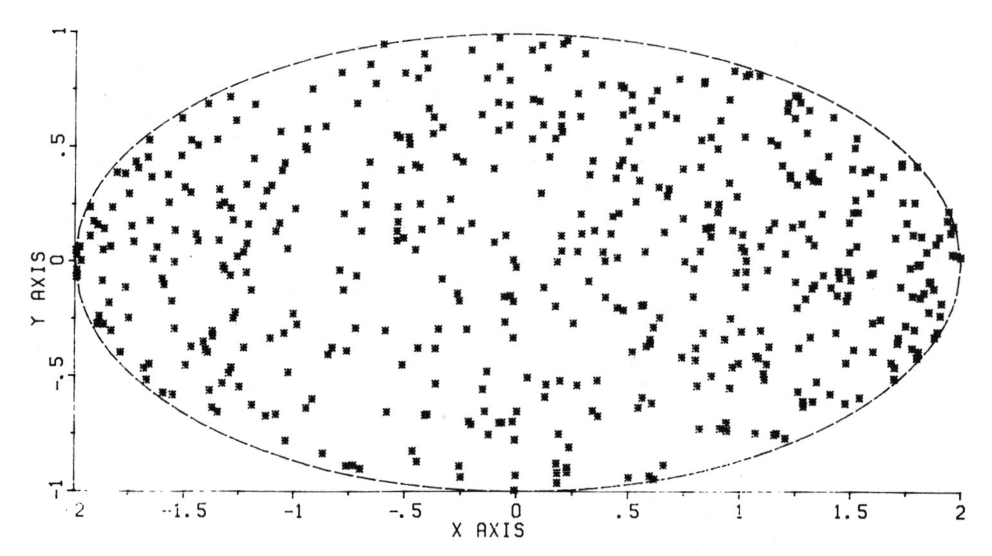

Figure 9.4 Nonuniformly distributed random points in the ellipse $x^2 + 4y^2 = 4$

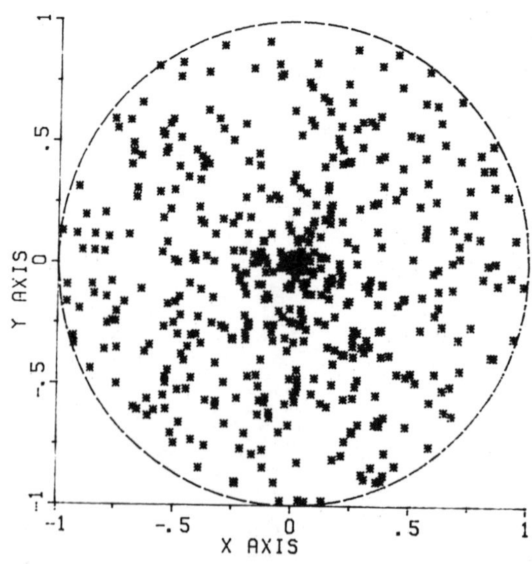

Figure 9.5 Nonuniformly distributed random points in the circle $x^2 + y^2 = 1$

Computer Problems 9.1

1. Generate 1000 random numbers x_i according to a uniform distribution in the interval (0, 1). Define a function f on (0, 1) as follows. $f(t)$ is the number of random numbers $x_1, x_2, \ldots, x_{1000}$ less than t. Compute $f(t)/1000$ for 200 points t uniformly distributed in (0, 1). What do you expect $f(t)/1000$ to be? Is this expectation borne out by the experiment? If a plotter is available, plot $f(t)/1000$.
2. Test the random number generator on your computer system in the following way. Generate 1000 random numbers $x_1, x_2, \ldots, x_{1000}$.
   (a) In any interval of width h approximately $1000h$ of the x_i should lie in that interval. Count the number of random numbers in each of ten intervals $[0, 1/n]$, $n = 1, 2, \ldots, 9, 10$.
   (b) Inequality $x_i < x_{i+1}$ should occur approximately 500 times. Count them in your sample.
3. Write a subprogram to generate with each call a random vector of the form $\mathbf{x} = (x_1, \ldots, x_{20})^T$ where each x_i is an integer from 1 to 100 and no two components of $\mathbf{x}$ are the same.
4. Write a program to generate 1000 random points uniformly distributed in

   (a) the equilateral triangle in Fig. 9.6 and (b) the diamond in Fig. 9.7.

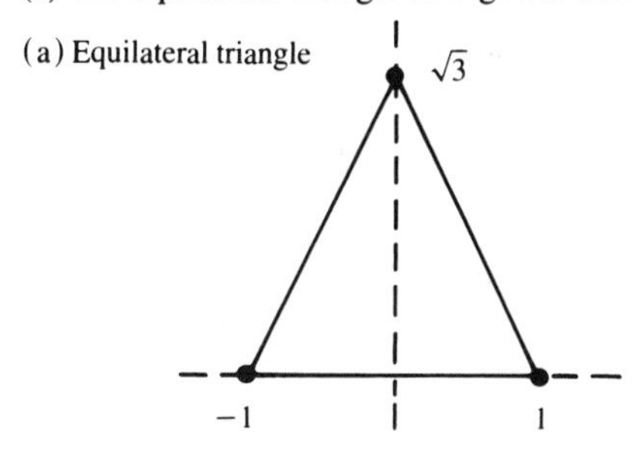

(b) Diamond

Figure 9.6 Figure 9.7

   Store the random points (x_i, y_i) in arrays X and Y.
5. If $x_1, x_2, x_3, \ldots$ is a random sequence of numbers uniformly distributed in the interval (0, 1), what proportion would you expect to satisfy the inequality $40x^2 + 7 > 43x$? Write a program to test this on 1000 random numbers produced by subroutine **RANDOM**.
6. Write a program to generate and print 1000 points uniformly and randomly distributed in the circle $(x - 3)^2 + (y + 1)^2 \leqq 9$.
7. Test the algorithms (1) and (2) on page 205 as random number generators.
8. Compute the average distance between two points in the circle $x^2 + y^2 = 1$. To solve this, generate N random pairs of points (x_i, y_i), (v_i, w_i) in the circle and compute

$$N^{-1} \sum_{i=1}^{N} [(x_i - v_i)^2 + (y_i - w_i)^2]^{1/2}$$

9. Generate in the computer 1000 random numbers in the interval (0, 1) and print the less significant digits of each. Examine them for evidence of nonrandom behavior.
10. Generate 1000 random numbers x_i $(1 \leqq i \leqq 1000)$ on your computer. Let n_i denote the eighth decimal digit in x_i. Count how many 0's, 1's, . . . , 9's there are among the 1000 numbers n_i. How many of each would you expect? This code can be written with nine Fortran statements.
11. (Continuation) Using random number generator **RANDOM**, generate 1000 random numbers with seed 256. Count how many times the digit i occurs in the jth decimal place. Print a table of these values — that is, frequency of digit versus decimal place. By examining the table, de-

termine which decimal place seems to produce the best uniform distribution of random digits. *Hint*: Use the subroutine from Computer Problem 6 of Sec. 1.1 to compute the arithmetic mean, variance, and standard deviations of the table entries.

12. Using random integers, write a short program to simulate five persons matching coin-flips. Print the percentage of match-ups (five of a kind) after 125 match-offs.

9.2 Estimation of Areas and Volumes by Monte Carlo Techniques

Now we turn to applications, the first being the approximation of a definite integral by the Monte Carlo method. If we select the first n elements $x_1, x_2, \ldots, x_n$ from a random sequence in the interval (0, 1), then

$$\int_0^1 f(x)\,dx \approx \frac{1}{n}\sum_{i=1}^{n} f(x_i)$$

That is, the integral is approximated by the average of n numbers $f(x_1), \ldots, f(x_n)$. When this is actually carried out, the error is of order $1/\sqrt{n}$, which is not at all competitive with good algorithms, such as the Romberg method. However, in higher dimensions, the Monte Carlo method can be quite attractive. For example,

$$\int_0^1\int_0^1\int_0^1 f(x, y, z)\,dx\,dy\,dz \approx \frac{1}{n}\sum_{i=1}^{n} f(x_i, y_i, z_i)$$

where (x_i, y_i, z_i) is a random sequence of n triplets in the cube $0 \leqq x \leqq 1, 0 \leqq y \leqq 1, 0 \leqq z \leqq 1$. In order to obtain random points in the cube, we assume that we have a random sequence in (0, 1) denoted by $\xi_1, \xi_2, \xi_3, \xi_4, \ldots$. To get our first random point p_1 in the cube, just let $p_1 = (\xi_1, \xi_2, \xi_3)$. The second is, of course, $p_2 = (\xi_4, \xi_5, \xi_6)$ and so on.

If the length of the interval (in a one-dimensional integral) is not one, remember that the average of f is not simply the integral but rather

$$\frac{1}{b-a}\int_a^b f(x)\,dx$$

which agrees with our intention that the function 1 have average one. Similarly, in higher dimensions, we must define the average of f over a region by integrating and dividing by the area, volume, or measure of that region. For instance,

$$\frac{1}{8}\int_1^3\int_{-1}^1\int_0^2 f(x, y, z)\,dx\,dy\,dz$$

is the average of f over the parallelepiped described by the inequalities $0 \leqq x \leqq 2$, $-1 \leqq y \leqq 1$, $1 \leqq z \leqq 3$.

So if (x_i, y_i) denote random points with appropriate uniform distribution, the following examples illustrate Monte Carlo techniques.

$$\int_0^5 f(x)\,dx \approx \frac{5}{n}\sum_{i=1}^{n} f(x_i)$$

$$\int_2^5\int_1^6 f(x, y)\,dx\,dy \approx \frac{15}{n}\sum_{i=1}^{n} f(x_i, y_i)$$

In each case, the random points should be uniformly distributed in the regions involved. Here we are using the fact that the average of a function on a set is equal to the integral of the function over the set divided by the measure of the set. Therefore,

$$\int_A f \approx (\text{Measure of } A) \times (\text{Average of } f \text{ on } A)$$

The volume of a complicated region in three-space can be computed by a Monte Carlo technique. Taking a simple case, let us determine the volume of the region whose points satisfy the inequalities

$$\begin{cases} 0 \leqq x \leqq 1, \quad 0 \leqq y \leqq 1, \quad 0 \leqq z \leqq 1 \\ x^2 + \sin y \leqq z \\ x - z + e^y \leqq 1 \end{cases}$$

The first line defines a cube whose volume is 1. The region defined by *all* the given inequalities is, therefore, a subset of this cube. If we generate n random points in the cube and determine that m of them satisfy the last two inequalities, the volume of the desired region must be approximately m/n. Here is a program that carries out this procedure.

```
      DATA  J,L,M,N/100,256,0,10/
      DO 3 K = 1,N
      DO 2 I = 1,J
      X = RANDOM(L)
      Y = RANDOM(L)
      Z = RANDOM(L)
      IF(X*X+SIN(Y) .GT. Z)  GO TO 2
      IF(X-Z+EXP(Y) .GT. 1.0)  GO TO 2
      M = M + 1
    2 CONTINUE
      V = FLOAT(M)/FLOAT(J*K)
    3 PRINT 4,V
    4 FORMAT(5X,F10.5)
      END
```

Observe that intermediate estimates are printed out when we reach 100, 200, . . . , 1000 points. An approximate value of 0.13 is determined for the volume of the region.

Continuing, let us consider the problem of obtaining the numerical value of the integral

$$\iint_\Omega \sin \sqrt{\ln (x + y + 1)}\; dx\, dy = \iint_\Omega f(x, y)\, dx\, dy$$

over the disk in xy space, defined by the inequality

$$\Omega : \left(x - \frac{1}{2}\right)^2 + \left(y - \frac{1}{2}\right)^2 \leqq \frac{1}{4}$$

A sketch of this domain, with a surface above it, is shown in Fig. 9.8.

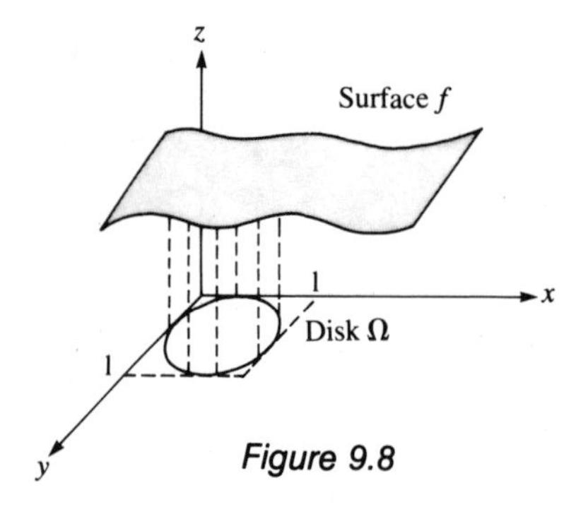

Figure 9.8

We proceed by generating random points in the square and discarding those that do not lie in the disk. We take $n = 1000$ points in the disk. If the points are $p_i = (x_i, y_i)$, then the integral is estimated to be

$$\iint_\Omega f\,dx\,dy \approx \text{(Area of base)} \times \text{(Average height)}$$
$$= (\pi r^2) \times \left[\frac{1}{n}\sum_{i=1}^{n} f(p_i)\right]$$
$$= \frac{\pi}{4n}\sum_{i=1}^{n} f(p_i)$$

The code for this example is given below. Intermediate estimates of the integral are printed for $n = 250, 500, 750$, and 1000. This gives us some idea of how the correct value is being approached by our averaging process.

```
      DATA  L,M,SUM/256,250,0.0/
      F(X,Y) = SIN(SQRT(ALOG(X + Y + 1.0)))
      PI4 = ATAN(1.0)
      DO 4 I = 1,4
      DO 3 J = 1,M
    2 X = RANDOM(L)
      Y = RANDOM(L)
      IF((X - 0.5)**2 + (Y - 0.5)**2 .GT. 0.25)  GO TO 2
    3 SUM = SUM + F(X,Y)
      VOL = PI4*SUM/FLOAT(I*M)
    4 PRINT 5,VOL
    5 FORMAT(5X,F10.5)
      END
```

We obtain an approximate numerical value of 0.57 for the integral.

Computer Problems 9.2

1. Write and test a program to evaluate the integral $\int_0^1 e^x\,dx$ by the Monte Carlo method, using $n = 25, 50, 100, 200, 400, 800$. Observe that 800 random numbers are needed and that the work in each case can be used in the next case. Print the exact answer.

2. Use the Monte Carlo method to approximate the integral

$$\int_{-1}^{1}\int_{-1}^{1}\int_{-1}^{1} (x^2 + y^2 + z^2)\, dx\, dy\, dz$$

Compare with the correct answer.

3. A Monte Carlo method for estimating $\int_a^b f(x)\,dx$ if $f(x) \geqq 0$ is as follows. Let $c \geqq \max_{a \leqq x \leqq b} f(x)$. Then generate n random points (x, y) in the rectangle $a \leqq x \leqq b, 0 \leqq y \leqq c$. Count the number k of these random points (x, y) that satisfy $y \leqq f(x)$. Then $\int_a^b f(x)\,dx \approx kc(b - a)/n$. Verify this and test the method on $\int_1^2 x^2\,dx$, $\int_0^1 (2x^2 - x + 1)\,dx$, and $\int_0^1 (x^2 + \sin 2x)\,dx$.
4. (Continuation) Modify the method outlined in Problem 3 to handle the case when f takes positive and negative values on $[a, b]$. Test the method on $\int_{-1}^{1} x^3\,dx$.
5. (Continuation) Use the method of Problem 3 to estimate $\pi = 4\int_0^1 \sqrt{1 - x^2}\,dx$. Generate random points in the unit square $0 \leqq x \leqq 1, 0 \leqq y \leqq 1$. Use $n = 100, 200, \ldots, 1000$ and try to determine whether the error is behaving like $1/\sqrt{n}$.
6. Another Monte Carlo method for evaluating $\int_a^b f(x)\,dx$ is as follows. Generate an odd number of random numbers in (a, b). Reorder these points so that $a < x_1 < x_2 < \ldots < x_n < b$. Now compute

$$f(x_1)(x_2 - a) + f(x_3)(x_4 - x_2) + f(x_5)(x_6 - x_4) + \cdots + f(x_n)(b - x_{n-1})$$

Test this method on

$$\int_0^1 (1 + x^2)^{-1}\,dx, \qquad \int_0^1 (1 - x^2)^{-1/2}\,dx, \qquad \int_0^1 x^{-1} \sin x\,dx$$

7. What is the expected value of the volume of a tetrahedron formed by four points chosen randomly inside the tetrahedron whose vertices are (0, 0, 0), (0, 1, 0), (0, 0, 1), (1, 0, 0)? (The precise answer is unknown!)
8. Run the programs given in this section and verify that they produce reasonable answers.
9. Write and test a short program to estimate the area under the curve $y = e^{-(x+1)^2}$ and inside the triangle having vertices (1, 0), (0, 1), (−1, 0).
10. Using the Monte Carlo approach, find the area of the irregular figure defined by

$$\begin{cases} 1 \leqq x \leqq 3 \\ -1 \leqq y \leqq 4 \\ x^3 + y^3 \leqq 29 \\ y \geqq e^x - 2 \end{cases}$$

11. Use the Monte Carlo method to estimate the volume of the solid whose points (x, y, z) satisfy

$$\begin{cases} 0 \leqq x \leqq 1 \\ 1 \leqq y \leqq 2 \\ -1 \leqq z \leqq 3 \\ e^x \leqq y \\ (\sin z)y \geqq 0 \end{cases}$$

12. Estimate the area of the region determined by the inequalities $0 \leqq x \leqq 1$, $10 \leqq y \leqq 13$, $y \geqq 12 \cos x$, $y \geqq 10 + x^3$, using a Monte Carlo technique. Print intermediate answers.

9.3 Simulation

We next illustrate the idea of *simulation*. We consider a physical situation in which an element of chance is present and try to imitate the situation on the computer. Statistical conclusions can be drawn if the experiment is performed many times.

In simulation problems we must often produce random variables with prescribed distribution. Suppose, for example, that we want to simulate the throw of a loaded die and that the probabilities of various outcomes have been determined as shown.

Outcome	1	2	3	4	5	6
Probability	.2	.14	.22	.16	.17	.11

If the random variable x is uniformly distributed in the interval $(0, 1)$, then by breaking this interval into six subintervals of lengths given by the table, we can simulate the throw of this loaded die. For example, we agree that if x is in $(0, 0.2)$, the die shows 1; if x is in $[0.2, 0.34)$, the die shows 2, and so on. A program to count the outcome of 1000 throws of this die might be written like this.

```
      DIMENSION  M(6),Y(6)
      DATA  L,N,(Y(I),I=1,6)/256,1000,0.2,0.34,0.56,0.72,0.89,1.0/
      DO 2 I = 1,6
    2 M(I) = 0
      DO 4 K = 1,N
      DO 3 I = 1,6
      IF(RANDOM(L) .LT. Y(I))  GO TO 4
    3 CONTINUE
    4 M(I) = M(I) + 1
      PRINT 5,(M(I),I = 1,6)
    5 FORMAT(5X,6I5)
      END
```

The next example of a simulation is a very old problem known as ***Buffon's needle problem***. Imagine that a needle of unit length is dropped onto a sheet of paper ruled by parallel lines one unit apart. What is the probability that the needle intersects one of the lines?

To make the problem precise, assume that the center of the needle lands between the lines at a random point. Assume further that the angular orientation of the needle is another random variable. Finally, assume that our random variables are drawn from a uniform distribution.

Figure 9.9 shows the geometry of the situation. Let the distance of the center of the needle from the nearest of the two lines be u and the angle from the horizontal be v. Here u and v are the two random variables. The needle intersects one of the lines if and only if $u \leqq \frac{1}{2} \sin v$. We perform the experiment many times, say 1000. Because of the problem's symmetry, we select u from a uniform random distribution on the interval $(0, \frac{1}{2})$, v from a uniform random distribution on the interval $(0, \pi/2)$, and determine the number of times that $2u \leqq \sin v$. We let $w = 2u$ and test $w \leqq \sin v$, where w is a random variable in $(0, 1)$. In the program the value of $\pi/2$ is needed. Since $\tan(\pi/4) = 1$, it follows that $\pi/4 = \text{Arctan}(1)$ and $\pi/2 = 2\,\text{Arctan}(1)$. Thus the statement `PI2=2.0*ATAN(1.0)` produces

$\pi/2$. Alternatively, the numerical value of $\pi/2$ can be inserted with a data statement. In this program intermediate answers are printed out so that their progression can be observed. Also, the theoretical answer, $t = 2/\pi \approx 0.63662$, is printed for comparison.

```
      DATA  L,M,N/256,0,1000/
      PI2 = 2.0*ATAN(1.0)
      T = 1.0/PI2
      DO 3 I = 1,10
      DO 2 J = 1,N
      W = RANDOM(L)
      V = PI2*RANDOM(L)
      IF(W .LE. SIN(V))  M = M + 1
    2 CONTINUE
      K = I*N
      X = FLOAT(M)/FLOAT(K)
    3 PRINT 4,K,X,T
    4 FORMAT(5X,I5,2F10.5)
      END
```

Our next example again has an analytic solution. This is advantageous for us, since we wish to compare the results of Monte Carlo simulations to theoretical solutions. Consider the experiment of tossing two dice. For an (unloaded) die, the numbers 1, 2, 3, 4, 5, 6 are equally likely to occur. We ask "What is the probability of throwing a 12 (i.e., 6 appearing on each die) in 24 throws of the dice?" The answer is $1 - (35/36)^{24} = 49.1\%$. If we simulate this process, a single experiment consists of throwing the dice 24 times, and this experiment must be repeated a large number of times, say 1000. For the outcome of the throw of a single die, we need random integers uniformly distributed in the set $\{1, 2, 3, 4, 5, 6\}$. It is easier to imagine the dice labeled $0, 1, \ldots, 5$ and regard as a success the appearance of two 5's. If x is a random variable in $(0, 1)$, then $6x$ is a random variable in $(0, 6)$ and the integer part of $6x$ is a random integer in $\{0, 1, 2, 3, 4, 5\}$. Here is a program.

```
      DATA  L,M,N/256,0,1000/
      DO 3 K = 1,N
      DO 2 J = 1,24
      I1 = INT(6.0*RANDOM(L))
      I2 = INT(6.0*RANDOM(L))
      IF(I1+I2 .LT. 10)  GO TO 2
      M = M + 1
      GO TO 3
    2 CONTINUE
    3 CONTINUE
      X = FLOAT(M)/FLOAT(N)
      PRINT 4,X
    4 FORMAT(5X,F10.5)
      END
```

This program computes the probability of throwing a 12 in 24 throws of the dice at "even money"—that is, 50.5%.

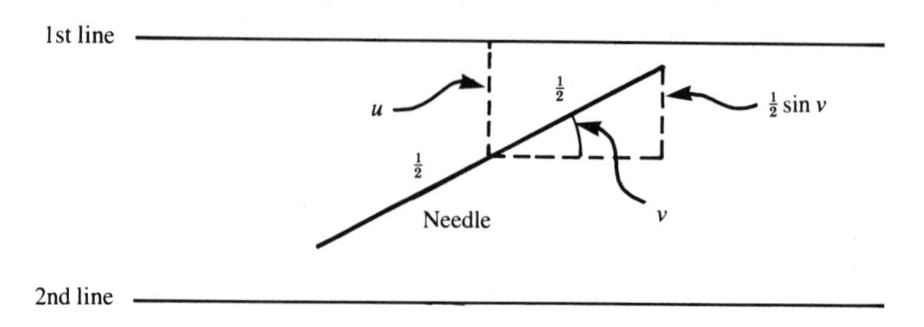

Figure 9.9

Our final example concerns neutron shielding. We take a simple model of neutrons passing into a lead wall. It is assumed that each neutron enters the lead wall at a right angle to the wall and travels a unit distance. Then it collides with a lead atom and rebounds in a random direction. Again it travels a unit distance before colliding with another lead atom. It rebounds in a random direction and so on. Assume that after eight collisions all the neutron's energy is spent. Assume also that the lead wall is five units thick. The question is "What percentage of neutrons can be expected to emerge from the other side of the lead wall?"

Let x be the distance measured from the initial surface where the neutron enters (Fig. 9.10).

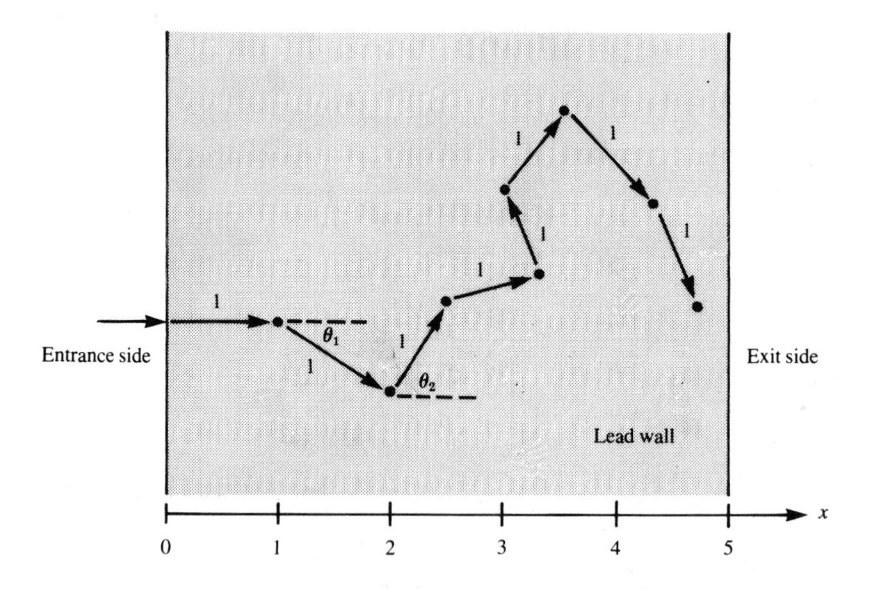

Figure 9.10

From trigonometry we recall that in a right triangle with hypotenuse 1 one side is $\cos\theta$. Also note that $\cos\theta \leqq 0$ when $\pi/2 \leqq \theta \leqq \pi$ (Fig. 9.11). The first collision occurs at a point where $x = 1$; the second at a point where $x = 1 + \cos\theta_1$. The third collision occurs at a point where $x = 1 + \cos\theta_1 + \cos\theta_2$ and so on. If $x \geqq 5$, the neutron has exited. If $x < 5$ for all eight collisions, the wall has shielded the area from that particular neutron. For a

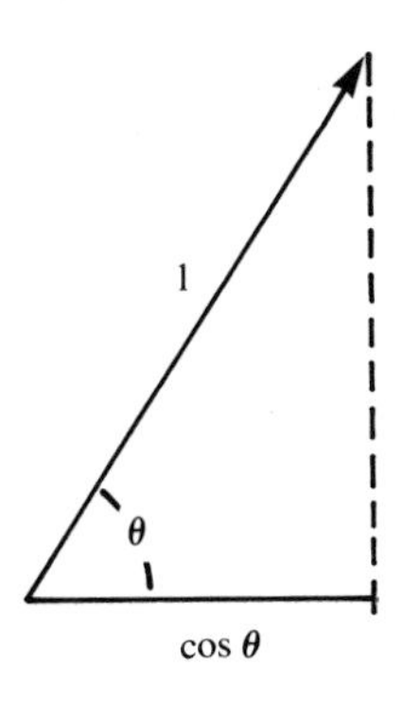

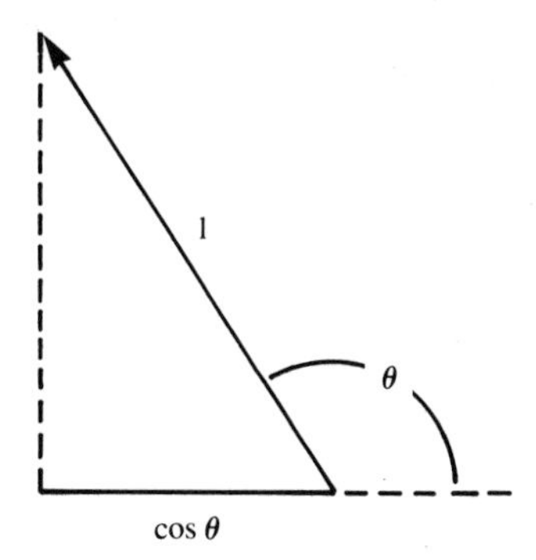

Figure 9.11

Monte Carlo simulation, we can use random angles θ_i in the interval $(0, \pi)$ because of symmetry. The simulation program is then

```
      DATA  L,M,N/256,0,100/
      PI = 4.0*ATAN(1.0)
      DO 4 I = 1,10
      DO 3 J = 1,N
      X = 1.0
      DO 2 K = 1,7
      Y = PI*RANDOM(L)
      X = X + COS(Y)
      IF(X .LE. 0.0)  GO TO 3
      IF(X .LT. 5.0)  GO TO 2
      M = M + 1
      GO TO 3
    2 CONTINUE
    3 CONTINUE
      PER = 100.0*FLOAT(M)/FLOAT(I*N)
    4 PRINT 5,PER
    5 FORMAT(5X,F10.5)
      END
```

Based on its results, we can say that approximately 1.9% of the neutrons can be expected to emerge from the lead wall.

Computer Problems 9.3

1. Suppose that a die is "loaded" so that the six faces are not equally likely to turn up when the die is rolled. The probabilities associated with the six faces are as follows.

1	2	3	4	5	6
.15	.2	.25	.15	.1	.15

   Write and run a program to simulate 1500 throws of such a die.
2. Write a program to simulate the following physical phenomenon. A particle is moving in the xy plane under the effect of a random force. It starts at $(0, 0)$. At the end of each second it moves

one unit in a random direction. We want to record in a table its position at the end of each second, taking altogether 1000 seconds.

3. (A Random Walk) On a windy night a drunkard begins walking at the origin of a two-dimensional coordinate system. His steps are one unit in length and are random in the following way. With probability $1/6$ he takes a step east, with probability $1/4$ he takes a step north, with probability $1/4$ he takes a step south, and with probability $1/3$ he takes a step west. What is the probability that after 50 steps he will be more than 20 units distant from the origin? Write a program to simulate this problem.
4. The lattice points in the plane are defined as those points whose coordinates are integers. A circle of diameter 1.5 is dropped on the plane in such a way that its center is a uniformly distributed random point in the square $0 \leqq x \leqq 1, 0 \leqq y \leqq 1$. What is the probability that two or more lattice points lie inside the circle? Use the Monte Carlo simulation to compute an approximate answer.
5. (Another Random Walk) Consider the lattice points (points with integer coordinates) in the square $0 \leqq x \leqq 6$, $0 \leqq y \leqq 6$. A particle starts at the point (4, 4) and moves in the following way. At each step it moves with equal probability to one of the four adjacent lattice points. What is the probability that when the particle first crosses the boundary of the square, it crosses the bottom side? Use Monte Carlo simulation.
6. What is the probability that within 20 generations the Kzovck family name will die out? Use the following data. In the first generation there is one male Kzovck. In each succeeding generation the probability that a male Kzovck will have exactly one male offspring is $4/11$, the probability that he will have exactly two is $1/11$, and the probability that he will have more than two is zero.
7. Write a program that simulates the random shuffle of a deck of 52 cards.
8. A merry-go-round with a total of 24 horses allows children to jump on at three gates and jump off at only one gate while it continues to turn slowly. If the children get on and off randomly (at most one per gate), how many revolutions go by before someone must wait longer than one revolution to ride? Assume a probability of $\frac{1}{2}$ that a child gets on or off.
9. Run the programs given in this section and determine whether the results are reasonable.
10. If two points are randomly chosen in the cube $\{(x, y, z) : 0 \leqq x \leqq 1, 0 \leqq y \leqq 1, 0 \leqq z \leqq 1\}$, what is the expected distance between them?

Supplementary Computer Problems (Chapter 9)

1. A particle breaks off from a random point on a rotating flywheel. Referring to Fig. 9.12, determine the probability of the particle hitting the window. Perform a Monte Carlo simulation to compute the probability in an experimental way.

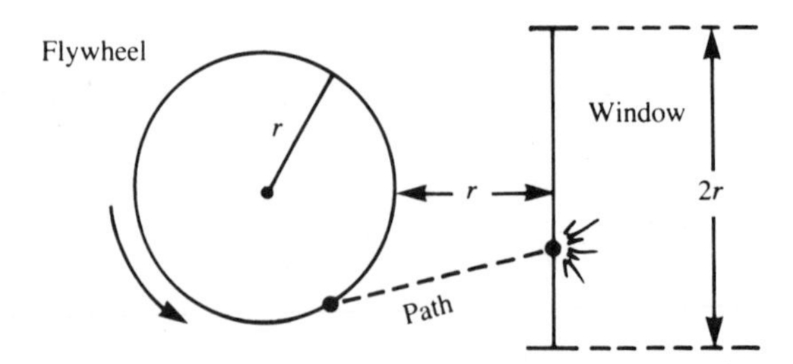

Figure 9.12

2. A point (a, b) is chosen at random in a rectangle defined by inequalities $|a| \leqq 1, |b| \leqq 2$. What is the probability that the resulting quadratic equation $ax^2 + bx + 1 = 0$ has *real* roots? Find the answer both analytically and by the Monte Carlo method.
3. Consider a circle of radius 1. A point is chosen at random inside the circle and a chord having the chosen point as midpoint is drawn. What is the probability that the chord will have length greater than 3/2? Solve the problem analytically and by the Monte Carlo method.
4. Write a program to generate 1600 random triples uniformly distributed in the sphere defined by $x^2 + y^2 + z^2 \leqq 1$. Count the number of random points in the first octant.
5. Write a program to simulate 1000 flips of three coins. Print the number of times that two of the three coins come up "heads."
6. Write a program to verify numerically that $\pi = \int_0^2 (4 - x^2)^{1/2}\, dx$. Use the Monte Carlo method and 2500 random numbers.
7. Write a program to estimate the probability that three random points on the edges of a square form an obtuse triangle (see Fig. 9.13).

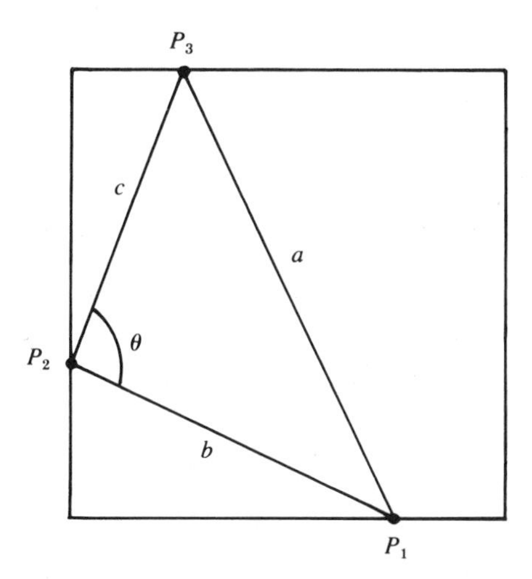

Figure 9.13

Hint: $\cos\theta = (b^2 + c^2 - a^2)/2bc$ (law of cosines).

8. A *histogram* is a graphical device for displaying frequencies by means of rectangles whose heights are proportional to frequencies. For example, in throwing two dice 3600 times, the resulting sums 2, 3, . . . , 12 should occur with frequencies close to those shown in the histogram in Fig. 9.14.

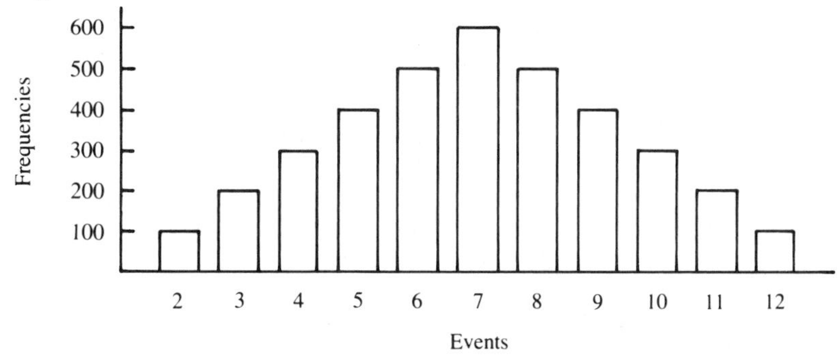

Figure 9.14

By means of a Monte Carlo simulation, determine a histogram for the frequency of digits 0, . . . , 9 appearing in 1000 random numbers.

9. Consider a neutron-shielding problem similar to the one in the text but modified as follows. Imagine the neutron beam impinging on the wall one unit above its base. The wall can be very high, and neutrons cannot escape from the top. But they can escape from the bottom as well as from the exit side. Find the percentage of escaping neutrons.
10. Consider a pair of loaded dice as described in Sec. 9.3. By a Monte Carlo simulation, determine the probability of throwing a 12 in 25 throws of the dice.
11. An integral can be estimated by the formula

$$\int_0^1 f(x)\,dx \approx \frac{1}{n}\sum_{i=1}^{n} f(x_i)$$

even if the x_i are not random numbers, and, in fact, some nonrandom sequences may be better. Use the sequence x_i = (fractional part of $i\sqrt{2}$) and test the corresponding numerical integration scheme. Test whether the estimates converge at the rate $1/n$ or $1/\sqrt{n}$, using some simple examples, such as $\int_0^1 e^x\,dx$, $\int_0^1 (1 + x^2)^{-1}\,dx$.

12. Consider a circular city of diameter 20 kilometers (see Fig. 9.15). Radiating from the center are 36 straight roads, spaced 10° apart in angle. There are also 20 circular roads spaced 1 kilometer apart. What is the average distance between road intersection points in the city measured along the roads?

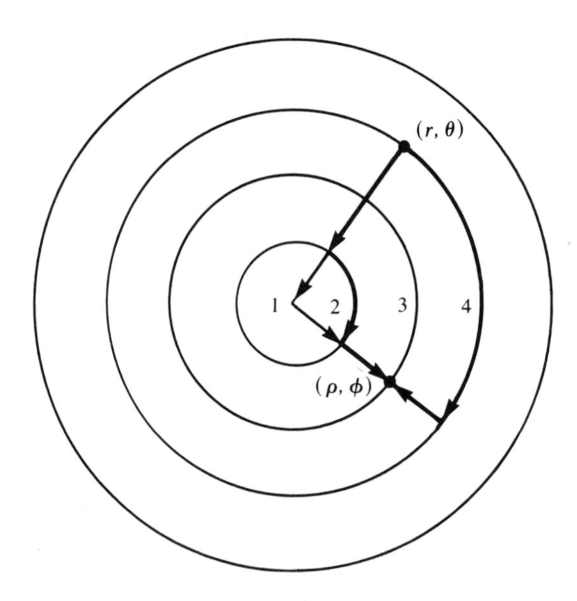

Figure 9.15

13. Write a program to estimate

$$\int_0^2 \int_3^6 \int_{-1}^1 (yx^2 + z\log y + e^x)\,dx\,dy\,dz$$

14. Consider the cardioid given by $(x^2 + y^2 + x)^2 = (x^2 + y^2)$. Write a program to find the average distance, *staying within the cardioid*, between two points randomly selected within the figure. Use 1000 points and print intermediate estimates.

15. Let R denote the region in the xy plane defined by inequalities

$$\begin{cases} \frac{1}{3} \leqq 3x \leqq 9 - y \\ \sqrt{x} \leqq y \leqq 3 \end{cases}$$

Estimate the integral

$$\iint_{R} (e^x + \cos xy)\, dx\, dy$$

16. Using a Monte Carlo technique, estimate the area of the region defined by inequalities $4x^2 + 9y^2 \leqq 36$, $y \leqq \text{Arctan}\,(x + 1)$.
17. Write a program to estimate the area of the region defined by inequalities

$$\begin{cases} x^2 + y^2 \leqq 4 \\ |y| \leqq e^x \end{cases}$$

18. Find the length of the lemniscate whose equation in polar coordinates is $r^2 = \cos 2\theta$. *Hint*: In polar coordinates $ds^2 = dr^2 + r^2\, d\theta^2$.
19. Two points are selected at random on the circumference of a circle. What is the average distance from the center of the circle to the center of gravity of the two points?
20. Let n_i $(1 \leqq i \leqq 1000)$ be a sequence of integers satisfying $0 \leqq n_i \leqq 9$. Write a program to test the given sequence for periodicity. (The sequence is *periodic* if there is an integer k such that $n_i = n_{i+k}$ for all i.)
21. Consider the ellipsoid

$$\frac{x^2}{4} + \frac{y^2}{16} + \frac{z^2}{4} = 1$$

(a) Write a program to generate and store 2500 random points uniformly distributed in the first octant of this ellipsoid.
(b) Write a program to estimate the volume of this ellipsoid in the first octant.

22. (The French Railroad System) Define the "distance" between two points (x_1, y_1) and (x_2, y_2) in the plane to be $\sqrt{(x_1 - x_2)^2 + (y_1 - y_2)^2}$ if the points are on a straight line through the origin but $\sqrt{x_1^2 + y_1^2} + \sqrt{x_2^2 + y_2^2}$ in all other cases. Draw a picture to illustrate. Compute the average "distance" between two points randomly selected in a unit circle centered at the origin.
23. (Repeated Birthdays Problem) Suppose that in a room of N persons each of the 365 days of the year is equally likely to be someone's birthday. From probability theory it can be shown that, contrary to intuition, only 23 persons need be present for the chances to be better than fifty-fifty that at least two of them will have the same birthday. With 56 persons, the probability is 0.988, or almost theoretically certain. Write a program to simulate this problem and print the probabilities for various values of N such as 23, 32, 40, and 56.

10

The Minimization of Multivariate Functions

An engineering design problem leads to a function

$$f(x, y) = (\cos x + e^y)^2 + 3(xy)^4$$

in which x and y are parameters to be selected and $f(x, y)$ is a function related to the cost of manufacturing and is to be minimized. Methods for locating optimal points (x, y) in such problems are developed in this chapter.

In calculus an important application concerns the problem of locating the local minima of a function. There, the principal technique for minimization is to differentiate the function whose minimum is sought, set the derivative equal to zero, and locate the points that satisfy the resulting equation. Problems of maximization are covered by the theory of minimization, since the maxima of f occur at points where $-f$ has its minima.

This technique can be used on functions of one or several variables. For example, if a minimum value of $f(x_1, x_2, x_3)$ is sought, we look for the points where all three partial derivatives are simultaneously zero:

$$\frac{\partial f}{\partial x_1} = \frac{\partial f}{\partial x_2} = \frac{\partial f}{\partial x_3} = 0$$

This procedure cannot be readily accepted as a *general-purpose* numerical method because it requires differentiation, followed by the solution of one or more equations in one or more variables. This task may be as difficult to accomplish as a direct frontal attack on the original problem.

The minimization problem has two forms: the unconstrained and the constrained. In an "unconstrained minimization" problem a function F is defined from an n-dimensional space $\mathbf{R}^n$ into the real line $\mathbf{R}$ and a point $\mathbf{z} \in \mathbf{R}^n$ is sought with the property that

$$F(\mathbf{z}) \leqq F(\mathbf{x}) \quad \text{for all } \mathbf{x} \in \mathbf{R}^n$$

It is convenient to write points in $\mathbf{R}^n$ simply as $\mathbf{x}$, $\mathbf{y}$, $\mathbf{z}$ and so on. If it is necessary to display the components of a point, we write $\mathbf{x} = (x_1, x_2, \ldots, x_n)^T$. In a "constrained minimization" problem a subset K in $\mathbf{R}^n$ is prescribed and a point $\mathbf{z} \in K$ is sought for which

$$F(\mathbf{z}) \leqq F(\mathbf{x}) \quad \text{for all } \mathbf{x} \in K$$

Such problems are usually more difficult because of the need to keep the points within the set K. Sometimes the set K is defined in a complicated way.

Consider the elliptic paraboloid $F(x_1, x_2) = x_1^2 + x_2^2 - 2x_1 - 2x_2 + 4$, which is sketched in Fig. 10.1. Clearly the unconstrained minimum occurs at $(1, 1)$. However, if $K = \{(x_1, x_2): x_1 \leqq 0, x_2 \leqq 0\}$, the constrained minimum is 4 at $(0, 0)$. Here we note that $F(x_1, x_2) = (x_1 - 1)^2 + (x_2 - 1)^2 + 2$.

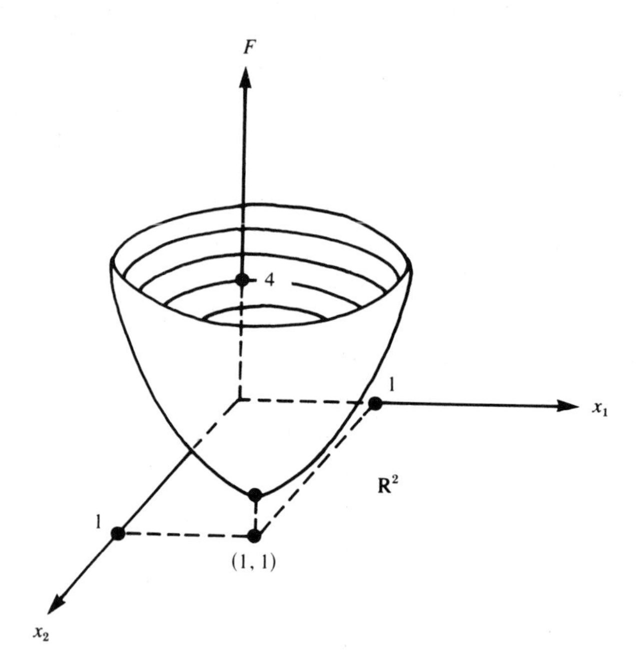

Figure 10.1

10.1 One-Variable Case

The special case in which a function F is defined on $\mathbf{R}$ is considered first because the more general problem with n variables is often solved by a sequence of one-variable problems.

Suppose that $F: \mathbf{R} \to \mathbf{R}$ and that we seek a point $z \in \mathbf{R}$ with the property that $F(z) \leqq F(x)$ for all $x \in \mathbf{R}$. Note that if no assumptions are made about F, this problem is insoluble in its general form. For instance, the function $f(x) = 1/(1 + x^2)$ has no minimum point. Even for relatively well-behaved functions, such as

$$F(x) = x^2 + \sin 53x$$

numerical methods may experience some difficulties because of the large number of purely local minima. Recall that a point z is a *local minimum* point of a function F if there is some neighborhood N of z in which all points $x \in N$ satisfy $F(z) \leqq F(x)$.

One reasonable assumption is that on some interval $[a, b]$ given to us in advance F has only a single local minimum. This property is often expressed by saying that F is

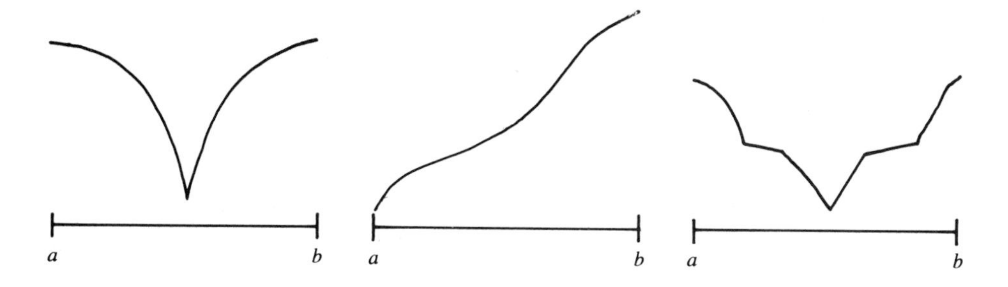

Figure 10.2 (a) Three unimodal functions.

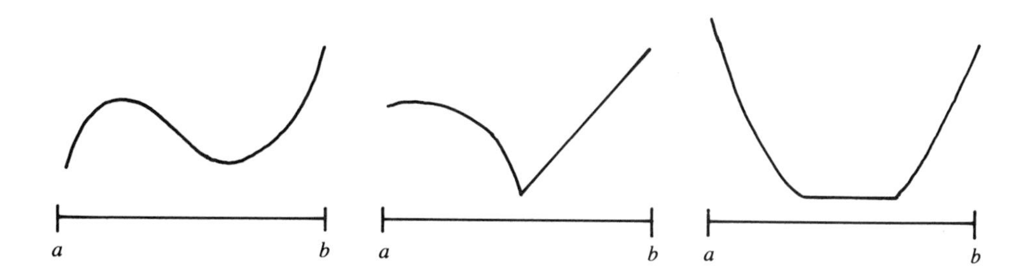

Figure 10.2 (b) Three functions that are not unimodal.

unimodal on $[a, b]$. (Caution: In statistics unimodal refers to a single local maximum.) Some unimodal functions are sketched in Fig. 10.2.

An important property of a continuous unimodal function, which might be surmised from Fig. 10.2, is that it is strictly decreasing up to the minimum point and strictly increasing thereafter. To be convinced of this, let x^* be the minimum point of F on $[a, b]$ and suppose, for instance, that F is not strictly decreasing on the interval $[a, x^*]$. Then points x_1 and x_2 satisfying $a \leqq x_1 < x_2 \leqq x^*$ and $F(x_1) \leqq F(x_2)$ must exist. Now let x^{**} be a minimum point of F on the interval $[a, x_2]$. (Recall that a continuous function on a closed finite interval attains its minimum value.) We can assume that $x^{**} \neq x_2$ because if x^{**} were initially chosen as x_2, it could be replaced by x_1 inasmuch as $F(x_1) \leqq F(x_2)$. But now we see that x^{**} is a local minimum point of F in the interval $[a, b]$, since it is a minimum point of F on $[a, x_2]$ but is not x^* itself. The presence of two local minimum points, of course, contradicts the unimodality of F.

Now we pose a problem concerning the search for the minimum point of a continuous unimodal function F on a given interval $[a, b]$. How accurately can the true minimum point x^* be computed with only n evaluations of F? With no evaluations of F, the best that can be said is that $x^* \in [a, b]$; taking the midpoint $\hat{x} = \frac{1}{2}(b + a)$ as the best estimate gives an error of $|x^* - \hat{x}| \leqq \frac{1}{2}(b - a)$. One evaluation by itself cannot be utilized, and so the best estimate and the error remain the same as in the previous case. Consequently, we need at least two function evaluations to obtain a better estimate.

Suppose that F is evaluated at a' and b' with the results shown in Fig. 10.3. If $F(a') \leqq F(b')$, then because F is increasing to the right of x^*, we can be sure that $x^* \in [a, b']$. On the other hand, similar reasoning for case $F(a') \geqq F(b')$ shows that $x^* \in [a', b]$. To make both "intervals of uncertainty" as small as possible, we move b' to the left and a' to the right. Thus F should be evaluated at two nearby points on either side of the midpoint, as shown in Fig. 10.4. Suppose that $a' = \frac{1}{2}(a + b) - 2\delta$ and $b' = \frac{1}{2}(a + b) + 2\delta$. Taking the midpoint of the appropriate subinterval $[a, b']$ or $[a', b]$ as the best estimate $\hat{x}$ of x^*, the error does not exceed $\frac{1}{4}(b - a) + \delta$. This the reader can easily verify.

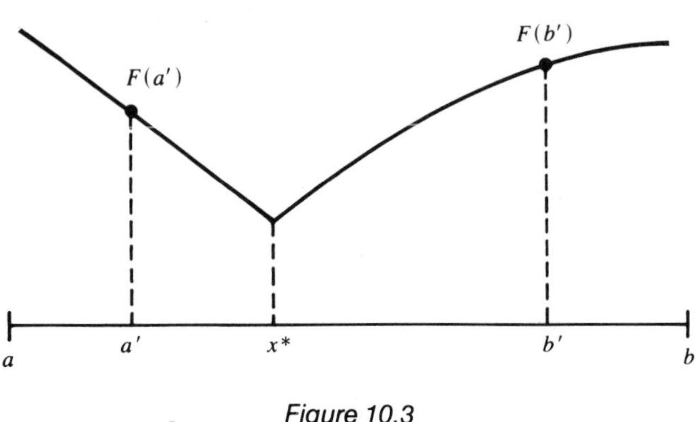

Figure 10.3

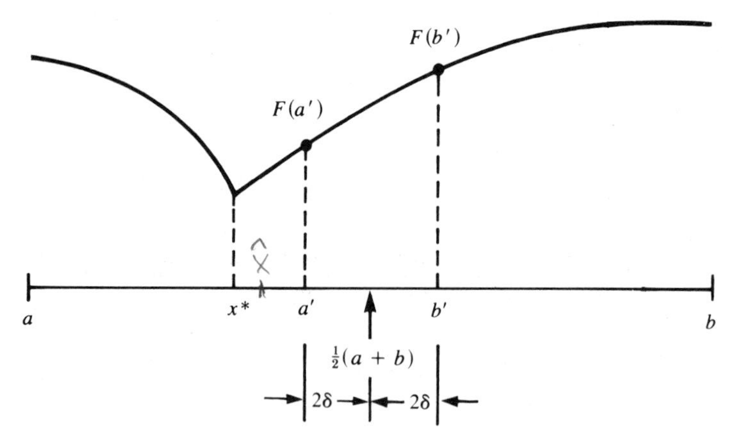

Figure 10.4

For $n = 3$, two evaluations are first made at the $\frac{1}{3}$ and $\frac{2}{3}$ points of the initial interval — that is, $a' = a + \frac{1}{3}(b - a)$ and $b' = a + \frac{2}{3}(b - a)$. From the two values $F(a')$ and $F(b')$ it can be determined whether $x^* \in [a, b']$ or $x^* \in [a', b]$. Next, relabel the new interval of uncertainty as $[a, b]$ by setting $a = a'$ if $F(a') \geqq F(b')$ or $b = b'$ if $F(a') \leqq F(b')$. We know the value of F at the midpoint of this interval, which is either a' or b', and we can use this information. The third evaluation is made at a point 2δ away from this point — that is, at either $b' = a' + 2\delta$ or $a' = b' - 2\delta$. The former case is shown in Fig. 10.5. Now if $F(a') \geqq F(b')$, then we set $a = a'$; otherwise set $b = b'$. Finally, $\hat{x} = \frac{1}{2}(a + b)$ is the best estimate of x^*. The error does not exceed $\frac{1}{6}(b - a) + \delta$ for the original interval $[a, b]$.

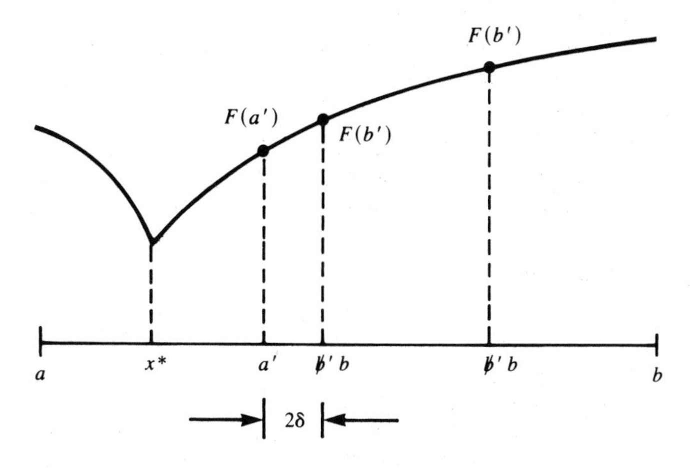

Figure 10.5

By continuing the search pattern as outlined, the result is that with n evaluations of F the estimate $\hat{x}$ of x^* is given with error not exceeding

$$\frac{\frac{1}{2}(b - a)}{\lambda_n} + \varepsilon \tag{1}$$

where ϵ is a designated small positive number and λ_n are integers forming the *Fibonacci sequence:*

$$\begin{cases} \lambda_0 = \lambda_1 = 1 \\ \lambda_n = \lambda_{n-1} + \lambda_{n-2} \qquad (n = 2, 3, 4, \ldots) \end{cases} \tag{2}$$

The first few elements $\lambda_0, \lambda_1, \lambda_2, \lambda_3, \lambda_4, \lambda_5, \lambda_6, \lambda_7, \ldots$ are 1, 1, 2, 3, 5, 8, 13, 21, ...

In the *Fibonacci search algorithm* we initially determine the number of steps n for a desired accuracy ε by selecting n to be the subscript of the smallest Fibonacci number greater than $(b - a)/\varepsilon$. We define a sequence of intervals, starting with the given interval $[a, b]$ of length $\ell = b - a$ and, for $k = n, n - 1, \ldots, 3$, using these formulas for updating:

$$\Delta = \frac{\lambda_{k-2}}{\lambda_k}(b - a) \tag{3}$$

$$a' = a + \Delta, \quad b' = b - \Delta$$

$$\begin{cases} a = a' & \text{if } F(a') \geqq F(b') \\ b = b' & \text{if } F(a') < F(b') \end{cases}$$

At the step $k = 2$ we set

$$\begin{cases} a' = \frac{1}{2}(a + b) - \varepsilon & \text{if } F(a) \geqq F(b) \\ b' = \frac{1}{2}(a + b) + \varepsilon & \text{if } F(a) \leqq F(b) \end{cases}$$

and we have the final interval $[a, b]$ from which we compute $\hat{x} = \frac{1}{2}(a + b)$. This algorithm requires only one function evaluation per step after the initial step.

To verify the algorithm, consider the situation shown in Fig. 10.6. Since $\lambda_k = \lambda_{k-1} + \lambda_{k-2}$, we have

$$\ell' = \ell - \Delta = \ell - \left(\frac{\lambda_{k-2}}{\lambda_k}\right)\ell = \left(\frac{\lambda_{k-1}}{\lambda_k}\right)\ell \tag{4}$$

and the length of the interval of uncertainty has been reduced by the factor $(\lambda_{k-1}/\lambda_k)$. The next step yields

$$\Delta' = \left(\frac{\lambda_{k-3}}{\lambda_{k-1}}\right)\ell' \tag{5}$$

and Δ' is actually the distance between a' and b'. Therefore one of the preceding points at which the function was evaluated is at one end or the other of ℓ'—that is,

$$\begin{aligned} b' - a' = \ell - 2\Delta &= \left(\frac{\lambda_k - 2\lambda_{k-2}}{\lambda_k}\right)\ell \\ &= \left(\frac{\lambda_{k-1} - \lambda_{k-2}}{\lambda_k}\right)\ell = \left(\frac{\lambda_{k-3}}{\lambda_k}\right)\ell \\ &= \left(\frac{\lambda_{k-3}}{\lambda_{k-1}}\right)\ell' = \Delta' \end{aligned}$$

by (2), (4), and (5).

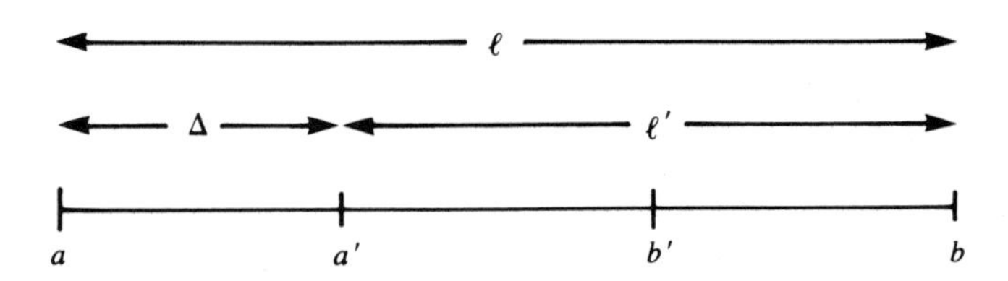

Figure 10.6

It is clear by (4) that after $n - 1$ function evaluations the next-to-last interval has length (λ_1/λ_n) times the initial interval $[a, b]$. So the final interval is $(b - a)(\lambda_1/\lambda_n) + 2\varepsilon$ wide and the maximum error (1) is established. The final step is similar to that outlined and F is evaluated at a point ε away from the midpoint of the next-to-last interval. Finally, set $\hat{x} = \frac{1}{2}(b + a)$ from the last interval $[a, b]$.

One disadvantage of the Fibonacci search is that the algorithm is somewhat complicated. Also, the desired precision must be given in advance and the number of steps to be computed for this precision determined before beginning the computation. Thus the initial evaluation points for the function F depend on n, the number of steps.

A similar algorithm that is free of these drawbacks is described next. It has been termed *Golden Section Search* because it depends on a ratio r known to the early Greeks as the Golden Section ratio:

$$r = \frac{1}{2}(\sqrt{5} - 1) \approx 0.6180339887$$

This number satisfies the equation $r^2 = 1 - r$. In each step of this iterative algorithm an interval $[a, b]$ is available from previous work. It is an interval known to contain the minimum point x^*, and our objective is to replace it by a smaller one that is also known to contain x^*. Also needed in each step are two values of F at two particular points in $[a, b]$:

$$\begin{cases} x = a + r(b - a), & u = F(x) \\ y = a + r^2(b - a), & v = F(y) \end{cases} \tag{6}$$

There are two cases to consider: either $u > v$ or $u \leqq v$. Let us take the first. Fig. 10.7

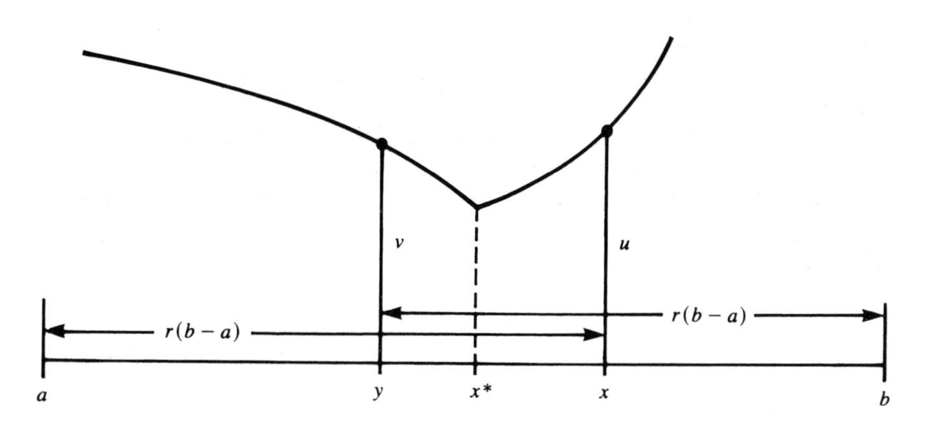

Figure 10.7

depicts this situation. Since F is assumed unimodal, the minimum of F must be in the interval $[a, x]$. This interval is the input interval at the beginning of the next step. Observe now that within the interval $[a, x]$ one evaluation of F is already available — namely, at y. Also note that $a + r(x - a) = y$ because $x - a = r(b - a)$. In the next step, therefore, y will play the role of x, and we shall need the value of F at the point $a + r^2(x - a)$. So what must be done in this step is to carry out the following replacements *in order*.

$$b \leftarrow x$$

$$x \leftarrow y$$

$$u \leftarrow v$$

$$y \leftarrow a + r^2(b - a)$$

$$v \leftarrow F(y)$$

The other case is similar. If $u \leqq v$, the picture might be as in Fig. 10.8. In this case, the minimum point must lie in $[y, b]$. Within this interval one value of F is available — namely, at x. Observe that $y + r^2(b - y) = x$ (see Problem 6). Thus x should now be given the role of y, and the value of F is to be computed at $y + r(b - y)$. The following replacements accomplish all of this.

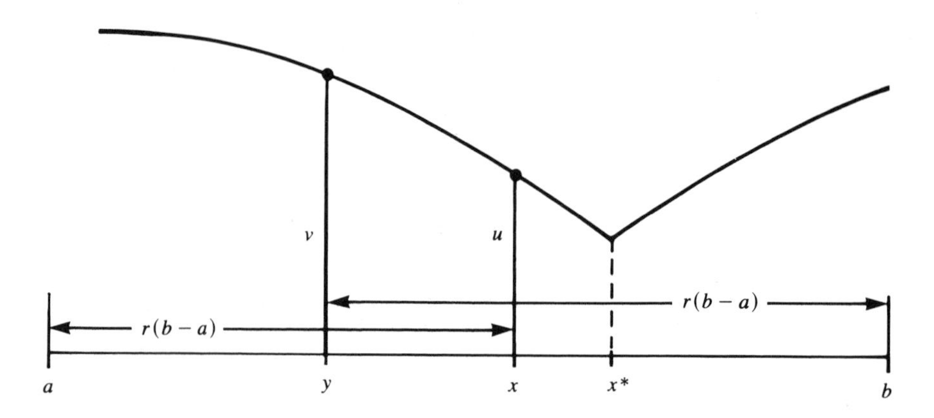

Figure 10.8

$$a \leftarrow y$$
$$y \leftarrow x$$
$$v \leftarrow u$$
$$x \leftarrow a + r(b - a)$$
$$u \leftarrow F(x)$$

In Problem 7 a shortcoming of this procedure is hinted at—this method is quite slow. Slowness refers, in this context, to the large number of function evaluations needed to achieve a reasonable precision. It can be surmised that this slowness is attributable to the extreme generality of the algorithm. No advantage has been taken of any smoothness that the function F may possess.

Suppose that F is represented by a Taylor series in the vicinity of the point x^*. Then

$$F(x) = F(x^*) + (x - x^*)F'(x^*) + \frac{1}{2}(x - x^*)^2 F''(x^*) + \cdots$$

Since x^* is a minimum point of F, $F'(x^*) = 0$. Thus we have

$$F(x) \approx F(x^*) + \frac{1}{2}(x - x^*)^2 F''(x^*)$$

This tells us that, in the neighborhood of x^*, $F(x)$ is approximated by a quadratic function whose minimum is also at x^*. Since we do not know x^* and do not want to involve derivatives in our algorithms, a natural stratagem is to interpolate F by a quadratic polynomial. Any three values $(x_i, F(x_i))$, $i = 1, 2, 3$, can be used for this purpose. The minimum point of the resulting quadratic function may be a better approximation to x^* than x_1, x_2, or x_3. Writing an algorithm that carries out this idea iteratively is not trivial. Many unpleasant cases must be handled. What should be done if the quadratic interpolant has a maximum instead of a minimum, for example?

Here is the outline of an algorithm for this procedure. At the beginning we have a function F whose minimum is sought. Two starting points x and y are given, as well as two control numbers δ and ε. Computing begins by evaluating the two numbers $u = F(x)$

and $v = F(y)$. If $u < v$, let $z = 2x - y$. If $v \leqq u$, let $z = 2y - x$. In either case, the number $w = F(z)$ is to be computed.

At this stage we have three points x, y, z, together with corresponding function values u, v, w. In the main iteration step of the algorithm one of these points and its accompanying function value are replaced by a new point and new function value. The process is repeated until a "success" or "failure" is reached.

In the main calculation a quadratic polynomial q is determined to interpolate F at the three current points x, y, z. The formulas are discussed below. Next, the point t where $q'(t) = 0$ is determined. Under ideal circumstances t is a *minimum* point of q and an *approximate minimum* point of F. So one of x, y, z should be replaced by t.

The *solution case* occurs if $q''(t) > 0$ and $\max\{|t - x|, |t - y|, |t - z|\} < \varepsilon$. The condition $q''(t) > 0$ indicates, of course, that q' is *increasing* in the vicinity of t so that t is indeed a minimum point of q. The second condition indicates that this estimate, t, of the minimum point of F is within distance ε of each of the three points x, y, z. In this case, t is accepted as a solution.

The *usual case* occurs if $q''(t) > 0$ and $\delta \geqq \max\{|t - x|, |t - y|, |t - z|\} \geqq \varepsilon$. These inequalities indicate that t is a minimum point of q but not near enough to the three initial points to be accepted as a solution. Also, t is not farther than δ units from each of x, y, z and can thus be accepted as a reasonable new point. The old point having the greatest function value is now replaced by t. For instance, if $u \geqq v$ and $u \geqq w$, then x is replaced by t and u by $F(t)$.

The *first bad case* occurs if $q''(t) > 0$ and $\max\{|t - x|, |t - y|, |t - z|\} > \delta$. Here t is a minimum point of q but is so remote that there is some danger in using it as a new point. We identify one of the original three points that is farthest from t—say, for example, x. Identify also the point closest to t, say z. Then replace x by $z + \delta\,\text{sgn}(t - z)$ and u by $F(x)$. Figure 10.9 shows this case. The curve is the graph of q.

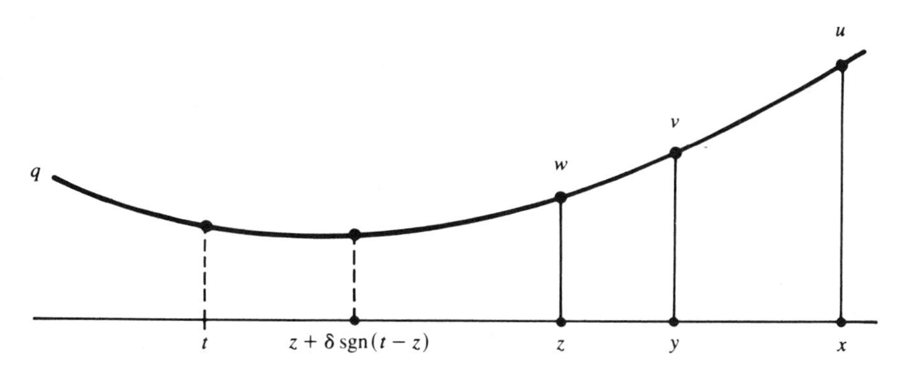

Figure 10.9

The *second bad case* occurs if $q''(t) < 0$, thus indicating that t is a maximum point of q. In this case, identify the greatest and the least among u, v, w. Suppose, for example, that $u \geqq v \geqq w$. Then replace x by $z + \delta\,\text{sgn}(z - x)$. An example is given in Fig. 10.10.

To complete the description of this algorithm, the formulas for t and $q''(t)$ must be given. They are obtained as follows.

$$\begin{cases} a = \dfrac{v - u}{y - x} \\ b = \dfrac{w - v}{z - y} \\ c = \dfrac{b - a}{z - x} \\ t = \dfrac{1}{2}[(x + y) - a/c] \\ q''(t) = 2c \end{cases} \tag{7}$$

Their derivation is outlined in Problem 9.

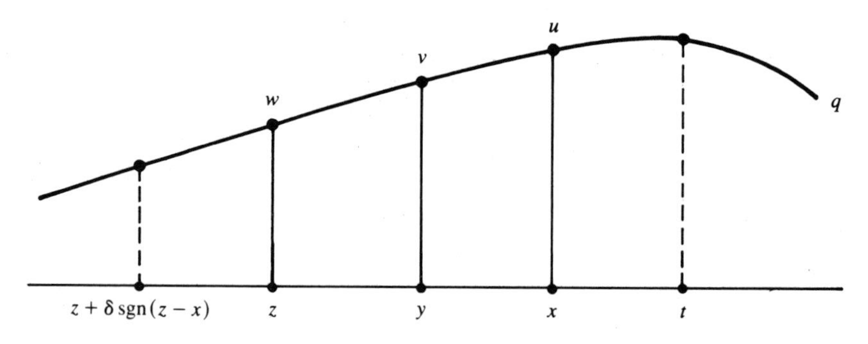

Figure 10.10

Problems 10.1

1. Find the unconstrained minimum point for the function $F(x_1, x_2, x_3) = x_1^2 + 3x_2^2 + 2x_3^2 - 4x_1 - 6x_2 + 8x_3$. Then find the constrained minimum over the set K defined by inequalities $x_1 \leqq 0$, $x_2 \leqq 0$, $x_3 \leqq 0$. Next, solve the same problem when K is defined by $x_1 \leqq 2$, $x_2 \leqq 0$, $x_3 \leqq -2$.
2. Find the unconstrained minimum of the function $F(x, y) = 13x^2 + 13y^2 - 10xy - 18x - 18y$. *Hint:* Try substituting $x = u + v$ and $y = u - v$.
3. If F is unimodal and continuous on the interval $[a, b]$, how many local maxima may F have on $[a, b]$?
4. For the Fibonacci search method, write expressions for $\hat{x}$ in the cases $n = 2, 3$.
5. The equation satisfied by Fibonacci numbers — namely, $\lambda_n - \lambda_{n-1} - \lambda_{n-2} = 0$ — is an example of a linear difference equation with constant coefficients. Solve it by postulating that $\lambda_n = \alpha^n$ and finding that $\alpha_1 = \frac{1}{2}(1 + \sqrt{5})$ or $\alpha_2 = \frac{1}{2}(1 - \sqrt{5})$ will serve for α. Initial conditions $\lambda_0 = \lambda_1 = 1$ can be met by a solution of the form $\lambda_n = c_1\alpha_1^n + c_2\alpha_2^n$. Find c_1 and c_2. Establish that $\lim_{n\to\infty}(\lambda_n/\lambda_{n-1}) = \alpha_1$. Show that this agrees with Eqs. (5) and (6) of Sec. 3.3.
6. Verify that $y + r^2(b - y) = x$ in the golden section algorithm. *Hint:* Use $r^2 + r = 1$.
7. If F is unimodal on an interval length ℓ, how many evaluations are necessary in the golden section algorithm to estimate the minimum point with an error of at most 10^{-k}?
8. (Continuation) In Problem 7 how large must n be if $\ell = 1$ and $k = 10$?

9. Using the divided-difference algorithm on the table

x	y	z
u	v	w

show that the quadratic interpolant in Newton form is

$$q(t) = u + a(t - x) + c(t - x)(t - y)$$

with a, b, c given by Eq. (7). Verify, then, the formulas for t and $q''(t)$ given in (7).

10. If subroutines can easily be written for F, F', and F'', how can Newton's method be used to locate the minimum point of F? Write down the formula that defines the iterative process. Does it involve F?
11. If subroutines are available for F and F', how can the secant method be used to minimize F?
12. Using the Fibonacci search algorithm, minimize $F(x) = x^2 - 6x + 2$ in the range $0 \leqq x \leqq 10$. Use $n = 4$ and $\varepsilon = \frac{1}{4}$.

Computer Problems 10.1

1. Write a subroutine to carry out the golden section algorithm for a given function and interval. The search should continue until a preassigned error bound is reached but not beyond 100 steps in any case.
2. (Continuation) Test the subroutine of Problem 1 on these examples.
   (a) $F(x) = \sin x$, $[0, \pi/2]$
   (b) $F(x) = [\text{Arctan } x]^2$, $[-1, 1]$
   (c) $F(x) = |\ln x|$, $[\frac{1}{2}, 4]$
   (d) $F(x) = |x|$, $[-1, 1]$
3. Write and test a subroutine for the following algorithm for approximating the minima of a function F of one variable over an interval $[a, b]$. The algorithm defines a sequence of quadruples $a < a' < b' < b$ by initially setting $a' = \frac{2}{3}a + \frac{1}{3}b$, $b' = \frac{1}{3}a + \frac{2}{3}b$ and repeatedly updating by $a = a'$, $a' = b'$, $b' = \frac{1}{2}(b + b')$ if $F(a') > F(b')$; $b = b'$, $a' = \frac{1}{2}(a + a')$, $b' = a$ if $F(a') < F(b')$; $a = a'$, $b = b'$, $a' = \frac{2}{3}a + \frac{1}{3}b$, $b' = \frac{1}{3}a + \frac{2}{3}b$, if $F(a') = F(b')$. *Note:* The construction ensures that $a < a' < b' < b$ and the minimum of F always occurs between a and b. Furthermore, only one new function value need be computed at each stage of the calculation after the first unless the case $F(a') = F(b')$ is obtained. The values of a, a', b', b tend to the same limit, which is a minimum point of F. Notice the similarity to the method of bisection.
4. Write and test a subroutine for the Fibonacci search algorithm. Verify that a partial algorithm for the Fibonacci search is as follows. Initially, set

$$\Delta = \frac{\lambda_{n-2}}{\lambda_n}(b - a)$$

$$a' = a + \Delta$$

$$b' = b - \Delta$$

$$u = F(a')$$

$$v = F(b')$$

Then loop on k from $n - 1$ downward to 3, updating

$$\begin{cases} \text{if } u \geqq v \\ a \leftarrow a' \\ a' \leftarrow b' \\ u \leftarrow v \\ \Delta \leftarrow \left(\dfrac{\lambda_{k-2}}{\lambda_k}\right)(b - a) \\ b' \leftarrow b - \Delta \\ v \leftarrow F(b') \end{cases} \qquad \begin{cases} \text{if } v > u \\ b \leftarrow b' \\ b' \leftarrow a' \\ v \leftarrow u \\ \Delta \leftarrow \left(\dfrac{\lambda_{k-2}}{\lambda_k}\right)(b - a) \\ a' \leftarrow a + \Delta \\ u \leftarrow F(a') \end{cases}$$

Add steps for $k = 2$.

5. (Berman Algorithm) Suppose that F is unimodal on $[a, b]$. Then if x_1, x_2 are any two points such that $a \leqq x_1 < x_2 \leqq b$, we have

$$F(x_1) > F(x_2) \text{ implies } x^* \in (x_1, b]$$

$$F(x_1) = F(x_2) \text{ implies } x^* \in [x_1, x_2]$$

$$F(x_1) < F(x_2) \text{ implies } x^* \in [a, x_2)$$

So by evaluating F at x_1 and x_2 and comparing function values, we are able to reduce the size of the interval known to contain x^*. This idea can easily be used to construct sequences that converge to x^*. The simplest approach is to start at the midpoint $x_0 = \frac{1}{2}(a + b)$ and if F is, say, decreasing for $x > x_0$, we test F at $x_0 + ih$, $i = 1, 2, \ldots, q$, with $h = (b - a)/(2q)$ until we find a point x_1 from which F begins to increase again (or until we reach b). Then we repeat this procedure starting at x and using a smaller step length h/q. Here q is the maximal number of evaluations at each step, say 4. Write a subroutine to perform the Berman algorithm and test it for evaluating the approximate minimization of one-dimensional functions. *Note:* The total number of evaluations of F needed for executing this algorithm up to some iterative step k depends on the location of x^*. If, for example, $x^* = b$, then clearly we need q evaluations at each iteration and hence kq evaluations. This number will decrease the closer x^* is to x_0, and it can be shown that with $q = 4$ the "expected" number of evaluations is three per step. It is interesting to compare the efficiency of the Berman algorithm ($q = 4$) with the Fibonacci search algorithm. The expected number of evaluations per step is three, and the uncertainty interval decreases by a factor $4^{-1/3} \approx 0.63$ per evaluation. In comparison, the Fibonacci search algorithm has a reduction factor of $2/(1 + \sqrt{5}) \approx 0.62$. Of course, the factor 0.63 in the Berman algorithm represents only an average and can be considerably lower but also as high as $4^{-1/4} \approx 0.87$.

10.2 Multivariate Case

Now we consider a real-valued function of n real variables $F: \mathbf{R}^n \to \mathbf{R}$. As before, a point $\mathbf{x}^*$ is sought such that $F(\mathbf{x}^*) \leqq F(\mathbf{x})$ for all $\mathbf{x}$. Some of the theory of multivariate functions must be developed in order to understand the rather sophisticated minimization algorithms in current use.

If the function F possesses partial derivatives of certain low orders (which is usually assumed in the development of these algorithms), then at any given point $\mathbf{x}$ a *gradient* vector $\mathbf{G}(\mathbf{x})$ is defined having components

$$G_i(\mathbf{x}) = \frac{\partial F(\mathbf{x})}{\partial x_i} \qquad (1 \leqq i \leqq n) \tag{1}$$

and a *Hessian* matrix $\mathbf{H}(\mathbf{x})$ is defined with components

$$H_{ij}(\mathbf{x}) = \frac{\partial^2 F(\mathbf{x})}{\partial x_i \, \partial x_j} \qquad (1 \leqq i, j \leqq n) \tag{2}$$

We interpret $\mathbf{G}(\mathbf{x})$ as an n-component vector and $\mathbf{H}(\mathbf{x})$ as an $n \times n$ matrix, both depending on $\mathbf{x}$.

Using the gradient and Hessian, the first few terms of the Taylor series for F can be written

$$F(\mathbf{x} + \mathbf{h}) = F(\mathbf{x}) + \sum_{i=1}^{n} G_i(\mathbf{x})h_i + \frac{1}{2}\sum_{i=1}^{n}\sum_{j=1}^{n} h_i H_{ij}(\mathbf{x})h_j + \cdots \tag{3}$$

Here $\mathbf{x}$ is the fixed point of expansion in $\mathbf{R}^n$ and $\mathbf{h}$ is the variable in $\mathbf{R}^n$ with components $h_1, h_2, \ldots, h_n$. Equation (3) can also be written in an elegant vector-matrix form.

$$F(\mathbf{x} + \mathbf{h}) = F(\mathbf{x}) + \mathbf{G}(\mathbf{x})^T\mathbf{h} + \frac{1}{2}\mathbf{h}^T\mathbf{H}(\mathbf{x})\mathbf{h} + \cdots \tag{4}$$

The three dots indicate higher-order terms in $\mathbf{h}$ that are not needed in this discussion.

A result in calculus states that the *order* in which partial derivatives are taken is immaterial if all partial derivatives occurring are continuous. In the special case of the Hessian matrix, if the second partial derivatives of F are all continuous, then $\mathbf{H}$ is a *symmetric* matrix.

$$H_{ij}(\mathbf{x}) = \frac{\partial^2 F}{\partial x_i \, \partial x_j} = \frac{\partial^2 F}{\partial x_j \, \partial x_i} = H_{ji}$$

To illustrate formula (4), let us compute the first three terms in the Taylor series for the function $F(x_1, x_2) = \cos \pi x_1 + \sin \pi x_2 + e^{x_1 x_2}$, taking $(1, 1)$ as the point of expansion. Partial derivatives are

$$\frac{\partial F}{\partial x_1} = -\pi \sin \pi x_1 + x_2 e^{x_1 x_2} \qquad \frac{\partial F}{\partial x_2} = \pi \cos \pi x_2 + x_1 e^{x_1 x_2}$$

$$\frac{\partial^2 F}{\partial x_1^2} = -\pi^2 \cos \pi x_1 + x_2^2 e^{x_1 x_2} \qquad \frac{\partial^2 F}{\partial x_2 \, \partial x_1} = (x_1 x_2 + 1)e^{x_1 x_2}$$

$$\frac{\partial^2 F}{\partial x_1 \, \partial x_2} = (x_1 x_2 + 1)e^{x_1 x_2} \qquad \frac{\partial^2 F}{\partial x_2^2} = -\pi^2 \sin \pi x_2 + x_1^2 e^{x_1 x_2}$$

At the particular point $\mathbf{x} = (1, 1)^T$ we have

$$F(\mathbf{x}) = -1 + e, \quad \mathbf{G}(\mathbf{x}) = \begin{pmatrix} e \\ -\pi + e \end{pmatrix}, \quad \mathbf{H}(\mathbf{x}) = \begin{pmatrix} \pi^2 + e & 2e \\ 2e & e \end{pmatrix}$$

So by (4)

$$F(1 + h_1, 1 + h_2) = -1 + e + (e, -\pi + e)\begin{pmatrix} h_1 \\ h_2 \end{pmatrix}$$

$$+ \frac{1}{2}(h_1, h_2)\begin{pmatrix} \pi^2 + e & 2e \\ 2e & e \end{pmatrix}\begin{pmatrix} h_1 \\ h_2 \end{pmatrix} + \cdots$$

or, equivalently, by (3)

$$F(1 + h_1, 1 + h_2) = -1 + e + eh_1 + (-\pi + e)h_2$$

$$+ \frac{1}{2}[(\pi^2 + e)h_1^2 + (2e)h_1h_2$$

$$+ (2e)h_2h_1 + eh_2^2] + \cdots$$

Note the "equality of cross derivatives"—that is, $\partial^2 F/\partial x_1\,\partial x_2 = \partial^2 F/\partial x_2\,\partial x_1$.

Another form of the Taylor series is useful. First, let $\mathbf{z}$ be the point of expansion and then let $\mathbf{h} = \mathbf{x} - \mathbf{z}$. Now from (4)

$$F(\mathbf{x}) = F(\mathbf{z}) + \mathbf{G}(\mathbf{z})^T(\mathbf{x} - \mathbf{z}) + \frac{1}{2}(\mathbf{x} - \mathbf{z})^T\mathbf{H}(\mathbf{z})(\mathbf{x} - \mathbf{z}) + \cdots \tag{5}$$

We illustrate with two special types of functions. First, the *linear* function. This has the form

$$F(\mathbf{x}) = c + \sum_{i=1}^{n} b_i x_i = c + \mathbf{b}^T\mathbf{x}$$

for appropriate coefficients $c, b_1, \ldots, b_n$. Clearly for $1 \leqq i \leqq n$, $G_i(\mathbf{z}) = b_i$ and $H_{ij}(\mathbf{z}) = 0$; so Eq. (5) yields

$$F(\mathbf{x}) = F(\mathbf{z}) + \sum_{i=1}^{n} b_i(x_i - z_i) = F(\mathbf{z}) + \mathbf{b}^T(\mathbf{x} - \mathbf{z})$$

Secondly, consider a general *quadratic* function. For simplicity, we take only two variables. The form of the function is

$$F(x_1, x_2) = c + (b_1x_1 + b_2x_2) + \frac{1}{2}(a_{11}x_1^2 + 2a_{12}x_1x_2 + a_{22}x_2^2) \tag{6}$$

which can be interpreted as the Taylor series for F when the point of expansion is (0, 0). To verify this assertion, the partial derivatives must be computed and evaluated at (0, 0).

$$\frac{\partial F}{\partial x_1} = b_1 + a_{11}x_1 + a_{12}x_2 \qquad \frac{\partial F}{\partial x_2} = b_2 + a_{22}x_2 + a_{12}x_1$$

$$\frac{\partial^2 F}{\partial x_1^2} = a_{11} \qquad \frac{\partial^2 F}{\partial x_1\,\partial x_2} = a_{12}$$

$$\frac{\partial^2 F}{\partial x_2\,\partial x_1} = a_{12} \qquad \frac{\partial^2 F}{\partial x_2^2} = a_{22}$$

Letting $\mathbf{z} = (0, 0)^T$, we obtain from Eq. (5)

$$F(\mathbf{x}) = c + (b_1, b_2)\begin{pmatrix} x_1 \\ x_2 \end{pmatrix} + \frac{1}{2}(x_1, x_2)\begin{pmatrix} a_{11} & a_{12} \\ a_{12} & a_{22} \end{pmatrix}\begin{pmatrix} x_1 \\ x_2 \end{pmatrix}$$

This is the matrix form of the original quadratic function of two variables. It can also be written as

$$F(\mathbf{x}) = c + \mathbf{b}^T\mathbf{x} + \frac{1}{2}\mathbf{x}^T\mathbf{A}\mathbf{x} \tag{7}$$

where c is a scalar, $\mathbf{b}$ a vector, and $\mathbf{A}$ a matrix. Equation (7) holds for a general quadratic function of n variables, with $\mathbf{b}$ an n-component vector and $\mathbf{A}$ an $n \times n$ matrix.

Returning to Eq. (3), we now write out the complicated double sum in more detail to assist in understanding it.

$$\mathbf{x}^T\mathbf{H}\mathbf{x} = \sum_{i=1}^{n}\sum_{j=1}^{n} x_i H_{ij} x_j = \left\{\begin{array}{l} \displaystyle\sum_{j=1}^{n} x_1 H_{1j} x_j \\ \displaystyle + \sum_{j=1}^{n} x_2 H_{2j} x_j \\ + \cdots \\ + \cdots \\ \displaystyle + \sum_{j=1}^{n} x_n H_{n,j} x_j \end{array}\right\}$$

Moreover, we have in complete detail

$$\mathbf{x}^T\mathbf{H}\mathbf{x} = \left\{\begin{array}{lllll} & x_1 H_{11} x_1 & + x_1 H_{12} x_2 & + \cdots & + x_1 H_{1n} x_n \\ + & x_2 H_{21} x_1 & + x_2 H_{22} x_2 & + \cdots & + x_2 H_{2n} x_n \\ + & \cdots & & & + \cdots \\ + & \cdots & & & + \cdots \\ + & x_n H_{n1} x_1 & + x_n H_{n2} x_2 & + \cdots & + x_n H_{nn} x_n \end{array}\right\}$$

Thus $\mathbf{x}^T\mathbf{H}\mathbf{x}$ can be interpreted as the sum of all n^2 terms in a square matrix of which the (i, j) element is $x_i H_{ij} x_j$.

A crucial property of the gradient vector $\mathbf{G}(\mathbf{x})$ is that it points in the direction of most rapid increase in the function F—the direction of "steepest ascent." Conversely, $-\mathbf{G}(\mathbf{x})$ points in the direction of "steepest descent." This fact is so important that it is worth a few words of justification. Suppose that $\mathbf{h}$ is a unit vector, $\sum_{i=1}^{n} h_i^2 = 1$. Now the rate of change of F (at $\mathbf{x}$), in the direction determined by $\mathbf{h}$, is defined naturally by

$$\left.\frac{d}{dt}F(\mathbf{x} + t\mathbf{h})\right|_{t=0}$$

This rate of change can be evaluated by using Eq. (4). From that equation, it follows that

$$F(\mathbf{x} + t\mathbf{h}) = F(\mathbf{x}) + t\mathbf{G}(\mathbf{x})^T\mathbf{h} + \frac{1}{2}t^2\mathbf{h}^T\mathbf{H}(\mathbf{x})\mathbf{h} + \cdots \tag{8}$$

Differentiation with respect to t leads to

$$\frac{d}{dt}F(\mathbf{x} + t\mathbf{h}) = \mathbf{G}(\mathbf{x})^T\mathbf{h} + t\mathbf{h}^T\mathbf{H}(\mathbf{x})\mathbf{h} + \cdots \tag{9}$$

By letting $t = 0$ here, we see that the rate of change of F in direction $\mathbf{h}$ is nothing else than $\mathbf{G}(\mathbf{x})^T\mathbf{h}$. Now we ask ''For what unit vector $\mathbf{h}$ is the rate of change a maximum?'' The simplest path to the answer is to invoke the powerful *Cauchy–Schwarz inequality*. It states that

$$\sum_{i=1}^{n} u_i v_i \leqq \left(\sum_{i=1}^{n} u_i^2 \sum_{i=1}^{n} v_i^2 \right)^{1/2} \tag{10}$$

with equality holding only if one of the vectors $\mathbf{u}$ and $\mathbf{v}$ is a nonnegative multiple of the other. Applying this to $\mathbf{G}(\mathbf{x})^T\mathbf{h} = \sum_{i=1}^{n} G_i(x)h_i$ and remembering that $\sum_{i=1}^{n} h_i^2 = 1$, we conclude that the maximum occurs when $\mathbf{h}$ is a positive multiple of $\mathbf{G}(\mathbf{x})$ — that is, when $\mathbf{h}$ points in the direction of $\mathbf{G}$.

Based on the foregoing discussion, a minimization procedure called *steepest descent* can be described. At any given point $\mathbf{x}$ the gradient vector $\mathbf{G}(\mathbf{x})$ is calculated. Then a one-dimensional minimization problem is solved by determining the value t^* for which the function

$$\phi(t) = F(\mathbf{x} + t\mathbf{G}(\mathbf{x}))$$

is a minimum. Then we replace $\mathbf{x}$ by $\mathbf{x} + t^*\mathbf{G}(\mathbf{x})$ and begin anew.

The method of steepest descent is not usually competitive with other methods, but it has the advantage of simplicity. One way of speeding it up is described in Computer Problem 2.

In understanding how methods work on functions of two variables, it is often helpful to draw *contour* diagrams. A *contour* of a function F is a set of the form $\{\mathbf{x} : F(\mathbf{x}) = c\}$, where c is a given constant. For example, the contours of function $F(\mathbf{x}) = 25x_1^2 + x_2^2$ are ellipses, as shown in Fig. 10.11. Contours are also called *level sets* by some authors. At any point on a contour the gradient of F is perpendicular to the curve. So the path of the steepest descent algorithm may look like Fig. 10.12.

To explain more advanced algorithms, we consider a general function F of n variables. Suppose that we have obtained the first three terms in the Taylor series of F in the vicinity of a point $\mathbf{z}$. How can they be used to guess the minimum point of F? Obviously we could ignore all terms beyond the quadratic terms and find the minimum of the resulting quadratic function.

$$F(\mathbf{z} + \mathbf{x}) = F(\mathbf{z}) + \mathbf{G}(\mathbf{z})^T\mathbf{x} + \frac{1}{2}\mathbf{x}^T\mathbf{H}(\mathbf{z})\mathbf{x} + \cdots \tag{11}$$

Here $\mathbf{z}$ is fixed and $\mathbf{x}$ is the variable. To find the minimum of this quadratic function of $\mathbf{x}$, we must compute the first partial derivatives and set them equal to zero. Denoting this quadratic function by Q and simplifying the notation slightly, we have

$$Q(\mathbf{x}) = F(\mathbf{z}) + \sum_{i=1}^{n} G_i x_i + \frac{1}{2} \sum_{i=1}^{n} \sum_{j=1}^{n} x_i H_{ij} x_j \tag{12}$$

from which it follows that (see Problem 8)

$$\frac{\partial Q}{\partial x_k} = G_k + \sum_{j=1}^{n} H_{kj} x_j \qquad (1 \leqq k \leqq n) \tag{13}$$

The point $\mathbf{x}$ that is sought is thus a solution of the system of n equations

$$\sum_{j=1}^{n} H_{kj} x_j = -G_k \qquad (1 \leqq k \leqq n) \tag{14}$$

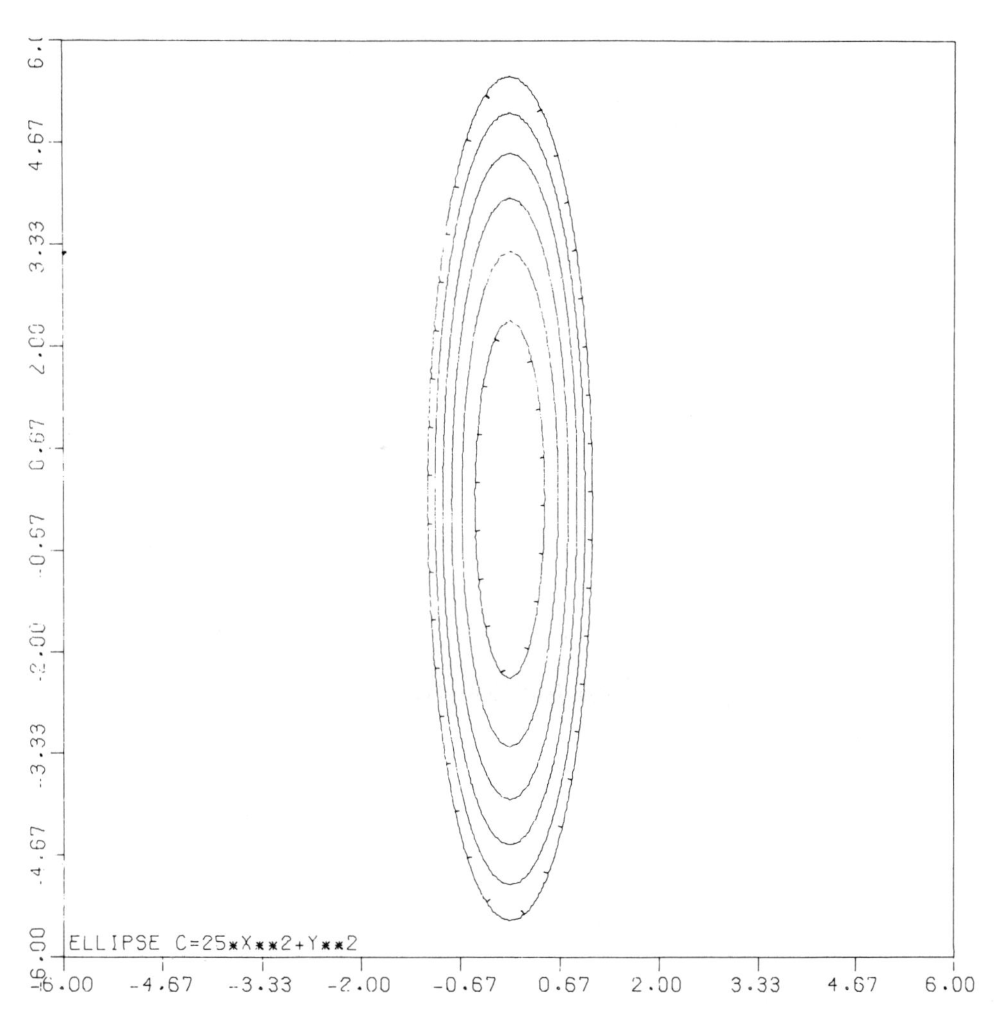

Figure 10.11 Contours of ellipse $25x^2 + y^2 = c$ for $c = 5, 10, 15, \ldots, 30$
(The small dashes point in the direction of negative gradient, *i.e.*, direction of steepest descent.)

or, equivalently,

$$\mathbf{Hx} = -\mathbf{G} \tag{15}$$

Here $\mathbf{H} = \mathbf{H}(\mathbf{z})$ and $\mathbf{G} = \mathbf{G}(\mathbf{z})$.

The preceding analysis suggests the following iterative procedure for locating a minimum point of a function F. Start with a point $\mathbf{z}$ that is a current estimate of the minimum point. Compute the gradient and Hessian of F at point $\mathbf{z}$. They can be denoted $\mathbf{G}$ and $\mathbf{H}$, respectively. Of course, $\mathbf{G}$ is an n-component vector of numbers and $\mathbf{H}$ is a $n \times n$ matrix of numbers. Then solve the matrix equation $\mathbf{Hx} = -\mathbf{G}$, obtaining an n-component vector $\mathbf{x}$. Replace $\mathbf{z}$ by $\mathbf{z} + \mathbf{x}$ and return to the beginning of the procedure.

There are many reasons for expecting trouble from the iterative procedure just outlined. One especially noisome aspect is that we can only expect to find a point where the first partial derivatives of F vanish; it need not be a minimum point. It is what we call a *stationary point*. Such points can be classified into three types: minimum point, maximum point, and saddlepoint. They can be illustrated by simple quadric surfaces familiar from analytic geometry. The graphs are shown in Fig. 10.13.

(a) minimum of $F(x, y) = x^2 + y^2$ at $(0, 0)$
(b) maximum of $F(x, y) = 1 - x^2 - y^2$ at $(0, 0)$
(c) saddlepoint of $F(x, y) = x^2 - y^2$ at $(0, 0)$

If $\mathbf{z}$ is a stationary point, then $\mathbf{G}(\mathbf{z}) = 0$. Moreover, a criterion ensuring that Q, as defined in (12), has a minimum point is this.

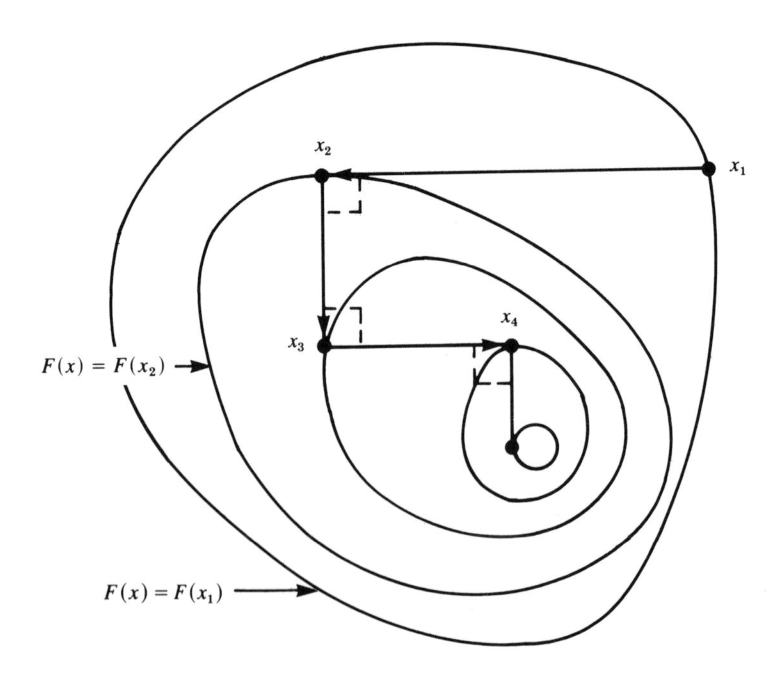

Figure 10.12

> If the matrix $\mathbf{H}$ has the property that $\mathbf{x}^T\mathbf{H}\mathbf{x} > 0$ for every nonzero vector $\mathbf{x}$, then the quadratic function Q has a minimum point.

(See Problem 9 to see why.) A matrix having this property is said to be *positive definite*. Notice that this criterion involves only second-degree terms in the quadratic function Q.

As examples of quadratic functions that do not have minima, consider

1. $-x_1^2 - x_2^2 + 13x_1 + 6x_2 + 12$
2. $x_1^2 - x_2^2 + 3x_1 + 5x_2 + 7$
3. $x_1^2 - 2x_1x_2 + x_1 + 2x_2 + 3$
4. $2x_1 + 4x_2 + 6$

In the first two let $x_1 = 0$ and $x_2 \to \infty$. In the third let $x_1 = x_2 \to \infty$. In the last let $x_1 = 0$ and $x_2 \to -\infty$. In each case, the function values approach $-\infty$, and no global minimum can exist.

The algorithms currently recommended for minimization are of a type called *quasi-Newton*. The principal example is an algorithm introduced in 1959 by Davidon and called the *variable metric algorithm*. Subsequently important modifications and improvements were made by others, notably R. Fletcher, M. J. D. Powell, C. G. Broyden, P. E. Gill, and W. Murray.

This algorithm proceeds iteratively, assuming in each step that a local quadratic approximation is known for function F whose minimum is sought. The minimum of this

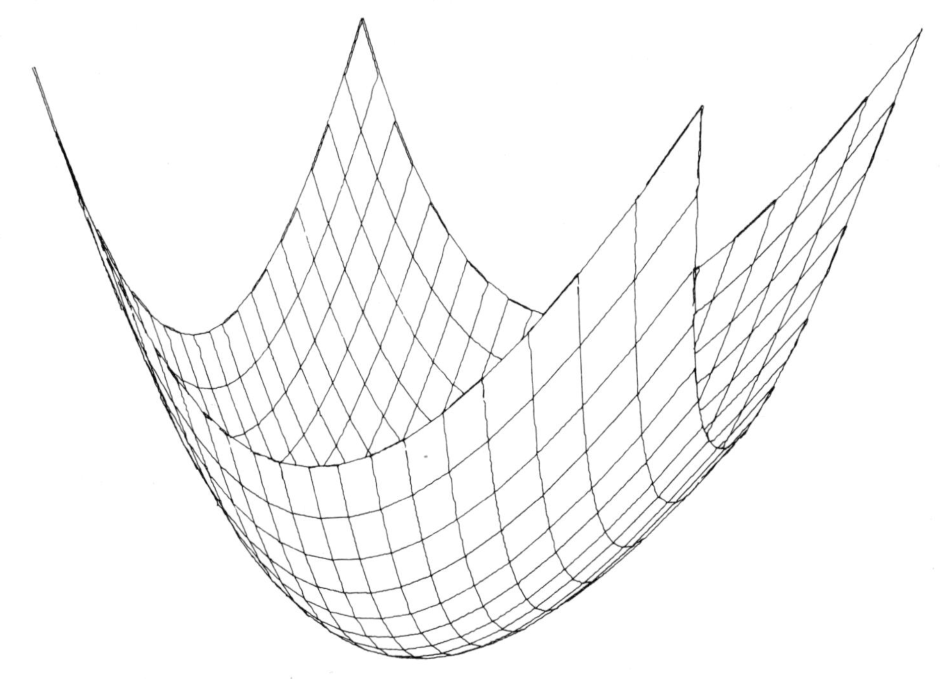

Figure 10.13 (a) $F(x,y) = x^2 + y^2$.

quadratic function provides either the new point directly or is used to determine a line along which a one-dimensional search is carried out. In implementing the algorithm, the gradient can either be provided in the form of a subroutine or computed numerically in the procedure. The Hessian **H** is not computed, but an estimate of its **LU** factorization is kept up to date as the process continues.

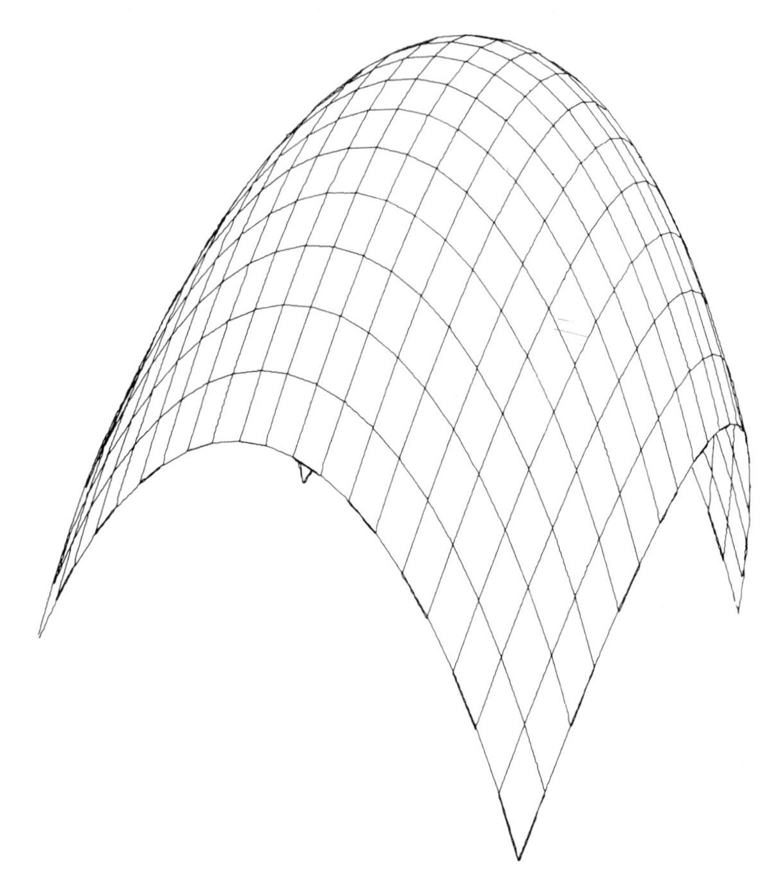

Figure 10.13 (b) $F(x, y) = 1 - x^2 - y^2$.

Problems 10.2 Q6

1. Determine whether these functions have minimum values in $\mathbf{R}^2$.
   (a) $x_1^2 - x_1x_2 + x_2^2 + 3x_1 + 6x_2 - 4$
   (b) $x_1^2 - 3x_1x_2 + x_2^2 + 7x_1 + 3x_2 + 5$
   (c) $2x_1^2 - 3x_1x_2 + x_2^2 + 4x_1 - x_2 + 6$
   (d) $ax_1^2 + 2bx_1x_2 + cx_2^2 + dx_1 + ex_2 + f \quad (a > 0,\ c > 0)$
   Hint: Use the method of completing the square.
2. Locate the minimum point of $3x^2 - 2xy + y^2 + 3x - 4y + 7$ by finding the gradient, Hessian, and solving the appropriate linear equations.
3. Write the first three terms of the Taylor series for function $F(x, y) = e^x \cos y - y \cdot \ln(x + 1)$, using $(0, 0)$ as the point of expansion.

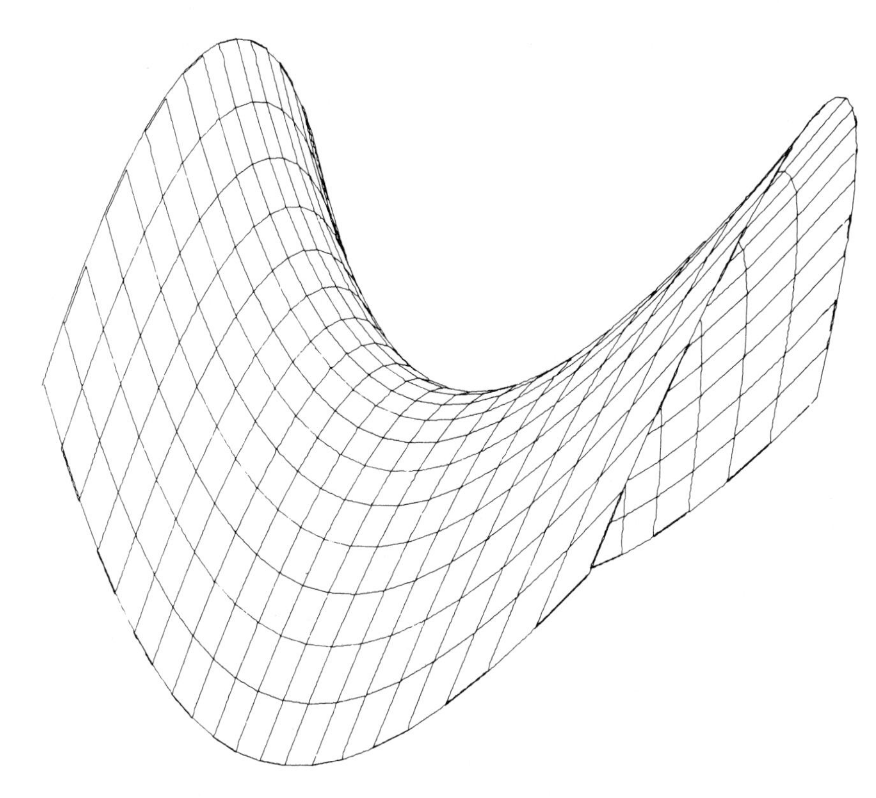

Figure 10.13 (c) $F(x, y) = x^2 - y^2$.

Figure 10.13c

4. Write the first three terms of the Taylor series for function $F(x, y) = 2x^2 - 4xy + 7y^2 - 3x + 5y$, using $(1, 1)$ as the point of expansion.
5. We want to find the minimum of function $f(x, y, z) = y^2z^2(1 + \sin^2 x) + (y + 1)^2(z + 3)^2$. The program to be used requires the gradient of the function. What formulas must we program for the gradient?
6. Let F be a function of two variables whose gradient at $(0, 0)$ is $(-5, 1)^T$ and whose Hessian is

$$\begin{pmatrix} 6 & -1 \\ -1 & 2 \end{pmatrix}$$

Make a reasonable guess as to the minimum point of F. Explain.
7. Write the function $F(x_1, x_2) = 3x_1^2 + 6x_1x_2 - 2x_2^2 + 5x_1 + 3x_2 + 7$ in the form of Eq. (7) with appropriate $\mathbf{A}$, $\mathbf{b}$, c. Show, in matrix form, the linear equations that must be solved in order to find a point where the first partial derivatives of F vanish. Finally, solve these equations to locate this point numerically.
8. Verify Eq. (13). In differentiating the double sum in Eq. (12), first write all terms containing x_k. Then differentiate and use the symmetry of the matrix $\mathbf{H}$.
9. Consider the quadratic function Q in Eq. (12). Show that if $\mathbf{H}$ is positive definite, then the stationary point is a minimum point.
10. (General Quadratic Equation) Generalize Eq. (6) to n variables. Show that a general quadratic function $Q(\mathbf{x})$ of n variables can be written in the matrix-vector form of Eq. (7) where $\mathbf{A}$ is an

$n \times n$ symmetric matrix, $\mathbf{b}$ a vector of length n, and c a scalar. Establish that the gradient and Hessian are $\mathbf{G}(\mathbf{x}) = \mathbf{Ax} + \mathbf{b}$ and $\mathbf{H}(\mathbf{x}) = \mathbf{A}$, respectively.

11. Let $\mathbf{A}$ be an $n \times n$ symmetric matrix and define an upper-triangular matrix $\mathbf{U} = (u_{ij})$ by putting

$$u_{ij} = \begin{cases} a_{ij} & \text{if } i = j \\ 2a_{ij} & \text{if } i < j \\ 0 & \text{if } i > j \end{cases}$$

Show that $\mathbf{x}^T\mathbf{Ux} = \mathbf{x}^T\mathbf{Ax}$ for all vectors $\mathbf{x}$.

12. Show that the general quadratic function $Q(\mathbf{x})$ of n variables can be written

$$Q(\mathbf{x}) = c + \mathbf{b}^T\mathbf{x} + \frac{1}{2}\mathbf{x}^T\mathbf{Ux}$$

where $\mathbf{U}$ is an upper-triangular matrix. Can this simplify the work of finding the stationary point of Q?

Computer Problems 10.2

1. Select a subprogram from the program library of your computer center for minimizing a function of many variables without the need to program derivatives. Test it on one or more of the following well-known functions. Starting values are written (x, y, z, w).
   (a) Rosenbrock's: $100(y - x^2)^2 + (1 - x)^2$. Start at $(-1.2, 1.0)$.
   (b) Powell's: $(x + 10y)^2 + 5(z - w)^2 + (y - 2z)^4 + 10(x - w)^4$. Start at $(3, -1, 0, 1)$.
   (c) Fletcher and Powell's: $100[z - 10\phi]^2 + [\sqrt{x^2 + y^2} - 1]^2 + z^2$ in which ϕ is an angle determined from (x, y) by $\cos 2\pi\phi = x/\sqrt{x^2 + y^2}$, $\sin 2\pi\phi = y/\sqrt{x^2 + y^2}$, $-\pi/2 < 2\pi\phi \leqq 3\pi/2$. Start at $(1, 1, 1)$.
   (d) Powell's: $x^2 + 2y^2 + 3z^2 + 4w^2 + (x + y + z + w)^4$. Start at $(1, -1, -1, 1)$.
   (e) Wood's: $100(x^2 - y)^2 + (1 - x)^2 + 90(z^2 - w)^2 + (1 - z)^2 + 10.1[(y - 1)^2 + (w - 1)^2] + 19.8(y - 1)(w - 1)$. Start at $(-3, -1, -3, -1)$.
2. (Accelerated Steepest Descent) This version of steepest descent is superior to the basic one. A sequence of points $\mathbf{x}_1, \mathbf{x}_2, \mathbf{x}_3, \ldots$ is generated as follows. Point $\mathbf{x}_1$ is specified as the starting point. Then $\mathbf{x}_2$ is obtained by one step of steepest descent from $\mathbf{x}_1$. In the general step, if $\mathbf{x}_1, \ldots, \mathbf{x}_m$ have been obtained, we find a point $\mathbf{z}$ by steepest descent from $\mathbf{x}_m$. Then $\mathbf{x}_{m+1}$ is taken as the minimum point on the line $\mathbf{x}_{m-1} + t(\mathbf{z} - \mathbf{x}_{m-1})$. Program and test this algorithm on one of the examples in Problem 1.

Supplementary Problems (Chapter 10)

1. Show that the gradient and Hessian satisfy the equation $\mathbf{H}(\mathbf{z})(\mathbf{x} - \mathbf{z}) = \mathbf{G}(\mathbf{x}) - \mathbf{G}(\mathbf{z})$ for a general quadratic function of n variables.
2. Using Taylor series show that a general quadratic function of n variables can be written in block form

$$\mathrm{Q}(\mathbf{x}) = \frac{1}{2}\mathcal{X}^T\mathcal{A}\mathcal{X} + \mathcal{B}^T\mathcal{X} + c$$

where

$$\mathcal{X} = \begin{pmatrix} \mathbf{x} \\ \mathbf{z} \end{pmatrix}, \quad \mathcal{A} = \begin{pmatrix} \mathbf{A} & -\mathbf{A} \\ \mathbf{A} & -\mathbf{A} \end{pmatrix}, \quad \mathcal{B} = \begin{pmatrix} \mathbf{b} \\ -\mathbf{b} \end{pmatrix}$$

Here $\mathbf{z}$ is the point of expansion.

3. The Taylor series expansion about zero can be written

$$F(\mathbf{x}) = F(0) + \mathbf{G}(0)^T\mathbf{x} + \frac{1}{2}\mathbf{x}^T\mathbf{H}(0)\mathbf{x} + \ldots$$

Show that the Taylor series about $\mathbf{z}$ can be written in a similar form by using matrix-vector notation—that is,

$$F(\mathbf{x}) = F(\mathbf{z}) + \mathcal{G}(\mathbf{z})^T \mathcal{X} + \frac{1}{2}\mathcal{X}^T \mathcal{H}(\mathbf{z})\mathcal{X} + \ldots$$

where

$$\mathcal{X} = \begin{pmatrix} \mathbf{x} \\ \mathbf{z} \end{pmatrix}, \quad \mathcal{G}(\mathbf{z}) = \begin{pmatrix} \mathbf{G}(\mathbf{z}) \\ -\mathbf{G}(\mathbf{z}) \end{pmatrix}, \quad \mathcal{H}(\mathbf{z}) = \begin{pmatrix} \mathbf{H}(\mathbf{z}) & -\mathbf{H}(\mathbf{z}) \\ \mathbf{H}(\mathbf{z}) & -\mathbf{H}(\mathbf{z}) \end{pmatrix}$$

4. Show that the gradient of $F(x, y)$ is perpendicular to the contour. *Hint*: Interpret the equation $F(x, y) = 0$ as defining y as a function of x. Then by the chain rule

$$\frac{\delta F}{\delta x} + \frac{\delta F}{\delta y}\frac{dy}{dx} = 0$$

From it obtain the slope of the tangent to the contour.

5. Consider the function

$$F(x_1, x_2, x_3) = 3e^{x_1 x_2} - x_3 \cos x_1 + x_2 \ln x_3$$

(a) Determine the gradient vector and Hessian matrix.
(b) Derive the first three terms of the Taylor series expanded about (0, 1, 1).
(c) What linear system should be solved for a reasonable guess as to the minimum point for F? What is the value of F at this point?

6. (Least Squares Problem) Consider the function

$$F(\mathbf{x}) = (\mathbf{b} - \mathbf{A}\mathbf{x})^T(\mathbf{b} - \mathbf{A}\mathbf{x}) + \alpha\,\mathbf{x}^T\mathbf{x}$$

where $\mathbf{A}$ is a real $m \times n$ matrix, $\mathbf{b}$ a real column vector of order m, and α a positive real number. We want the minimum point of F for given $\mathbf{A}$, $\mathbf{b}$, and α. Show that

$$F(\mathbf{x} + \mathbf{h}) - F(\mathbf{x}) = (\mathbf{A}\mathbf{h})^T(\mathbf{A}\mathbf{h}) + \alpha\,\mathbf{h}^T\mathbf{h} \geqq 0$$

for $\mathbf{h}$ a vector of order n, provided that

$$(\mathbf{A}^T\mathbf{A} + \alpha\mathbf{I})\mathbf{x} = \mathbf{A}^T\mathbf{b}$$

This means that any solution of this linear system minimizes $F(\mathbf{x})$; that is, this is the "normal equation."

7. We want to find the minimum of $F(x, y, z) = z^2 \cos x + x^2y^2 + x^2e^z$, using a computer program that requires subprograms for the gradient of F together with F. Write the subprograms needed.

8. Assume that subroutine `XMIN(F, GRAD, N, X, G)` is available to compute the minimum value of a function of two variables. Suppose that this routine requires not only the function but also its gradient in array `G`. If we are going to use this routine with the function $f(x, y) = e^x \cos^2(xy)$, what subprograms will be needed? Write the appropriate code.

9. The golden section ratio $r = \frac{1}{2}(1 + \sqrt{5})$ has many mystical properties—for example,

(a) $$r = 1 + \cfrac{1}{1 + \cfrac{1}{1 + \cfrac{1}{1 + \cdots}}}$$

(b) $r = \sqrt{1 + \sqrt{1 + \sqrt{1 + \sqrt{1 + \cdots}}}}$

(c) $r^n = r^{n-1} + r^{n-2}$

(d) $r = r^{-1} + r^{-2} + r^{-3} + \cdots$

Establish these properties.

10. For $f(x, y) = (y - x)^{-1}$ the Taylor series can be written as

$$f(x + h, y + k) = Af + Bf^2 + Cf^3 + \cdots$$

where $f = f(x, y)$. Determine the coefficients A, B, and C.

11

Linear Programming

In studying how the U.S. economy is affected by changes in the supply and cost of energy, it has been found appropriate to use a "linear programming" model. This is a large system of linear inequalities governing the variables in the model, together with a linear function of these variables to be maximized. Typically the variables are the activity levels of various processes in the economy, such as the number of barrels of oil pumped per day or the number of men's shirts produced per day. A model containing reasonable detail could easily involve thousands of variables and thousands of linear inequalities. Such problems are discussed in this chapter, and some guidance is offered on how to use existing software.

11.1 Standard Forms and Duality

Linear programming is a branch of mathematics that deals with finding extreme values of linear functions when the variables are constrained by linear inequalities. Any problem of this type can be put into the following standard form by simple manipulations (to be discussed later).

FIRST PRIMAL FORM: Given data c_j, a_{ij}, b_i $(1 \leqq j \leqq n,\ 1 \leqq i \leqq m)$, we wish to determine the x_j $(1 \leqq j \leqq n)$ that maximize the linear function

$$\sum_{j=1}^{n} c_j x_j$$

subject to the constraints

$$\begin{cases} \displaystyle\sum_{j=1}^{n} a_{ij} x_j \leqq b_i & (1 \leqq i \leqq m) \\ x_j \geqq 0 & (1 \leqq j \leqq n) \end{cases}$$

In matrix notation the linear programming problem in first primal form looks like this:

$$\begin{cases} \text{maximize: } \mathbf{c}^T\mathbf{x} \\ \text{subject to: } \mathbf{A}\mathbf{x} \leqq \mathbf{b},\ \mathbf{x} \geqq 0 \end{cases} \tag{1}$$

Here $\mathbf{c}$ and $\mathbf{x}$ are n-component vectors, $\mathbf{b}$ is an m-component vector, and $\mathbf{A}$ an $m \times n$ matrix. A *vector* inequality $\mathbf{u} \leqq \mathbf{v}$ means that $\mathbf{u}$ and $\mathbf{v}$ are vectors with the same number of components and that *all* the individual components satisfy the inequality $u_i \leqq v_i$. The linear function $\mathbf{c}^T\mathbf{x}$ is called the *objective function*.

In a linear programming problem the set of all vectors satisfying the constraints is called the *feasible* set and its elements are the *feasible* points. So in the preceding notation the feasible set is

$$\mathcal{K} = \{\mathbf{x} \in \mathbf{R}^n : \mathbf{x} \geqq 0 \text{ and } \mathbf{A}\mathbf{x} \leqq \mathbf{b}\}$$

A more precise statement of the linear programming problem, then, is: determine $\mathbf{x}^* \in \mathcal{K}$ such that $\mathbf{c}^T\mathbf{x}^* \geqq \mathbf{c}^T\mathbf{x}$ for all $\mathbf{x} \in \mathcal{K}$.

To get an idea of the type of practical problem that can be solved by linear programming, consider a simple example of optimization. Suppose that a certain factory uses two raw materials to produce two products. Suppose also that

1. Each unit of the first product uses 5 units of the first raw material and 3 of the second.
2. Each unit of the second product uses 3 units of the first raw material and 6 of the second.
3. On hand are 15 units of the first raw material and 18 of the second.
4. The profits on sales of the products are 2 per unit and 3 per unit, respectively.

How should the raw materials be used to realize a maximum profit? In order to answer this question, variables x_1 and x_2 are introduced to represent the number of units of the two products to be manufactured. In terms of these variables, the profit is

$$2x_1 + 3x_2 \tag{2}$$

The process uses up $5x_1 + 3x_2$ units of the first raw material and $3x_1 + 6x_2$ units of the second. The limitations in 3 above, then, are expressed by these inequalities:

$$\begin{cases} 5x_1 + 3x_2 \leqq 15 \\ 3x_1 + 6x_2 \leqq 18 \end{cases} \tag{3}$$

Of course, $x_1 \geqq 0$ and $x_2 \geqq 0$. Thus the solution to the problem is a vector $\mathbf{x}$ that maximizes the linear function in (2) while satisfying the constraints (3). More precisely, among all vectors $\mathbf{x}$ in the set

$$\mathcal{K} = \{\mathbf{x} : \mathbf{x} \geqq 0,\ 5x_1 + 3x_2 \leqq 15,\ 3x_1 + 6x_2 \leqq 18\}$$

we want the one that makes $2x_1 + 3x_2$ as large as possible.

Because the number of variables in this example is only 2, the problem is amenable to graphical solution. To locate the solution, we begin by graphing the set $\mathcal{K}$. This is the shaded region in Fig. 11.1. Then we draw some of the lines $2x_1 + 3x_2 = \alpha$, where α is given various values. These lines are dashed in the figure and labeled with the value of α. Finally, we select one of these lines with a maximum α that intersects $\mathcal{K}$. That intersection is the solution point and a vertex of $\mathcal{K}$. It is obtained numerically by solving

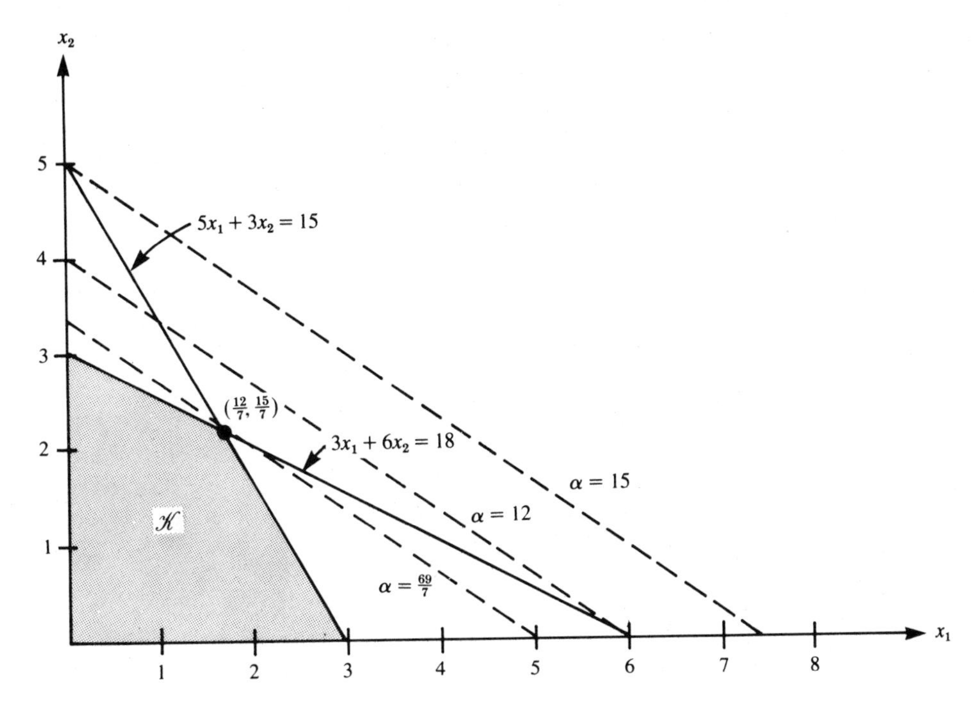

Figure 11.1 Graphical Method

simultaneously the equations $5x_1 + 3x_2 = 15$ and $3x_1 + 6x_2 = 18$. Thus $x = (12/7, 15/7)$, and the corresponding profit from Eq. (2) is $2(12/7) + 3(15/7) = 69/7$.

Note in this example that the units used, whether dollars, pesos, pounds, or kilograms, do not matter for the mathematical method as long as they are used consistently. Notice also that x_1 and x_2 are permitted to be arbitrary real numbers. The problem would be quite different if only integer values were acceptable as a solution. This situation would occur if the products being produced consisted of indivisible units, such as a manufactured article. If the integer constraint is imposed, only points with integer coordinates inside $\mathscr{K}$ are acceptable. Of them, the best is (0, 3). Observe particularly that we *cannot* simply round off the solution (1.71, 2.14) to the nearest integers to solve the problem with integer constraints. The point (2, 2) lies just outside $\mathscr{K}$. So if the company could alter the constraints slightly by increasing the amount of the first raw material to 16, the integer solution (2, 2) would be allowable. Special programs for *integer* linear programming are available but are outside the scope of this book.

Observe how the solution would be altered if our profit or objective function were $2x_1 + x_2$. In this case, the dashed lines in the figure would have a different slope (namely, -2) and a different vertex of the shaded region would have occurred as the solution—namely, (3, 0). A characteristic feature of linear programming problems is that the solutions always can be found among the vertices if any exist.

A linear programming problem not already in the first primal form can be put into that form by some standard techniques.

1. If the original problem calls for the minimization of the linear function $\mathbf{c}^T\mathbf{x}$, this is the same as maximizing $(-\mathbf{c})^T\mathbf{x}$.
2. If the original problem contains a constraint like $\mathbf{a}^T\mathbf{x} \geqq \beta$, it can be replaced by the constraint $(-\mathbf{a})^T\mathbf{x} \leqq -\beta$.
3. If the objective function contains a constant, this fact has no effect on the solution. For example, the maximum of $\mathbf{c}^T\mathbf{x} + \lambda$ occurs for the same $\mathbf{x}$ as the maximum of $\mathbf{c}^T\mathbf{x}$.
4. If the original problem contains equality constraints, each can be replaced by two inequality constraints. Thus the equation $\mathbf{a}^T\mathbf{x} = \beta$ is equivalent to $\mathbf{a}^T\mathbf{x} \leqq \beta$ and $\mathbf{a}^T\mathbf{x} \geqq \beta$.
5. If the original problem does not require a variable (say x_i) to be nonnegative, we can replace x_i by the difference of two nonnegative variables, say $x_i = u_i - v_i$, where $u_i \geqq 0$ and $v_i \geqq 0$.

Here is an example that illustrates all five techniques. Consider the linear programming problem

$$\text{minimize:} \quad 2x_1 + 3x_2 - x_3 + 4$$

$$\text{constraints:} \quad \begin{cases} x_1 - x_2 + 4x_3 \geqq 2 \\ x_1 + x_2 + x_3 = 15 \\ x_2 \geqq 0 \geqq x_3 \end{cases}$$

It is equivalent to the following problem in first primal form.

$$\text{maximize:} \quad -2u + 2v - 3z - w$$

$$\text{constraints:} \quad \begin{cases} -u + v + z + 4w \leqq -2 \\ u - v + z - w \leqq 15 \\ -u + v - z + w \leqq -15 \\ u \geqq 0,\ v \geqq 0,\ z \geqq 0,\ w \geqq 0 \end{cases}$$

Corresponding to a given linear programming problem in first primal form is another problem called its *dual*. It is obtained from the original problem

$$\text{(P)} \quad \begin{cases} \text{maximize: } \mathbf{c}^T\mathbf{x} \\ \text{subject to: } \mathbf{A}\mathbf{x} \leqq \mathbf{b},\ \mathbf{x} \geqq 0 \end{cases}$$

by defining the dual to be the problem

$$\text{(D)} \quad \begin{cases} \text{minimize: } \mathbf{b}^T\mathbf{y} \\ \text{subject to: } \mathbf{A}^T\mathbf{y} \geqq \mathbf{c},\ \mathbf{y} \geqq 0 \end{cases}$$

For example, the dual of this problem

$$\text{maximize:} \quad 2x_1 + 3x_2$$

$$\text{subject to:} \quad \begin{cases} 4x_1 + 5x_2 \leqq 6 \\ 7x_1 + 8x_2 \leqq 9 \\ 10x_1 + 11x_2 \leqq 12 \\ x_1 \geqq 0,\ x_2 \geqq 0 \end{cases}$$

is

$$\begin{aligned}
&\text{minimize:} \quad 6y_1 + 9y_2 + 12y_3 \\
&\text{subject to:} \quad \begin{cases} 4y_1 + 7y_2 + 10y_3 \geqq 2 \\ 5y_1 + 8y_2 + 11y_3 \geqq 3 \\ y_1 \geqq 0,\ y_2 \geqq 0,\ y_3 \geqq 0 \end{cases}
\end{aligned}$$

Note that, in general, the dual problem has different dimensions from that of the original problem. Thus the number of *inequalities* in the original problem becomes the number of *variables* in the dual problem.

An elementary relationship between the original primal problem (P) and its dual (D) is as follows.

> If $\mathbf{x}$ satisfies the constraints of (P) and $\mathbf{y}$ satisfies the constraints of (D), then $\mathbf{c}^T\mathbf{x} \leqq \mathbf{b}^T\mathbf{y}$. Consequently, if $\mathbf{c}^T\mathbf{x} = \mathbf{b}^T\mathbf{y}$, then $\mathbf{x}$ and $\mathbf{y}$ are solutions of (P) and (D), respectively.

It is easily verified: by the assumptions made, $\mathbf{x} \geqq 0$, $\mathbf{Ax} \leqq \mathbf{b}$, $\mathbf{y} \geqq 0$, and $\mathbf{A}^T\mathbf{y} \geqq \mathbf{c}$. Consequently,

$$\mathbf{c}^T\mathbf{x} \leqq (\mathbf{A}^T\mathbf{y})^T\mathbf{x} = \mathbf{y}^T\mathbf{Ax} \leqq \mathbf{y}^T\mathbf{b} = \mathbf{b}^T\mathbf{y}$$

This relationship can be used to estimate the number $\lambda = \max\{\mathbf{c}^T\mathbf{x} : \mathbf{x} \geqq 0 \text{ and } \mathbf{Ax} \leqq \mathbf{b}\}$. (This number is often termed the *value* of the linear programming problem.) To estimate λ, take any $\mathbf{x}$ and $\mathbf{y}$ satisfying $\mathbf{x} \geqq 0$, $\mathbf{y} \geqq 0$, $\mathbf{Ax} \leqq \mathbf{b}$, $\mathbf{A}^T\mathbf{y} \geqq \mathbf{c}$. Then $\mathbf{c}^T\mathbf{x} \leqq \lambda \leqq \mathbf{b}^T\mathbf{y}$.

The importance of the dual problem stems from the fact that the extreme values in the primal and dual problems are the same. Formally stated,

> DUALITY THEOREM: If the original problem has a solution x^*, then the dual problem has a solution $\mathbf{y}^*$, and furthermore, $\mathbf{c}^T\mathbf{x}^* = \mathbf{b}^T\mathbf{y}^*$.

This result is nicely illustrated by the numerical example from the beginning of this section. The dual to that problem is

$$\begin{aligned}
&\text{minimize:} \quad 15y_1 + 18y_2 \\
&\text{constraints:} \quad \begin{cases} 5y_1 + 3y_2 \geqq 2 \\ 3y_1 + 6y_2 \geqq 3 \\ y_1 \geqq 0,\ y_2 \geqq 0 \end{cases}
\end{aligned}$$

The graph of this problem is given in Fig. 11.2. Moving the line $15y_1 + 18y_2 = \alpha$, we see that vertex $(1/7, 3/7)$ is the minimum point. Notice that $15(1/7) + 18(3/7) = 69/7$ so that values of the objective functions are indeed identical. The solution $\mathbf{x} = (12/7, 15/7)$ and $\mathbf{y} = (1/7, 3/7)$ can be related, but we shall not cover this.

Returning to the general problem in the first primal form, we introduce additional nonnegative variables $x_{n+1}, x_{n+2}, \ldots, x_{n+m}$, known as *slack variables*, so that some of the inequalities can be written as equalities. Using this device, we can put the original problem into the following standard form.

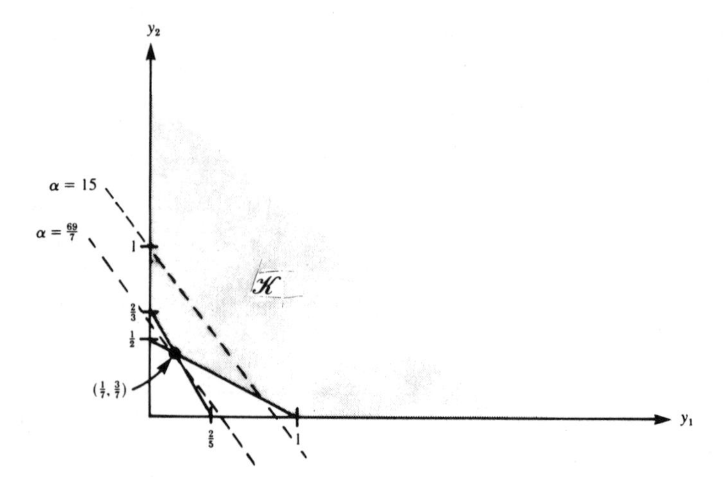

Figure 11.2 Graphical method of dual problem.

SECOND PRIMAL FORM: Maximize the linear function

$$\sum_{j=1}^{n} c_j x_j$$

subject to the constraints

$$\begin{cases} \sum_{j=1}^{n} a_{ij} x_j + x_{n+i} = b_i & (1 \leqq i \leqq m) \\ x_i \geqq 0 & (1 \leqq i \leqq m) \end{cases}$$

Using the simpler matrix notation, we have

$$\begin{cases} \text{maximize: } \mathbf{c}^T\mathbf{x} \\ \text{constraints: } \mathbf{Ax} = \mathbf{b},\ \mathbf{x} \geqq 0 \end{cases}$$

Here it is assumed that the $m \times n$ matrix **A** contains an $m \times m$ identity matrix in its last m columns and that the last m entries of c are zero. Also, note that when a problem in first primal form is changed to second primal form, we increase the number of variables and thus alter the quantities n, **x**, **c**, and **A**. That is, a problem in the first primal form with n variables would contain $n + m$ variables in the second.

To illustrate the transformation of a problem from first to second primal form, consider the example introduced at the beginning of this section:

$$\text{maximize: } \quad 2x_1 + 3x_2$$

$$\text{constraints: } \quad \begin{cases} 5x_1 + 3x_2 \leqq 15 \\ 3x_1 + 6x_2 \leqq 18 \\ x_1 \geqq 0,\ x_2 \geqq 0 \end{cases}$$

Two slack variables x_3 and x_4 are introduced to take up the "slack" in two of the inequalities. The new problem in second primal form is then

$$\text{maximize:} \quad 2x_1 + 3x_2 + 0x_3 + 0x_4$$

$$\text{constraints:} \quad \begin{cases} 5x_1 + 3x_2 + x_3 = 15 \\ 3x_1 + 6x_2 + x_4 = 18 \\ x_1 \geqq 0,\ x_2 \geqq 0,\ x_3 \geqq 0,\ x_4 \geqq 0 \end{cases}$$

Problems involving absolute values of the variables or absolute values of linear expressions can often be turned into linear programming problems. To illustrate, consider the problem of minimizing $|x - y|$ subject to linear constraints on x and y. We can introduce a new variable $z \geqq 0$ and impose constraints $x - y \leqq z$, $-x + y \leqq z$. Then we seek to minimize the linear form $0x + 0y + 1z$.

Problems 11.1

1. Put the following problem into first primal form. Minimize $|x_1 + 2x_2 - x_3|$ subject to

$$\begin{cases} x_1 + 3x_2 - x_3 \leqq 8 \\ 2x_1 - 4x_2 - x_3 \geqq 1 \\ |4x_1 + 5x_2 + 6x_3| \leqq 12 \\ x_1 \geqq 0,\ x_2 \geqq 0,\ x_3 \geqq 0 \end{cases}$$

   Hint: $|A| \leqq B$ can be written as $-B \leqq A \leqq B$.

2. A program is available for solving linear programming problems in first primal form. Put the following problem into that form. Minimize $5x_1 + 6x_2 - 2x_3 + 8$ subject to

$$\begin{cases} 2x_1 - 3x_2 \geqq 5 \\ x_2 + x_1 \leqq 15 \\ x_3 + 2x_1 \leqq 25 + x_2 \\ x_1 + x_2 - x_3 \geqq 1 \\ x_1 \geqq 0,\ x_2 \geqq 0,\ x_3 \geqq 0 \end{cases}$$

3. Consider the linear problems

   (a) maximize: $2x_1 + 3x_2$

$$\text{constraints:} \quad \begin{cases} x_1 + 2x_2 \geqq -6 \\ -x_1 + 3x_2 \leqq 3 \\ |2x_1 - 5x_2| \leqq 5 \\ x_1 \geqq 0,\ x_2 \geqq 0 \end{cases}$$

   (b) minimize: $7x_1 + x_2 - x_3 + 4$

$$\text{constraints:} \quad \begin{cases} x_1 - x_2 + x_3 \geqq 2 \\ x_1 + x_2 + x_3 \leqq 10 \\ -2x_1 + 4 \leqq x_2 \\ x_1 \geqq 0,\ x_2 \geqq 0 \end{cases}$$

   Rewrite each problem in first primal form and give the dual problem.

4. Sketch the feasible region for the following constraints.

$$\begin{cases} x - y \leqq 2 \\ x + y \leqq 3 \\ 2x + y \leqq 3 \\ x \geqq 0,\ y \geqq 0 \end{cases}$$

(a) By substituting the vertices into the objective function

$$z(x, y) = x + 2y$$

determine the minimum value of this function on the feasible region.

(b) Let

$$z(x, y) = (x - \tfrac{1}{2})^2 + (y - \tfrac{1}{2})^2$$

Show that the minimum value of z over the feasible region does not occur at a vertex.

5. Solve each of the linear programming problems by the graphical method. Determine $\mathbf{x}$ to

$$\begin{cases} \text{maximize: } \mathbf{c}^T\mathbf{x} \\ \text{constraints: } \mathbf{A}\mathbf{x} \leqq \mathbf{b},\ \mathbf{x} \geqq 0 \end{cases}$$

Here nonunique and unbounded "solutions" may be obtained.

(a) $\mathbf{c} = (2, -4)^T$

$$\mathbf{A} = \begin{pmatrix} -3 & -5 \\ 4 & 9 \end{pmatrix}$$

$\mathbf{b} = (-15, 36)$

(b) $\mathbf{c} = (2, \tfrac{1}{2})^T$

$$\mathbf{A} = \begin{pmatrix} 6 & 5 \\ 4 & 1 \end{pmatrix}$$

$\mathbf{b} = (30, 12)^T$

(c) $\mathbf{c} = (3, 2)^T$

$$\mathbf{A} = \begin{pmatrix} -3 & 2 \\ -4 & 9 \end{pmatrix}$$

$\mathbf{b} = (6, 36)^T$

(d) $\mathbf{c} = (2, -3)^T$

$$\mathbf{A} = \begin{pmatrix} -1 & 1 \\ 0 & 1 \end{pmatrix}$$

$\mathbf{b} = (0, 5)^T$

(e) $\mathbf{c} = (-4, 11)^T$

$$\mathbf{A} = \begin{pmatrix} -3 & 4 \\ -4 & 11 \end{pmatrix}$$

$\mathbf{b} = (12, 44)^T$

(f) $\mathbf{c} = (-3, 4)^T$

$$\mathbf{A} = \begin{pmatrix} 2 & 3 \\ -4 & -5 \end{pmatrix}$$

$\mathbf{b} = (6, -20)^T$

(g) $\mathbf{c} = (2, 1)^T$

$$\mathbf{A} = \begin{pmatrix} 1 & 1 \\ 1 & 2 \end{pmatrix}$$

$\mathbf{b} = (0, -2)^T$

(h) $\mathbf{c} = (3, 1)^T$

$$\mathbf{A} = \begin{pmatrix} 2 & 4 \\ 5 & 3 \end{pmatrix}$$

$\mathbf{b} = (21, 18)^T$

6. Solve the following linear programming problem by hand, using a graph for help.

$$\text{maximize: } 4x + 4y + z$$

$$\text{constraints: } \begin{cases} 3x + 2y + z = 12 \\ 7x + 7y + 2z \leqq 144 \\ 7x + 5y + 2z \leqq 80 \\ 11x + 7y + 3z \leqq 132 \\ x \geqq 0,\ y \geqq 0 \end{cases}$$

Hint: Use the equation to eliminate z from all other expressions. Solve the resulting two-dimensional problem.

7. Put this linear programming problem into second primal form. You may want to make changes of variables. If so, include a "dictionary" relating new and old variables.

$$\text{minimize:} \quad \varepsilon_1 + \varepsilon_2 + \varepsilon_3$$

$$\text{constraints:} \quad \begin{cases} |3x + 4y + 6| \leqq \varepsilon_1 \\ |2x - 8y - 4| \leqq \varepsilon_2 \\ |-x - 3 \quad + 5| \leqq \varepsilon_3 \\ \varepsilon_1 > 0,\ \varepsilon_2 > 0,\ \varepsilon_3 > 0,\ x > 0,\ y > 0 \end{cases}$$

Solve the resulting problem.

8. Consider the linear programming problem

$$\text{maximize:} \quad c_1x_1 + c_2x_2$$

$$\text{constraints:} \quad \begin{cases} a_1x_1 + a_2x_2 \leqq b \\ x_1 \geqq 0, \quad x_2 \geqq 0 \end{cases}$$

In the special case when all data are positive, show that the dual problem has the same extreme value as the original problem.

9. Suppose that a linear programming problem in first primal form has the property that $\mathbf{c}^T\mathbf{x}$ is not bounded on the feasible set. What conclusion can be drawn about the dual problem?

Computer Problems 11.1

1. A.Western shop wishes to purchase 300 felt and 200 straw cowboy hats. Bids have been received from three wholesalers. Texas Hatters has agreed to supply not more than 200 hats, Lone Star Hatters not more than 250, and Lariat Ranch Wear not more than 150. The owner of the shop has estimated that his profit per hat sold from Texas Hatters would be \$3/felt and \$4/straw, from Lone Star Hatters \$3.80/felt and \$3.50/straw, from Lariat Ranch Wear \$4/felt and \$3.60/straw. Set up a linear programming problem to maximize the owner's profits. Solve by using a program from the computer center program library.
2. The ABC Drug Company makes two types of liquid painkiller with brand names Relieve (R) and Ease (E) and containing a different mixture of three basic drugs A, B, and C produced by the company. Each bottle of R requires $\frac{7}{9}$ units of drug A, $\frac{1}{2}$ units of drug B, and $\frac{3}{4}$ units of drug C. Each bottle of E requires $\frac{4}{9}$ units of drug A, $\frac{5}{2}$ units of drug B, and $\frac{1}{4}$ units of drug C. The company is able to produce each day only 5 units of drug A, 7 units of drug B, and 9 units of C. Moreover, Federal Drug Administration regulations stipulate that the number of bottles of R cannot exceed twice the number of bottles of E. The profit margin for each bottle of E and R is \$7 and \$3, respectively. Set up the linear programming problem in first primal form in order to determine the number of bottles of the two painkillers that the company should produce each day so as to maximize their profits. Solve by using a subroutine from the computer center program library.
3. Suppose that the university student government wishes to charter planes to transport at least 750 students to the bowl game. Two airlines A1 and A2 agree to supply aircraft for the trip. A1 has five aircraft available carrying 75 passengers each and A2 has three aircraft available carrying 250 passengers each. The cost per aircraft is \$900 and \$3250 for the trip from A1 and A2, respectively. The student government wants to charter at most six aircraft. How many of each type should be chartered to minimize the cost of the airlift? How much should the student government charge in order to make 50¢ profit per student? Solve by the graphical method and verify by using a routine from your computer center program library.
4. Rework Problem 3 in the following two possibly different ways:
   (a) the number of students going on the airlift is maximized.
   (b) the cost per student going is minimized.

5. (A Diet Problem) A university dining hall wishes to provide at least 5 units of vitamin C and 3 units of vitamin E per serving. Three foods are available containing these vitamins. Food f_1 contains 2.5 and 1.25 units per ounce of C and E, respectively, whereas food f_2 contains just the opposite amounts. The third food f_3 contains an equal amount of each vitamin at one unit per ounce. Food f_1 costs 25¢ per ounce, food f_2 costs 56¢ per ounce, and food f_3 costs 10¢ per ounce. The dietician wishes to provide the meal at a minimum cost per serving that satisfies the minimum vitamin requirements. Set up this linear programming problem in second primal form. Solve with the aid of a code from your computer center program library.

11.2 The Simplex Method

The principal algorithm used in solving linear programming problems is the *simplex method*. Here enough of the background of this method is described so that the reader can use available computer programs incorporating it.

Consider a linear programming problem in second primal form:

$$\text{maximize:} \quad \mathbf{c}^T\mathbf{x}$$

$$\text{constraints:} \begin{cases} \mathbf{A}\mathbf{x} = \mathbf{b} \\ \mathbf{x} \geqq 0 \end{cases}$$

It is assumed that $\mathbf{c}$ and $\mathbf{x}$ are n-component vectors, $\mathbf{b}$ an m-component vector, and $\mathbf{A}$ an $m \times n$ matrix. Also, it is assumed that $\mathbf{b} \geqq 0$ and that $\mathbf{A}$ contains an $m \times m$ identity matrix in its last m columns.

As before, we define the set of *feasible* points as

$$\mathcal{K} = \{\mathbf{x} \in \mathbf{R}^n : \mathbf{A}\mathbf{x} = \mathbf{b},\ \mathbf{x} \geqq 0\}$$

The points of $\mathcal{K}$ are exactly the points that are competing to maximize $\mathbf{c}^T\mathbf{x}$.

The set $\mathcal{K}$ is a polyhedral set in $\mathbf{R}^n$, and the algorithm to be described proceeds from vertex to vertex in $\mathcal{K}$, always increasing the value of $\mathbf{c}^T\mathbf{x}$ as it goes from one to another. Let us give a precise definition of ''vertex.'' A point $\mathbf{x}$ in $\mathcal{K}$ is called a *vertex* if it is impossible to express it as $\mathbf{x} = \frac{1}{2}(\mathbf{u} + \mathbf{v})$ with both $\mathbf{u}$ and $\mathbf{v}$ in $\mathcal{K}$ and $\mathbf{u} \neq \mathbf{v}$. In other words, $\mathbf{x}$ is not the midpoint of any line segment whose endpoints lie in $\mathcal{K}$.

We denote by $\mathbf{a}^{(1)}, \mathbf{a}^{(2)}, \ldots, \mathbf{a}^{(n)}$ the column vectors constituting the matrix $\mathbf{A}$. The following statement relates the columns of $\mathbf{A}$ to the vertices of $\mathcal{K}$.

> Let $\mathbf{x} \in \mathcal{K}$ and define $\mathcal{I}(\mathbf{x}) = \{i : x_i > 0\}$. Then the following are equivalent.
> (i) $\mathbf{x}$ is a vertex of $\mathcal{K}$.
> (ii) The set $\{\mathbf{a}^{(i)} : i \in \mathcal{I}(\mathbf{x})\}$ is linearly independent.

If (i) is false, then we can write $\mathbf{x} = \frac{1}{2}(\mathbf{u} + \mathbf{v})$ with $\mathbf{u} \in \mathcal{K}$, $\mathbf{v} \in \mathcal{K}$, and $\mathbf{u} \neq \mathbf{v}$. For every index i not in the set $\mathcal{I}(\mathbf{x})$, we have $x_i = 0$, $u_i \geqq 0$, $v_i \geqq 0$, and $x_i = \frac{1}{2}(u_i + v_i)$. This forces u_i and v_i to be zero. Thus all the nonzero components of $\mathbf{u}$ and $\mathbf{v}$ correspond to indices i in $\mathcal{I}(\mathbf{x})$. Since $\mathbf{u}$ and $\mathbf{v}$ belong to $\mathcal{K}$,

$$\mathbf{b} = \mathbf{A}\mathbf{u} = \sum_{i=1}^{n} u_i \mathbf{a}^{(i)} = \sum_{i \in \mathcal{J}(\mathbf{x})} u_i \mathbf{a}^{(i)}$$

and

$$\mathbf{b} = \mathbf{A}\mathbf{v} = \sum_{i=1}^{n} v_i \mathbf{a}^{(i)} = \sum_{i \in \mathcal{J}(\mathbf{x})} v_i \mathbf{a}^{(i)}$$

Hence,

$$\sum_{i \in \mathcal{J}(\mathbf{x})} (u_i - v_i)\mathbf{a}^{(i)} = 0$$

showing the linear dependence of the set $\{\mathbf{a}^{(i)} : i \in \mathcal{J}(\mathbf{x})\}$. Thus (ii) is false. Consequently, (ii) implies (i).

For the converse, assume that (ii) is false. From the linear dependence of column vectors $\mathbf{a}^{(i)}$ for $i \in \mathcal{J}(\mathbf{x})$, we have

$$\sum_{i \in \mathcal{J}(\mathbf{x})} y_i \mathbf{a}^{(i)} = 0 \quad \text{with} \quad \sum_{i \in \mathcal{J}(\mathbf{x})} |y_i| \neq 0,$$

for appropriate coefficients y_i. For each $i \notin \mathcal{J}(\mathbf{x})$, let $y_i = 0$. Form the vector $\mathbf{y}$ with components y_i for $i = 1, 2, \ldots, n$. Then, for any λ, we see that because $\mathbf{x} \in \mathcal{K}$

$$\mathbf{A}(\mathbf{x} + \lambda \mathbf{y}) = \sum_{i=1}^{n} (x_i \pm \lambda y_i)\mathbf{a}^{(i)} = \sum_{i=1}^{n} x_i \mathbf{a}^{(i)} \pm \lambda \sum_{i \in \mathcal{J}(\mathbf{x})} y_i \mathbf{a}^{(i)} = \mathbf{A}\mathbf{x} = \mathbf{b}$$

Now select the real number λ positive but so small that $\mathbf{x} + \lambda\mathbf{y} \geqq 0$ and $\mathbf{x} - \lambda\mathbf{y} \geqq 0$. [To see that it is possible, consider separately the components for $i \in \mathcal{J}(\mathbf{x})$ and $i \notin \mathcal{J}(\mathbf{x})$.] The resulting vectors, $\mathbf{u} = \mathbf{x} + \lambda\mathbf{y}$ and $\mathbf{v} = \mathbf{x} - \lambda\mathbf{y}$, belong to $\mathcal{K}$. They differ and obviously $\mathbf{x} = \frac{1}{2}(\mathbf{u} + \mathbf{v})$. Thus $\mathbf{x}$ is not a vertex of $\mathcal{K}$; that is, (i) is false. So (i) implies (ii).

Given a linear programming problem, there are three possibilities.

1. There are no feasible points — that is, the set $\mathcal{K}$ is empty.
2. $\mathcal{K}$ is not empty and $\mathbf{c}^T\mathbf{x}$ is not bounded on $\mathcal{K}$.
3. $\mathcal{K}$ is not empty and $\mathbf{c}^T\mathbf{x}$ is bounded on $\mathcal{K}$.

It is true (but not obvious) that in case 3 there is a point $\mathbf{x}$ in $\mathcal{K}$ such that $\mathbf{c}^T\mathbf{x} \geqq \mathbf{c}^T\mathbf{y}$ for all $\mathbf{y}$ in $\mathcal{K}$. We have assumed that our problem is in the second primal form so that possibility 1 cannot occur. Indeed, $\mathbf{A}$ contains an $m \times m$ identity matrix and so has the form

$$\mathbf{A} = \begin{bmatrix} a_{11} & a_{12} & \cdots & a_{1k} & 1 & 0 & \cdots & 0 \\ a_{21} & a_{22} & \cdots & a_{2k} & 0 & 1 & \cdots & 0 \\ \vdots & \vdots & & \vdots & \vdots & \vdots & & \vdots \\ a_{m1} & a_{m2} & \cdots & a_{mk} & 0 & 0 & \cdots & 1 \end{bmatrix}$$

where $k = n - m$. Consequently, we can *construct* a feasible point $\mathbf{x}$ easily by setting $x_1 = x_2 = \cdots = x_k = 0$ and $x_{k+1} = b_1$, $x_{k+2} = b_2$, and so on. It is then clear that $\mathbf{A}\mathbf{x} = \mathbf{b}$. The inequality $\mathbf{x} \geqq 0$ follows from our initial assumption that $\mathbf{b} \geqq 0$.

Here is a brief outline of the simplex method for solving linear programming problems. It involves a sequence of exchanges so that the trial solution proceeds systematically from one vertex to another in $\mathcal{K}$. This procedure is stopped when the value of $\mathbf{c}^T\mathbf{x}$ is no longer increased as a result of the exchange.

Simplex Method. Select a small positive value for ε. In each step we have a set of m indices $\{k_1, k_2, \ldots, k_m\}$.

1. Put columns $\mathbf{a}^{(k_1)}, \mathbf{a}^{(k_2)}, \ldots, \mathbf{a}^{(k_m)}$ into $\mathbf{B}$ and solve $\mathbf{Bx} = \mathbf{b}$.
2. If $x_i > 0$ for $1 \leqq i \leqq m$, continue. Otherwise exit, since the algorithm has failed.
3. Set $\mathbf{e} = (c_{k_1}, c_{k_2}, \ldots, c_{k_m})^T$ and solve $\mathbf{B}^T\mathbf{y} = \mathbf{e}$.
4. Choose any s in $\{1, 2, \ldots, n\}$ but not in $\{k_1, k_2, \ldots, k_m\}$ for which $c_s - \mathbf{y}^T\mathbf{a}^{(s)}$ is greatest.
5. If $c_s - \mathbf{y}^T\mathbf{a}^{(s)} < \varepsilon$, exit, since $\mathbf{x}$ is the solution.
6. Solve $\mathbf{Bz} = \mathbf{a}^{(s)}$.
7. If $z_i \leqq \varepsilon$ for $1 \leqq i \leqq m$, then exit because the objective function is unbounded on $\mathcal{K}$.
8. Among the ratios x_i/z_i having $z_i > 0$ for $1 \leqq i \leqq m$, let x_r/z_r be the smallest. In case of a tie, let r be the first occurrence.
9. Replace k_r by s and go to step 1.

A few remarks on this algorithm are in order. In the beginning select the indices $k_1, k_2, \ldots, k_m$ so that $\mathbf{a}^{(k_1)}, \mathbf{a}^{(k_2)}, \ldots, \mathbf{a}^{(k_m)}$ forms an $m \times m$ identity matrix. At step 5 where we say "$\mathbf{x}$ is a solution" we mean that the vector $\mathbf{v} = (v_i)$ given by $v_{k_i} = x_i$ for $1 \leqq i \leqq n$ and $v_i = 0$ for $i \notin \{k_1, k_2, \ldots, k_m\}$ is the solution. A convenient choice for the tolerance ε that occurs in steps 5 and 7 might be 10^{-6}.

In any reasonable implementation of the simplex method advantage must be taken of the fact that succeeding occurrences of step 1 are very similar. In fact, only one column of $\mathbf{B}$ changes at a time. Similar remarks hold for steps 3 and 6.

We do not recommend that the reader attempt to program the simplex algorithm. Efficient codes, refined over many years of experience, are usually available in the program library of computing centers. One feature that many possess is that they can provide solutions to a given problem *and* to its dual with very little additional computing. Sometimes this feature can be exploited to decrease the execution time of a problem. To see why, consider a linear programming problem in first primal form:

$$\text{(P)} \qquad \begin{cases} \text{maximize: } \mathbf{c}^T\mathbf{x} \\ \text{subject to: } \mathbf{Ax} \leqq \mathbf{b},\ \mathbf{x} \geqq 0 \end{cases}$$

As usual, we assume that $\mathbf{x}$ is an n vector and that $\mathbf{A}$ is an $m \times n$ matrix. When the simplex algorithm is applied to (P), it performs an iterative process on an $m \times m$ matrix denoted by $\mathbf{B}$ in the preceding description. If the number of inequality constraints m is very large relative to n, then the dual problem may be easier to solve, since the $\mathbf{B}$ matrices for it will be of dimension $n \times n$. Indeed, the dual problem is

$$\text{(D)} \qquad \begin{cases} \text{minimize: } \mathbf{b}^T\mathbf{y} \\ \text{subject to: } \mathbf{A}^T\mathbf{y} \geqq \mathbf{c},\ \mathbf{y} \geqq 0 \end{cases}$$

and the number of inequality constraints here is n. An example of this technique appears in the next section.

Problems 11.2

1. Show that the linear programming problem

$$\begin{cases} \text{maximize: } \mathbf{c}^T\mathbf{x} \\ \text{subject to: } \mathbf{A}\mathbf{x} \leqq \mathbf{b} \end{cases}$$

can be put into first primal form by increasing the number of variables by just one. *Hint:* Replace x_j by $y_j - y_0$.
2. Show that the set $\mathcal{K}$ can have only a finite number of vertices.
3. Suppose that $\mathbf{u}$ and $\mathbf{v}$ are solution points for a linear programming problem and that $\mathbf{x} = \frac{1}{2}(\mathbf{u} + \mathbf{v})$. Show that $\mathbf{x}$ is also a solution.
4. Using the simplex method as described, solve the illustrative problem of Sec. 11.1.
5. Using standard manipulations, put the dual problem (D) into first and second primal forms.
6. Show how a program for solving a linear programming problem in first primal form can be used to solve a system of n linear equations in n variables.

Computer Problems 11.2

Select a linear programming code from your computing center library and use it to solve these problems.

1. minimize: $8x_1 + 6x_2 + 6x_3 + 9x_4$

constraints:
$$\begin{cases} x_1 + 2x_2 + x_4 \geqq 2 \\ 3x_1 + x_2 + x_4 \geqq 4 \\ x_3 + x_4 \geqq 1 \\ x_1 + x_3 \geqq 1 \\ x_1 \geqq 0,\ x_2 \geqq 0,\ x_3 \geqq 0,\ x_4 \geqq 0 \end{cases}$$

2. minimize: $10x_1 - 5x_2 - 4x_3 + 7x_4 + x_5$

constraints:
$$\begin{cases} 4x_1 - 3x_2 - x_3 + 4x_4 + x_5 = 1 \\ -x_1 + 2x_2 + 2x_3 + x_4 + 3x_5 = 4 \\ x_1 \geqq 0,\ x_2 \geqq 0,\ x_3 \geqq 0,\ x_4 \geqq 0,\ x_5 \geqq 0 \end{cases}$$

3. maximize: $2x_1 + 4x_2 + 3x_3$

constraints:
$$\begin{cases} 4x_1 + 2x_2 + 3x_3 \leqq 15 \\ 3x_1 + 2x_2 + x_3 \leqq 7 \\ x_1 + x_2 + 2x_3 \leqq 6 \\ x_1 \geqq 0,\ x_2 \geqq 0,\ x_3 \geqq 0 \end{cases}$$

11.3 Approximate Solution of Inconsistent Linear Systems

Linear programming can be used for the approximate solution of systems of linear equations that are inconsistent. An $m \times n$ system of equations

$$\sum_{j=1}^{n} a_{ij}x_j = b_i \qquad (1 \leqq i \leqq m)$$

is said to be ***inconsistent*** if there is no vector $\mathbf{x} = (x_1, x_2, \ldots, x_n)^T$ that simultaneously satisfies all m equations in the system. For instance, the system

$$\begin{cases} 2x_1 + 3x_2 = 4 \\ x_1 - x_2 = 2 \\ x_1 + 2x_2 = 7 \end{cases} \tag{1}$$

is inconsistent, as can be seen by attempting to carry out the Gaussian elimination process.

Since no $\mathbf{x}$ vector can solve an inconsistent system of equations, the ***residuals***

$$r_i = \sum_{j=1}^{n} a_{ij}x_j - b_i \qquad (1 \leqq i \leqq m) \tag{2}$$

cannot be made to vanish simultaneously. Hence $\sum_{i=1}^{m} |r_i| > 0$. Now it is natural to ask for an $\mathbf{x}$ vector that renders the expression $\sum_{i=1}^{m} |r_i|$ as small as possible. This problem is called the ***ℓ_1 problem*** for this system of equations. Other criteria, leading to different "approximate solutions," might be to minimize $\sum_{i=1}^{m} r_i^2$ or $\max_{1 \leqq i \leqq m} |r_i|$. Chapter 12 discusses in detail the problem of minimizing $\sum_{i=1}^{m} r_i^2$.

The minimization of $\sum_{i=1}^{n} |r_i|$ by appropriate choice of the $\mathbf{x}$ vector is a problem for which special algorithms have been designed; see Barrodale and Roberts [1974]. However, if one of these special programs is not available or if the problem is of small scope, linear programming can be used.

A simple, direct restatement of the problem is

$$\text{minimize:} \quad \sum_{i=1}^{m} \varepsilon_i$$

$$\text{constraints:} \quad \begin{cases} \displaystyle\sum_{j=1}^{n} a_{ij}x_j - b_i \leqq \varepsilon_i \\ \displaystyle -\sum_{j=1}^{n} a_{ij}x_j + b_i \leqq \varepsilon_i \end{cases} \tag{3}$$

If a linear programming code is at hand in which the variables are not required to be nonnegative, then it can be used on problem (3). If the variables must be nonnegative, the following technique can be applied. Introduce a variable y_{n+1} and write $x_j = y_j - y_{n+1}$. Then define $a_{i,n+1} = -\sum_{j=1}^{n} a_{ij}$. This step creates an additional column in the $\mathbf{A}$ array. Now consider the linear programming problem

$$
\begin{aligned}
&\text{maximize:} \quad -\sum_{i=1}^{m} \varepsilon_i \\
&\text{constraints:} \begin{cases} \displaystyle\sum_{j=1}^{n+1} a_{ij} y_j - \varepsilon_i \leqq b_i & (1 \leqq i \leqq m) \\ \displaystyle -\sum_{j=1}^{n+1} a_{ij} y_j - \varepsilon_i \leqq b_i & (1 \leqq i \leqq m) \\ y \geqq 0, \ \varepsilon \geqq 0 \end{cases}
\end{aligned} \tag{4}
$$

which is in first primal form with $m + n + 1$ variables and $2m$ inequality constraints.

It is not hard to verify that (4) is equivalent to Problem (3). The main point is that

$$
\begin{aligned}
\sum_{j=1}^{n+1} a_{ij} y_j &= \sum_{j=1}^{n} a_{ij}(x_j + y_{n+1}) + a_{i,n+1} y_{n+1} \\
&= \sum_{j=1}^{n} a_{ij} x_j + y_{n+1} \sum_{j=1}^{n} a_{ij} + y_{n+1} \left(- \sum_{j=1}^{n} a_{ij} \right) \\
&= \sum_{j=1}^{n} a_{ij} x_j
\end{aligned}
$$

Another technique can be used to replace the $2m$ inequality constraints in (4) by a set of m equality constraints. We write

$$
\varepsilon_i = |r_i| = u_i + v_i
$$

where $u_i = r_i$ and $v_i = 0$ if $r_i \geqq 0$, but $v_i = -r_i$ and $u_i = 0$ if $r_i < 0$. The resulting linear programming problem is

$$
\begin{aligned}
&\text{maximize:} \quad -\sum_{i=1}^{m} u_i - \sum_{i=1}^{m} v_i \\
&\text{constraints:} \begin{cases} \displaystyle\sum_{j=1}^{n+1} a_{ij} y_j - u_i + v_i = b_i & (1 \leqq i \leqq m) \\ u \geqq 0, \ v \geqq 0, \ y \geqq 0 \end{cases}
\end{aligned} \tag{5}
$$

Using the preceding formulas, we have

$$
\begin{aligned}
r_i &= \sum_{j=1}^{n} a_{ij} x_j - b_i = \sum_{j=1}^{n} a_{ij}(y_j - y_{n+1}) - b_i \\
&= \sum_{j=1}^{n} a_{ij} y_j - y_{n+1} \sum_{j=1}^{n} a_{ij} - b_i \\
&= \sum_{j=1}^{n+1} a_{ij} y_j - b_i = u_i - v_i
\end{aligned}
$$

From it we conclude that $r_i + v_i = u_i \geqq 0$. Now v_i and u_i should be as small as possible, consistent with this restriction, because we are attempting to minimize $\sum_{i=1}^{m} (u_i + v_i)$. So if $r_i \geqq 0$, we take $v_i = 0$ and $u_i = r_i$, whereas if $r_i < 0$, we take $v_i = -r_i$ and $u_i = 0$. In either case, $|r_i| = u_i + v_i$. Thus minimizing $\sum_{i=1}^{m} (u_i + v_i)$ is the same as minimizing $\sum_{i=1}^{m} |r_i|$.

The example of the inconsistent linear system given by (1) could be solved in the ℓ_1 sense by solving the linear programming problem

$$\text{minimize:} \quad u_1 + v_1 + u_2 + v_2 + u_3 + v_3$$

$$\text{constraints:} \quad \begin{cases} 2y_1 + 3y_2 - 5y_3 - u_1 + v_1 = 4 \\ y_1 - y_2 - u_2 + v_2 = 2 \\ y_1 + 2y_2 - 3y_3 - u_3 + v_3 = 7 \\ y_1, y_2, y_3 \geqq 0, \; u_1, u_2, u_3 \geqq 0, \; v_1, v_2, v_3 \geqq 0 \end{cases}$$

The solution is

$$u_1 = 0, \quad u_2 = 0, \quad u_3 = 0$$

$$v_1 = 0, \quad v_2 = 0, \quad v_3 = 5$$

$$y_1 = 2, \quad y_2 = 0, \quad y_3 = 0$$

From it we recover the ℓ_1 solution of system (1) in the form

$$x_1 = y_1 - y_3 = 2 \qquad r_1 = u_1 - v_1 = 0$$

$$x_2 = y_2 - y_3 = 0 \qquad r_2 = u_2 - v_2 = 0$$

$$r_3 = u_3 - v_3 = -5$$

Consider again a system of m linear equations in n unknowns

$$\sum_{j=1}^{n} a_{ij}x_j = b_i \qquad (1 \leqq i \leqq m)$$

Supposing that the system is inconsistent, we know that for any $\mathbf{x}$ vector the residuals $r_i = \sum_{j=1}^{n} a_{ij}x_j - b_i$ cannot all be zero. So the quantity $\varepsilon = \max_{1 \leqq i \leqq m} |r_i|$ is positive. The problem of making ε a minimum is called the ℓ_∞ *problem* for the system of equations. An equivalent linear programming problem is

$$\text{minimize:} \quad \varepsilon$$

$$\text{constraints:} \quad \begin{cases} \displaystyle\sum_{j=1}^{n} a_{ij}x_j - \varepsilon \leqq b_i & (1 \leqq i \leqq m) \\ \displaystyle -\sum_{j=1}^{n} a_{ij}x_j - \varepsilon \leqq -b_i & (1 \leqq i \leqq m) \end{cases} \tag{6}$$

If a linear programming code is available in which the variables need not be greater than or equal to zero, then it can be used to solve the ℓ_∞ problem as formulated above. If the variables must be nonnegative, we first introduce a variable y_{n+1} so large that the quantities $y_j = x_j + y_{n+1}$ are positive. Next, we solve the linear programming problem

$$\text{minimize:} \quad \varepsilon$$

$$\text{constraints:} \begin{cases} \sum_{j=1}^{n+1} a_{ij} y_j - \varepsilon \leqq b_i & (1 \leqq i \leqq m) \\ -\sum_{j=1}^{n+1} a_{ij} y_j - \varepsilon \leqq -b_i & (1 \leqq i \leqq m) \\ \varepsilon \geqq 0, \; y_j \geqq 0 & (1 \leqq j \leqq n+1) \end{cases} \tag{7}$$

Here we have again defined $a_{i,n+1} = -\sum_{j=1}^{n} a_{ij}$.

For our example, the solution that minimizes the quantity max $\{|2x_1 - 3x_2 - 4|, |x_1 - x_2 - 2|, |x_1 + 2x_2 - 7|\}$ is obtained from the linear programming problem

$$\text{minimize:} \quad \varepsilon$$

$$\text{constraint:} \begin{cases} 2y_1 + 3y_2 - 5y_3 - \varepsilon \leqq 4 \\ y_1 - y_2 - \varepsilon \leqq 2 \\ y_1 + 2y_2 - 3y_3 - \varepsilon \leqq 7 \\ -2y_1 - 3y_2 + 5y_3 - \varepsilon \leqq -4 \\ -y_1 + y_2 - \varepsilon \leqq -2 \\ -y_1 - 2y_2 + 3y_3 - \varepsilon \leqq -7 \\ y_1, y_2, y_3 \geqq 0, \; \varepsilon \geqq 0 \end{cases} \tag{8}$$

The solution is

$$y_1 = \frac{8}{9}, \quad y_2 = \frac{5}{3}, \quad y_3 = 0, \quad \varepsilon = \frac{25}{9}$$

From it the ℓ_∞ solution of (1) is recovered as follows.

$$x_1 = y_1 - y_3 = \frac{8}{9} \qquad x_2 = y_2 - y_3 = \frac{15}{9}$$

In problems like (7) m is often much larger than n. Thus in accordance with remarks made in Sec. 11.2, it may be preferable to solve the *dual* of Problem (7), for it would have $2m$ variables but only $n + 2$ inequality constraints. To illustrate, the dual of Problem (8) is

$$\text{maximize:} \quad 4u_1 + 2u_2 + 7u_3 - 4u_4 - 2u_5 - 7u_6$$

$$\text{constraints:} \begin{cases} 2u_1 + u_2 + u_3 - 2u_4 - u_5 - u_6 \geqq 0 \\ 3u_1 - u_2 + 2u_3 - 3u_4 + u_5 - 2u_6 \geqq 0 \\ -5u_1 - 3u_3 + 5u_4 + 3u_6 \geqq 0 \\ -u_1 - u_2 - u_3 - u_4 - u_5 - u_6 \geqq 1 \\ u_i \geqq 0 \quad (1 \leqq i \leqq 4) \end{cases} \tag{9}$$

The three types of "approximate solution" that have been discussed (for an overdetermined system of linear equations) are useful in different situations. Broadly speaking, an ℓ_∞ solution is preferred when the data are known to be accurate. An ℓ_2 solution is preferred when the data are contaminated with errors that are believed to conform to the normal probability distribution. The ℓ_1 solution is often used when the data are suspected of containing "wild" points — points that result from gross errors, such as incorrect placement of a decimal point. Additional information can be found in Rice and White [1964]. (The ℓ_2 problem is discussed in Chapter 12.)

Problems 11.3

1. Consider the inconsistent linear system

$$\begin{cases} 5x_1 + 2x_2 = 6 \\ x_1 + x_2 + x_3 = 2 \\ 7x_2 - 5x_3 = 11 \\ 6x_1 + 9x_3 = 9 \end{cases}$$

Write, with nonnegative variables,
(a) the equivalent linear programming problem for solving the system in the ℓ_1 sense.
(b) the equivalent linear programming problem for solving the system in the ℓ_∞ sense.

2. (Continuation) Do the same for the system

$$\begin{cases} 3x + y = 7 \\ x - y = 11 \\ x + 6y = 13 \\ -x + 3y = -12 \end{cases}$$

3. We want to find a polynomial p of degree n that approximates a function f as well as possible *from below*. That is, we want $0 \leqq f - p \leqq \varepsilon$ for minimum ε. Show how p could be obtained with reasonable precision by solving a linear programming problem.

Computer Problems 11.3

1. Obtain numerical answers to parts (a) and (b) of Problem 1 in the preceding set.
2. (Continuation) Repeat for Problem 2.
3. Find a polynomial of degree 4 that represents the function e^x in the following sense. Select 20 equally spaced points x_i in interval $[0, 1]$ and require the polynomial to minimize the expression $\max_{1 \leqq i \leqq 20} |e^{x_i} - p(x_i)|$. *Hint:* This is the same as solving 20 equations in five variables in the ℓ_∞ sense. The ith equation is $A + Bx_i + Cx_i^2 + Dx_i^3 + Ex_i^4 = e^{x_i}$ and the unknowns are A, B, C, D, and E.

Supplementary Problems (Chapter 11)

1. Consider the following linear programming problem.

$$\text{maximize:} \quad 2x_1 + 2x_2 - 6x_3 - x_4$$

$$\text{constraints:} \begin{cases} 3x_1 + x_4 = 25 \\ x_1 + x_2 + x_3 + x_4 = 20 \\ 4x_1 + 6x_3 \geqq 5 \\ 2x_1 + 3x_3 + 2x_4 \geqq 0 \\ x_1 \geqq 0, \ x_2 \geqq 0, \ x_3 \geqq 0, \ x_4 \geqq 0, \end{cases}$$

Reformulate Problem 1 in (a) second primal form and (b) the dual problem.

2. Solve the following linear programming problem graphically.

$$\text{maximize:} \quad 3x_1 + 5x_2$$

$$\text{constraints:} \quad \begin{cases} x_1 \leqq 4 \\ x_2 \leqq 6 \\ 3x_1 + 2x_2 \leqq 18 \\ x_1 \geqq 0, \quad x_2 \geqq 0 \end{cases}$$

3. Solve the dual problem of Problem 2.
4. Show that the dual problem may be written

$$\begin{cases} \text{maximize: } \mathbf{b}^T\mathbf{y} \\ \text{constraints: } \mathbf{y}^T\mathbf{A} \geqq \mathbf{c}^T, \ \mathbf{y} \geqq 0 \end{cases}$$

5. Using standard techniques, put problem (P) on p. 258 into first primal form; then take the dual of it. What is the result?
6. Describe how the problem of minimizing the expression max $\{|x - y - 3|, |2x + y + 4|, |x + 2y - 7|\}$ can be solved by using a linear programming code.
7. Show how this problem can be solved by linear programming.

$$\text{minimize:} \quad |x - y|$$

$$\text{constraints:} \quad \begin{cases} x \leqq 3y \\ x \geqq y \\ y \leqq x - 2 \end{cases}$$

8. Consider the linear programming problem

$$\text{minimize:} \quad x_1 + x_4 + 25$$

$$\text{constraints:} \quad \begin{cases} 2x_1 + 2x_2 + x_3 < 7 \\ 2x_1 - 3x_2 + x_4 = 4 \\ x_2 - x_4 > 1 \\ 3x_2 - 8x_3 + x_4 = 5 \\ x_1, x_2, x_3, x_4 \geqq 0 \end{cases}$$

Write in matrix-vector form (a) the dual problem and (b) the second primal problem.

12

Smoothing of Data and the Method of Least Squares

Surface tension S in a liquid is known to be a linear function of temperature T. For a particular liquid, measurements have been made of the surface tension at certain temperatures. The results were

T	0	10	20	30	40	80	90	100
S	68.0	67.1	66.4	65.6	64.6	61.8	61.0	60.0

How can the most probable values of the constants in the equation

$$S = aT + b$$

be determined? Methods for solving such problems are developed in this chapter.

12.1 Method of Least Squares

In experimental, social, and behavioral sciences an experiment or survey often produces a mass of data. To interpret the data, the investigator may resort to graphical methods. For instance, an experiment in physics might produce a numerical table of the form

x	x_1	x_2	x_3	$\cdots$	x_m
y	y_1	y_2	y_3	$\cdots$	y_m

(1)

and from it m points on a graph could be plotted. Suppose that the resulting graph looked like Fig. 12.1. A reasonable tentative conclusion is that the underlying function is *linear* and that the failure of the points to fall *precisely* on a straight line is due to experimental errors. Proceeding on this assumption — or if theoretical reasons existed for believing that the function is linear — the next step is to determine the correct function. Assuming that $y = ax + b$, what are the coefficients a and b? Thinking geometrically, we ask

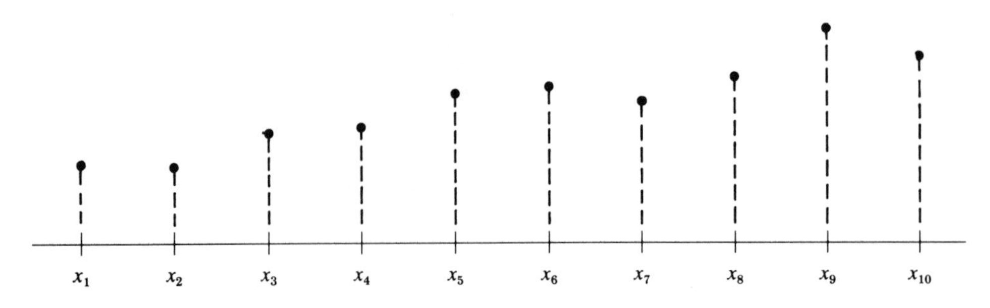

Figure 12.1

"What line most nearly passes through the ten points plotted?"

To answer this question, suppose that a guess is made about the correct values of a and b, which is equivalent to deciding on a specific line to represent the data. In general, the data points will not fall on the line $y = ax + b$. If by chance the kth datum falls on the line, then $ax_k + b - y_k = 0$. If it does not, then there is a discrepancy or *error* of magnitude $|ax_k + b - y_k|$. The total error for all m points is therefore $\sum_{k=1}^{m} |ax_k + b - y_k|$. This is a function of a and b, and it would be reasonable to choose a and b so that the function assumes its minimum value. This problem is an example of ℓ_1 *approximation* and can be solved by the techniques of linear programming, a subject dealt with in Chapter 11.

In practice, it is common to minimize a different function of a and b:

$$\phi(a, b) = \sum_{k=1}^{m} (ax_k + b - y_k)^2 \tag{2}$$

This function is suitable because of statistical considerations. Explicitly, if the errors follow a *normal probability distribution,* then the minimization of ϕ produces a best estimate of a and b. This is called an ℓ_2 - problem.

Let us try to make $\phi(a, b)$ a minimum. By calculus, the conditions

$$\frac{\partial \phi}{\partial a} = 0 \qquad \frac{\partial \phi}{\partial b} = 0$$

(partial derivatives of ϕ with respect to a and b, respectively) are *necessary* at the minimum. Taking the derivatives in (2), we obtain

$$\begin{cases} \displaystyle\sum_{k=1}^{m} 2(ax_k + b - y_k)x_k = 0 \\ \displaystyle\sum_{k=1}^{m} 2(ax_k + b - y_k) \quad = 0 \end{cases}$$

This is a pair of simultaneous linear equations in the unknowns a and b. Called the *normal equations,* they can be written

$$\begin{cases} \left(\sum_{k=1}^{m} x_k^2\right) a + \left(\sum_{k=1}^{m} x_k\right) b = \sum_{k=1}^{m} y_k x_k \\ \left(\sum_{k=1}^{m} x_k\right) a + \qquad mb = \sum_{k=1}^{m} y_k \end{cases} \tag{3}$$

Here, of course, $\sum_{k=1}^{m} 1 = m$. The explicit solution of Eq. (3) is easily written down. (See Problem 8.) This is an example of the *least squares* procedure in a simple case.

As a concrete example, the table

x	1.0	2.0	2.5	3.0
y	3.7	4.1	4.3	5.0

leads to the system of two equations

$$\begin{cases} 20.25a + 8.5b = 37.65 \\ 8.5a + \quad 4b = 17.1 \end{cases}$$

whose solution is $a = 0.6$ and $b = 3.0$. The value of ϕ, which ideally should be zero, is 0.1.

The method of least squares is not restricted to linear (first-degree) polynomials, nor to any specific functional form. Suppose, for instance, that we want to fit a table of values (x_k, y_k), $k = 1, 2, \ldots, m$, by a function of the form

$$a \ln x + b \cos x + ce^x$$

in the least squares sense. We consider the function

$$\phi(a, b, c) = \sum_{k=1}^{m} (a \ln x_k + b \cos x_k + ce^{x_k} - y_k)^2$$

and set $\partial\phi/\partial a = 0$, $\partial\phi/\partial b = 0$, $\partial\phi/\partial c = 0$. This results in the following three normal equations.

$$a \sum_{k=1}^{m} (\ln x_k)^2 + b \sum_{k=1}^{m} (\ln x_k)(\cos x_k) + c \sum_{k=1}^{m} (\ln x_k)e^{x_k} = \sum_{k=1}^{m} y_k \ln x_k$$

$$a \sum_{k=1}^{m} (\ln x_k)(\cos x_k) + b \sum_{k=1}^{m} (\cos x_k)^2 + c \sum_{k=1}^{m} (\cos x_k)e^{x_k} = \sum_{k=1}^{m} y_k \cos x_k$$

$$a \sum_{k=1}^{m} (\ln x_k)e^{x_k} + b \sum_{k=1}^{m} (\cos x_k)e^{x_k} + c \sum_{k=1}^{m} (e^{x_k})^2 = \sum_{k=1}^{m} y_k e^{x_k}$$

For the table

x	0.24	0.65	0.95	1.24	1.73	2.01	2.23	2.52	2.77	2.99
y	0.23	−0.26	−1.10	−0.45	0.27	0.10	−0.29	0.24	0.56	1.00

we obtain the 3 × 3 system

$$\begin{cases} 6.79410a - 5.34749b + 63.25889c = 1.61627 \\ -5.34749a + 5.10842b - 49.00859c = -2.38271 \\ 63.25889a - 49.00859b + 1002.50650c = 26.77277 \end{cases}$$

which has the solution $a = -1.04103$, $b = -1.26132$, $c = 0.03073$. So the curve

$$y = -1.04103 \ln x - 1.26132 \cos x + 0.03073e^x$$

has the required form and fits the table in the least squares sense. The value of $\phi(a, b, c)$ is 0.92557.

The principle of least squares, illustrated in these two simple cases, can be extended without involving any new ideas to general linear families of functions. Suppose that the data (1) are thought to conform to a relationship like

$$y = \sum_{j=1}^{n} c_j g_j(x) \tag{4}$$

in which the functions $g_1, \ldots, g_n$ are known and held fixed. The coefficients $c_1, \ldots, c_n$ are to be determined according to the principle of least squares. In other words, we define the expression

$$\phi(c_1, c_2, \ldots, c_n) = \sum_{k=1}^{m} \left[\sum_{j=1}^{n} c_j g_j(x_k) - y_k \right]^2 \tag{5}$$

and select the coefficients to make it as small as possible. The expression $\phi(c_1, \ldots, c_n)$ is, of course, the sum of the squares of the errors associated with each entry (x_k, y_k) in the given table.

Proceeding as before, we write down as necessary conditions for the minimum the n equations

$$\frac{\partial \phi}{\partial c_i} = 0 \qquad (1 \leqq i \leqq n)$$

These partial derivatives are obtained from Eq. (5). Indeed,

$$\frac{\partial \phi}{\partial c_i} = \sum_{k=1}^{m} 2 \left[\sum_{j=1}^{n} c_j g_j(x_k) - y_k \right] g_i(x_k) \qquad (1 \leqq i \leqq n)$$

When set equal to zero, the resulting equations can be rearranged as

$$\sum_{j=1}^{n} \left[\sum_{k=1}^{m} g_i(x_k) g_j(x_k) \right] c_j = \sum_{k=1}^{m} y_k g_i(x_k) \qquad (1 \leqq i \leqq n) \tag{6}$$

They are the *normal equations* in this situation and serve to determine the best values of the parameters $c_1, \ldots, c_n$. The normal equations are linear in $c_1, \ldots, c_n$, and thus, in principle, they can be solved by the method of Gaussian elimination (Chapter 6).

In practice, the normal equations may be difficult to solve if care is not taken in choosing the basic functions $g_1, \ldots, g_n$. Not only should these functions be appropriate to the problem at hand, but they should also be well conditioned for numerical work. We shall elaborate on this aspect of the problem in the next section.

Problems 12.1

1. Using the method of least squares, find the constant function that best fits the data shown.

x	1	2	3
y	$1\frac{1}{4}$	$1\frac{1}{3}$	$\frac{5}{12}$

2. Determine the *constant* function c that is produced by the least squares theory applied to table (1). Does the resulting formula involve points x_k in any way? Apply your general formula to Problem 1.
3. Find an equation of form $y = ae^{x^2} + bx^3$ that best fits the points $(-1, 0)$, $(0, 1)$, $(1, 2)$ in the least squares sense.
4. Suppose that the x points in table (1) are symmetrically situated about zero on the x axis. In this case, there is an especially simple formula for the line that best fits the points. Find it.
5. Find the equation of a parabola of form $y = ax^2 + b$ that best represents these data.

x	-1	0	1
y	3.1	0.9	2.9

Use the method of least squares.

6. Suppose that table (1) is known to conform to a function like $y = x^2 - x + c$. What value of c is obtained by the least squares theory?
7. Suppose that table (1) is thought to be represented by a function $y = c \log x$. If so, what value for c emerges from least squares theory?
8. Show that the solution of Eq. (3) is given by

$$a = \frac{1}{d}\left(m \sum_{k=1}^{m} x_k y_k - \sum_{k=1}^{m} x_k \sum_{k=1}^{m} y_k\right)$$

$$b = \frac{1}{d}\left(\sum_{k=1}^{m} x_k^2 \sum_{k=1}^{m} y_k - \sum_{k=1}^{m} x_k \sum_{k=1}^{m} x_k y_k\right)$$

$$d = m \sum_{k=1}^{m} x_k^2 - \left(\sum_{k=1}^{m} x_k\right)^2$$

9. (Continuation) How do we know that divisor d is not zero? In fact, show that d is positive for $m \geqq 2$. *Hint:* Show that

$$d = \sum_{k=2}^{m} \sum_{l=1}^{k-1} (x_k - x_l)^2$$

by induction on m. The Cauchy-Schwarz inequality can also be used to prove that $d > 0$.

10. (Continuation) Show that a and b can also be computed as follows.

$$\hat{x} = \frac{1}{m} \sum_{k=1}^{m} x_k \qquad \hat{y} = \frac{1}{m} \sum_{k=1}^{m} y_k$$

$$c = \sum_{k=1}^{m} (x_k - \hat{x})^2 \qquad a = \frac{1}{c} \sum_{k=1}^{m} (x_k - \hat{x})(y_k - \hat{y})$$

$$b = \hat{y} - a\hat{x}$$

Hint: Show that $d = mc$.

11. How do we know that coefficients $c_1, \ldots, c_n$ that satisfy the equations obtained by setting the partial derivatives to zero do not lead to a *maximum* in the function defined by Eq. (5)?
12. If table (1) is thought to conform to a relationship $y = \log(cx)$, what is the value of c obtained by the method of least squares?

Computer Problems 12.1

1. Write a subroutine that sets up the normal equations (6). Using that subroutine and other subroutines, such as GAUSS and SOLVE from Chapter 6, verify the solution given for the example involving $\ln x$, $\cos x$, and e^x.
2. Write a subroutine that fits a straight line to table (1).

12.2 Orthogonal Systems and Chebyshev Polynomials

Once functions $g_1, \ldots, g_n$ of Sec. 12.1 are chosen, the least squares problem can be interpreted as follows. The set of all functions g that can be expressed as linear combinations of $g_1, \ldots, g_n$ is a vector space $\mathcal{G}$. (Familiarity with vector spaces is not essential to understanding the discussion here.) In symbols,

$$\mathcal{G} = \left\{ g : \text{there exist } c_1, \ldots, c_n \text{ such that } g(x) = \sum_{j=1}^{n} c_j g_j(x) \right\}$$

The function sought in the least squares problem is thus an element of the vector space $\mathcal{G}$. The functions $g_1, \ldots, g_n$ form a *basis* for $\mathcal{G}$ if they are not linearly dependent. However, a given vector space has many different bases, and they can differ drastically in their numerical properties.

Let us turn our attention away from the given basis $\{g_1, \ldots, g_n\}$ to the vector space $\mathcal{G}$ generated by that basis. Without changing $\mathcal{G}$, we ask "What basis for $\mathcal{G}$ should be chosen for numerical work?" In the present problem the principal numerical task is to solve the normal equations

$$\sum_{j=1}^{n} \left[\sum_{k=1}^{m} g_i(x_k) g_j(x_k) \right] c_j = \sum_{k=1}^{m} y_k g_i(x_k) \qquad (1 \leqq i \leqq n) \tag{1}$$

The nature of this system obviously depends on the basis $\{g_1, \ldots, g_n\}$. We want these equations to be *easily* solved or to be capable of being *accurately* solved. The ideal situation is when the coefficient matrix in Eq. (1) is the identity matrix. This occurs if the basis $\{g_1, \ldots, g_n\}$ has the property of *orthonormality:*

$$\sum_{k=1}^{m} g_i(x_k) g_j(x_k) = \delta_{ij} = \begin{cases} 1 & \text{if } i = j \\ 0 & \text{if } i \neq j \end{cases}$$

In the presence of this property, Eq. (1) simplifies dramatically to

$$c_j = \sum_{k=1}^{m} y_k g_j(x_k) \qquad (1 \leqq j \leqq n)$$

which is no longer a system of equations to be solved but rather an explicit formula for the coefficients c_j.

Under rather general conditions the space $\mathcal{G}$ has a basis that is orthonormal in the sense just described. A procedure known as the *Gram–Schmidt process* can be used to obtain such a basis. There are some situations in which the effort of obtaining an orthonormal basis is justified, but simpler procedures will often suffice. We shall describe one procedure now.

Remember that our goal is to make Eq. (1) well-disposed for numerical solution. We want to avoid any matrix of coefficients that involves the difficulties encountered in connection with the Hilbert matrix (see Problem 1, Sec. 6.2). This objective is met if the basis for the space $\mathcal{G}$ is well-chosen.

Consider one of the important examples of the least squares theory – when the space $\mathcal{G}$ consists of all polynomials of degree $\leqq n$. It may seem natural to use the following $n + 1$ functions as a basis for $\mathcal{G}$:

$$g_0(x) = 1, \quad g_1(x) = x, \quad g_2(x) = x^2, \ldots, g_n(x) = x^n$$

Using this basis, we write a typical element of the space $\mathcal{G}$ in the form

$$g(x) = \sum_{j=0}^{n} c_j g_j(x) = \sum_{j=0}^{n} c_j x^j = c_0 + c_1 x + c_2 x^2 + \cdots + c_n x^n$$

This basis, however natural, is almost always a *poor* choice for numerical work. For many purposes, the Chebyshev polynomials (suitably defined for the interval involved) do form a *good* basis. Figure 12.2(a) gives an indication of why the simple polynomials x^n do not form a good basis for numerical work. These functions are too much alike. If a function g is given, and if we wish to analyze it into components, $g(x) = \sum_{j=0}^{n} c_j x^j$, it is difficult to determine the coefficients c_j precisely. Figure 12.2(b) shows a few of the Chebyshev polynomials; they are quite different from one another.

For simplicity, assume that the points in our least squares problem have the property

$$-1 = x_1 < x_2 < x_3 < \ldots < x_m = 1$$

Then the *Chebyshev polynomials* for the interval $[-1, 1]$ can be used. The traditional notation is

$$\begin{aligned} T_0(x) &= 1 \\ T_1(x) &= x \\ T_2(x) &= 2x^2 - 1 \\ T_3(x) &= 4x^3 - 3x \\ T_4(x) &= 8x^4 - 8x^2 + 1 \end{aligned}$$

etc.

A recursive formula for these polynomials is

$$T_j(x) = 2xT_{j-1}(x) - T_{j-2}(x) \qquad (j \geqq 2) \tag{2}$$

This formula, together with the equations $T_0(x) = 1$ and $T_1(x) = x$, provides a formal definition of the Chebyshev polynomials.

Linear combinations of Chebyshev polynomials are easy to evaluate because a special nested multiplication algorithm applies. In order to describe this procedure, consider an arbitrary linear combination of $T_0, T_1, \ldots, T_n$:

$$g(x) = \sum_{j=0}^{n} c_j T_j(x)$$

An algorithm to compute $g(x)$ for any given x goes as follows.

$$\begin{cases} w_{n+2} = w_{n+1} = 0 \\ w_k = c_k + 2xw_{k+1} - w_{k+2} \qquad (k=n,\ n-1,\ n-2,\ \ldots,\ 0) \\ g(x) = w_0 - xw_1 \end{cases} \tag{3}$$

In order to see that this algorithm actually produces $g(x)$, we write down the series for g, shift some indices, and use formulas (2) and (3).

$$g(x) = \sum_{k=0}^{n} c_k T_k(x)$$

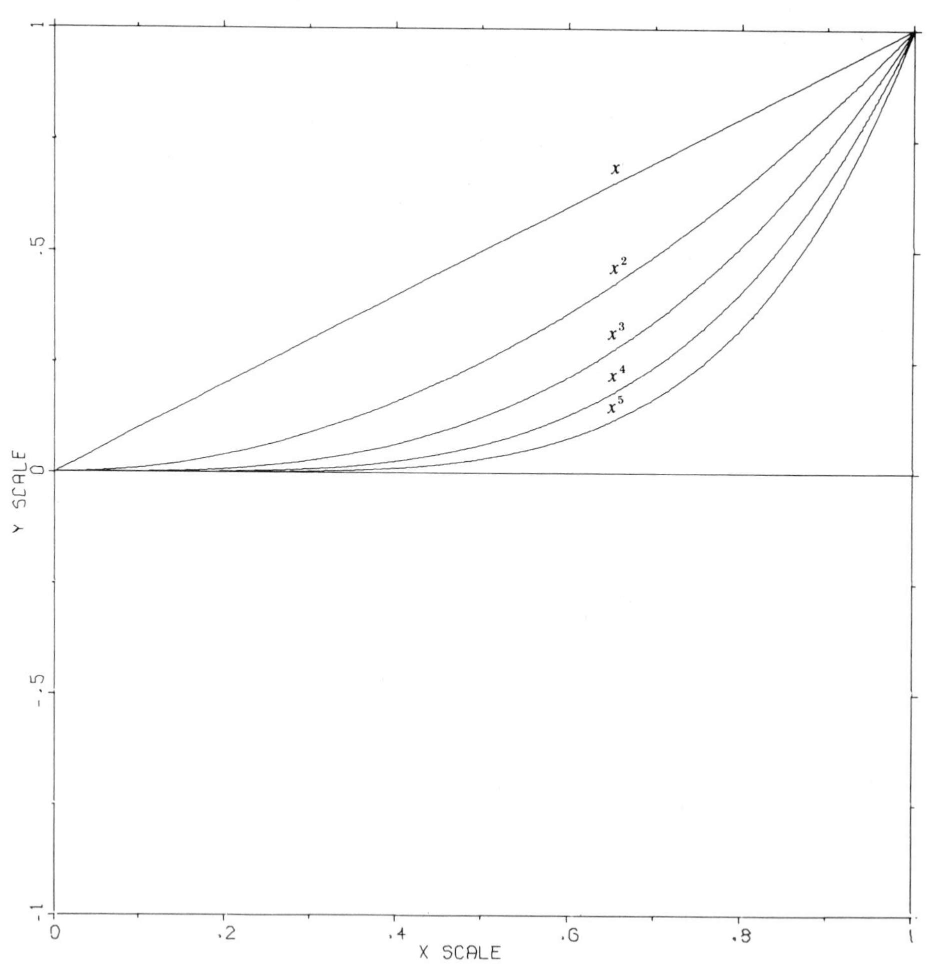

Figure 12.2(a) Polynomials x^n.

$$= \sum_{k=0}^{n} (w_k - 2xw_{k+1} + w_{k+2})T_k$$

$$= \sum_{k=0}^{n} w_k T_k - 2x \sum_{k=0}^{n} w_{k+1} T_k + \sum_{k=0}^{n} w_{k+2} T_k$$

$$= \sum_{j=0}^{n} w_j T_j - 2x \sum_{j=1}^{n+1} w_j T_{j-1} + \sum_{j=2}^{n+2} w_j T_{j-2}$$

$$= \sum_{j=0}^{n} w_j T_j - 2x \sum_{j=1}^{n} w_j T_{j-1} + \sum_{j=2}^{n} w_j T_{j-2}$$

$$= w_0 T_0 + w_1 T_1 + \sum_{j=2}^{n} w_j T_j - 2xw_1 T_0 - 2x \sum_{j=2}^{n} w_j T_{j-1} + \sum_{j=2}^{n} w_j T_{j-2}$$

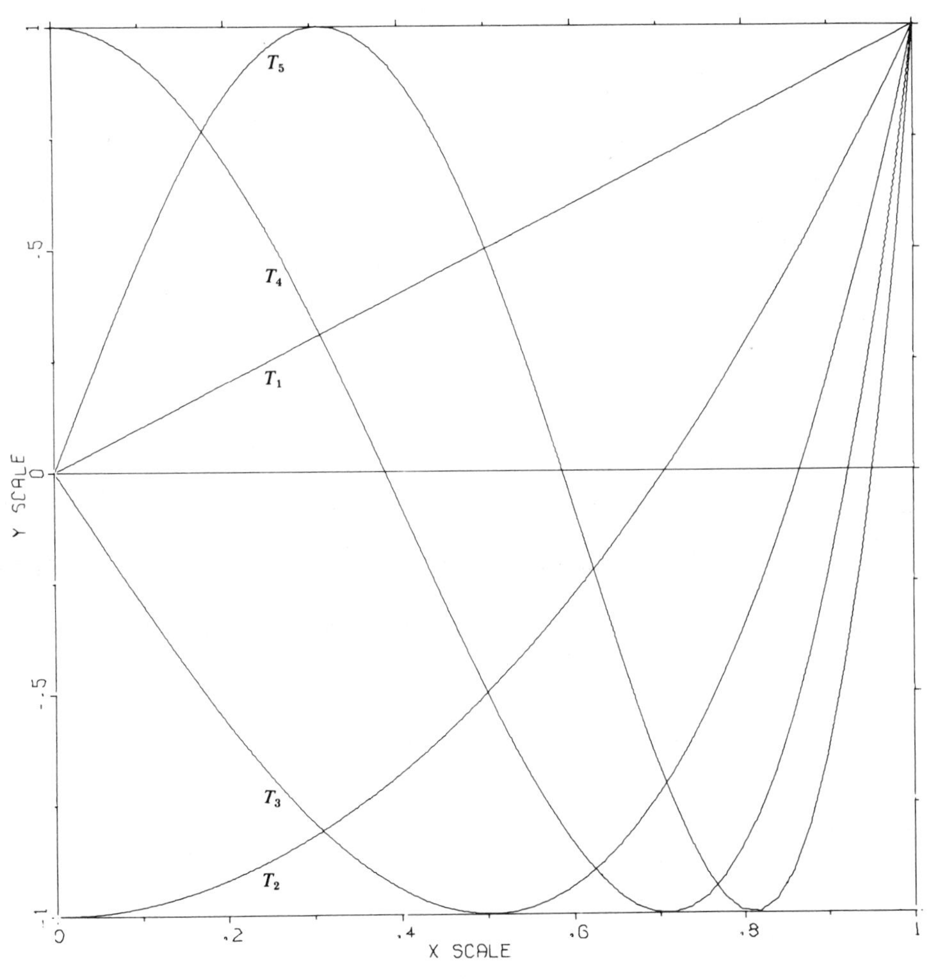

Figure 12.2(b) Chebyshev polynomials.

$$= w_0 + xw_1 - 2xw_1 + \sum_{j=2}^{n} w_j(T_j - 2xT_{j-1} + T_{j-2})$$

$$= w_0 - xw_1$$

In general, it is best to arrange the data so that all the abscissas $\{x_i\}$ lie in the interval $[-1, 1]$. Then if the first few Chebyshev polynomials are used as a basis for the polynomials, the normal equations should be reasonably "well conditioned." We have not given a technical definition of this term; it can be interpreted informally to mean that Gaussian elimination with pivoting produces an accurate solution to the normal equations.

If the original data do not satisfy the conditions min $\{x_k\} = -1$ and max $\{x_k\} = 1$ but lie instead in another interval $[a, b]$, then the change of variable $x = \frac{1}{2}(b - a)z + \frac{1}{2}(a + b)$ produces a variable z that traverses $[-1, 1]$ as x traverses $[a, b]$.

Here is an outline of a procedure, based on the preceding discussion, that produces a polynomial of degree $\leqq n$ that best fits a given table of values (x_k, y_k) $(1 \leqq k \leqq m)$. Here m is usually much greater than n.

1. Find the smallest interval $[a, b]$ containing all the x_k. Thus let $a = \min\{x_k\}$ and $b = \max\{x_k\}$.
2. Make a transformation to interval $[-1, 1]$ by defining $z_k = (2x_k - a - b)/(b - a)$, $1 \leqq k \leqq m$.
3. Decide on the value of n to be used. In this situation, 8 or 10 would be a large value for n.
4. Using Chebyshev polynomials as a basis, generate the $(n + 1) \times (n + 1)$ normal equations

$$\sum_{j=0}^{n} \left[\sum_{k=1}^{m} T_i(z_k)T_j(z_k) \right] c_j = \sum_{k=1}^{m} y_k T_i(z_k) \qquad (0 \leqq i \leqq n) \tag{4}$$

5. Use an equation-solving subroutine to solve the normal equations for coefficients c_0, $c_1, \ldots, c_n$ in the function $g = \sum_{j=0}^{n} c_j T_j$.
6. The polynomial sought is $g((2x - a - b)/(b - a))$.

The details of step 4 are as follows. Begin by introducing a doubly subscripted variable $d_{jk} = T_{j-1}(z_k)$, where $1 \leqq k \leqq m$ and $1 \leqq j \leqq n + 1$. The matrix $\mathbf{D} = (d_{jk})$ can be computed efficiently by using the recursive definition of the Chebyshev polynomials, Eq. (2), as in the following segment of code.

```
      DO 2 K = 1,M
      ZK = Z(K)
      D(1,K) = 1.0
      D(2,K) = ZK
      DO 2 J = 3,N+1
    2 D(J,K) = 2.0*ZK*D(J-1,K) - D(J-2,K)
```

The normal equations have a coefficient matrix $\mathbf{A} = (a_{ij})$ and a right-hand side $\mathbf{b} = (b_i)$ given by

$$a_{ij} = \sum_{k=1}^{m} T_{i-1}(z_k)T_{j-1}(z_k) = \sum_{k=1}^{m} d_{ik}d_{jk} \qquad (1 \leqq i, j \leqq n + 1)$$

$$b_i = \sum_{k=1}^{m} y_k T_{i-1}(z_k) = \sum_{k=1}^{m} y_k d_{ik} \qquad (1 \leqq i \leqq n + 1) \tag{5}$$

The code to calculate **A** and **b** is as follows.

```
      DO 5 I = 1,N+1
      S = 0.0
      DO 3 K = 1,M
    3 S = S + Y(K)*D(I,K)
      B(I) = S
      DO 5 J = I,N+1
      S = 0.0
      DO 4 K = 1,M
    4 S = S + D(I,K)*D(J,K)
      A(I,J) = S
    5 A(J,I) = S
```

To fit data with polynomials, other methods exist that employ systems of polynomials tailor-made for a given set of abscissae. The method outlined above is, however, simple and direct.

Problems 12.2

1. Let $g_1, g_2, \ldots, g_n$ be a set of functions such that $\sum_{k=1}^{m} g_i(x_k)g_j(x_k) = 0$ if $i \neq j$. What linear combination of these functions best fits the data of table (1) in Sec. 12.1?
2. Consider a system of polynomials $g_0, g_1, \ldots, g_n$ defined by $g_0(x) = 1$, $g_1(x) = x - 1$, $g_j(x) = 3xg_{j-1}(x) + 2g_{j-2}(x)$. Develop an efficient algorithm for computing values of function $f(x) = \sum_{j=0}^{n} c_j g_j(x)$.
3. Show that $\cos n\theta = 2 \cos \theta \cos (n - 1)\theta - \cos (n - 2)\theta$. *Hint:* Use the familiar identity $\cos (A \mp B) = \cos A \cos B \pm \sin A \sin B$.
4. (Continuation) Show that if $f_n(x) = \cos (n \arccos x)$, then $f_0(x) = 1$, $f_1(x) = x$, and $f_n(x) = 2xf_{n-1}(x) - f_{n-2}(x)$.
5. (Continuation) Show that an alternate definition of Chebyshev polynomials is $T_n(x) = \cos (n \arccos x)$ for $-1 \leqq x \leqq 1$.
6. (Continuation) Give a one-line proof that $T_n(T_m(x)) = T_{nm}(x)$.
7. (Continuation) Show that $|T_n(x)| \leqq 1$ for x in the interval $[-1, 1]$.
8. Define $g_k(x) = T_k(\frac{1}{2}x + \frac{1}{2})$. What recursive relation do these functions satisfy?
9. Show that $T_0, T_2, T_4, \ldots$ are even and that $T_1, T_3, \ldots$ are odd functions. Recall that an even function satisfies the equation $f(x) = f(-x)$; an odd function satisfies the equation $f(x) = -f(-x)$.
10. Count the number of operations involved in the algorithm used to compute $g(x) = \sum_{j=0}^{n} c_j T_j(x)$.
11. Show that the algorithm for computing $g(x) = \sum_{j=0}^{n} c_j T_j(x)$ can be modified to read

$$\begin{cases} w_{n-1} = c_{n-1} + 2xc_n \\ w_k = c_k + 2xw_{k+1} - w_{k+2} \qquad (k = n - 2, n - 3, \ldots, 1) \\ g(x) = c_0 + xw_1 - w_2 \end{cases}$$

thus making w_{n+2}, w_{n+1}, and w_0 unnecessary.

12. (Continuation) Count the operations for this algorithm.

Computer Problems 12.2

1. Carry out an experiment in data smoothing as follows. Start with a polynomial of modest degree, say 7. Compute 100 values of this polynomial at random points in the interval $[-1, 1]$. Perturb these values by adding random numbers chosen from a small interval, say $[-\frac{1}{8}, \frac{1}{8}]$. Try to recover the polynomial from these perturbed values by using the method of least squares.
2. Write `FUNCTION CHEB(N,X)` for evaluating $T_n(x)$. Use the recursive formula satisfied by Chebyshev polynomials. Do not use a subscripted variable. Test the program on these 15 cases: $n = 0, 1, 3, 6, 12$ and $x = 0, -1, 0.5$. [What is $T_n(0.5)$?]
3. Write `FUNCTION TCHEB(N,X,Y)` to calculate $T_0(x), T_1(x), \ldots, T_n(x)$ and store these numbers in the **Y** array. Use your subroutine, together with suitable plotting routines, to obtain graphs of T_0, T_1, T_2, T_8 on $[-1, 1]$.
4. Write `FUNCTION FX(N,C,X)` for evaluating $f(x) = \sum_{j=0}^{n} c_j T_j(x)$. Test your subroutine by means of the known formula $\sum_{n=0}^{\infty} t^n T_n(x) = (1 - tx)/(1 - 2tx + t^2)$, valid for $|t| < 1$. If $|t| \leqq \frac{1}{2}$, then only a few terms of the series are needed to give full machine precision. Add terms in ascending order of magnitude.
5. Obtain a graph of T_n for some reasonable value of n by means of the following idea. Generate 100 equally spaced angles θ_i in interval $[0, \pi]$. Define $x_i = \cos\theta_i$ and $y_i = T_n(x_i) = \cos(n \arccos x_i) = \cos n\theta_i$. Give the table of (x_i, y_i) to a suitable plotting routine.
6. Program a subroutine to carry out the procedure outlined in the text for fitting a table by a linear combination of Chebyshev polynomials. Test it in the manner of Computer Problem 1, first by using an unperturbed polynomial. Find out experimentally how large n can be in this process before roundoff errors become serious.
7. Select a modest value of n, say $5 \leqq n \leqq 20$. Select $m > 2n$ and define $x_k = \cos[(2k - 1)\pi/(2m)]$. Compute and print the matrix **A** whose elements are

$$a_{ij} = \sum_{k=1}^{m} T_{i-1}(x_k)T_{j-1}(x_k) \qquad (1 \leqq i, j \leqq n)$$

Interpret the results in terms of the least squares polynomial fitting problem.

12.3 Other Examples of the Least Squares Principle

The principle of least squares is also used in other situations. In one of these we attempt to "solve" an inconsistent system of linear equations of the form

$$\sum_{j=1}^{n} a_{kj}x_j = b_k \qquad (1 \leqq k \leqq m) \tag{1}$$

in which $m > n$. Here there are m equations but only n unknowns. If a given n tuple $(x_1, \ldots, x_n)$ is substituted on the left, the discrepancy between the two sides of the kth equation is termed the kth *residual*. Ideally, of course, all residuals should be zero. If it is not possible to select $(x_1, \ldots, x_n)$ so as to make all residuals zero, system (1) is said to be *inconsistent* or *incompatible*. In this case, an alternative is to minimize the sum of the squares of the residuals. So we are led to minimize the expression

$$\phi(x_1, \ldots, x_n) = \sum_{k=1}^{m} \left(\sum_{j=1}^{n} a_{kj}x_j - b_k \right)^2 \tag{2}$$

by making an appropriate choice of $(x_1, \ldots, x_n)$. Proceeding as before, we take partial derivatives with respect to x_i and set them equal to zero, arriving thereby at the normal equations

$$\sum_{j=1}^{n} \left(\sum_{k=1}^{m} a_{ki} a_{kj} \right) x_j = \sum_{k=1}^{m} b_k a_{ki} \qquad (1 \leqq i \leqq n) \tag{3}$$

This is a linear system of just n equations involving unknowns $x_1, \ldots, x_n$. It can be shown that this sytem is consistent, provided that the column vectors in the original coefficient array are linearly independent. System (3) can be solved, for instance, by Gaussian elimination. The solution of system (3) is then a best approximate solution of Eq. (1) in the least squares sense.

Special methods have been devised for the problem just discussed. Generally they gain in precision over the simple approach outlined here. One such algorithm for solving the least squares problem $\mathbf{Ax} = \mathbf{b}$ begins by factoring $\mathbf{A} = \mathbf{QR}$, where $\mathbf{Q}$ is an $m \times n$ matrix satisfying $\mathbf{Q}^T\mathbf{Q} = \mathbf{I}$ and $\mathbf{R}$ is an $n \times n$ matrix satisfying $r_{ii} > 0$, $r_{ij} = 0$ for $j < i$. Then the least squares solution is obtained by an algorithm called *the modified Gram–Schmidt process*. The reader is referred to Stewart [1973] for details.

Another important example of the principle of least squares occurs in fitting or approximating functions on *intervals* rather than *discrete* sets. For example, a given function f defined on an interval $[a, b]$ may be required to be approximated by a function like

$$g(x) = \sum_{j=1}^{n} c_j g_j(x)$$

It is natural, then, to attempt to minimize the expression

$$\phi(c_1, \ldots, c_n) = \int_a^b [g(x) - f(x)]^2 \, dx \tag{4}$$

by choosing coefficients appropriately. In some applications, it is desirable to force functions g and f into better agreement in certain parts of the interval. For this purpose, we can modify Eq. (4) by including a positive *weight* function $w(x)$, which can, of course, be one if all parts of the interval are to be treated the same. The result is

$$\phi(c_1, \ldots, c_n) = \int_a^b [g(x) - f(x)]^2 w(x) \, dx$$

The minimum of ϕ is again sought by differentiating with respect to each c_i and setting the partial derivatives equal to zero. The result is a system of normal equations.

$$\sum_{j=1}^{n} \left[\int_a^b g_i(x) g_j(x) w(x) \, dx \right] c_j = \int_a^b f(x) g_i(x) w(x) \, dx \qquad (1 \leqq i \leqq n) \tag{5}$$

Again it is a system of n linear equations in n unknowns $c_1, \ldots, c_n$ and can be solved by Gaussian elimination. Earlier remarks about choosing a good basis apply here also. The ideal situation is to have functions $g_1, \ldots, g_n$ that have the "orthogonality" property

$$\int_a^b g_i(x) g_j(x) w(x) \, dx = 0 \qquad (i \neq j) \tag{6}$$

Many such orthogonal systems have been developed over the years. Chebyshev polynomials form one such system — namely,

$$\int_{-1}^{1} T_{i-1}(x)T_{j-1}(x)(1 - x^2)^{-1/2}\,dx = \begin{cases} 0 & \text{if } i \neq j \\ \dfrac{\pi}{2} & \text{if } i = j > 1 \\ \pi & \text{if } i = j = 1 \end{cases}$$

The weight function $(1 - x^2)^{-1/2}$ assigns heavy weight to the ends of the interval $[-1, 1]$.

If a sequence of nonzero functions $g_1, g_2, \ldots$ is orthogonal according to Eq. (6), then the sequence $c_1^{-1}g_1, c_2^{-1}g_2, \ldots$ will be orthonormal for appropriate positive real numbers c_j — namely,

$$c_j = \left\{ \int_a^b [g_j(x)]^2 w(x)\,dx \right\}^{1/2}$$

As another example of the least squares principle, here is a nonlinear problem. Suppose that a table of points (x_k, y_k) is to be fitted by a function of the form

$$y = e^{cx}$$

Proceeding as before leads to the problem of minimizing the function

$$\phi(c) = \sum_{k=1}^{m} (e^{cx_k} - y_k)^2$$

The minimum occurs for a value of c such that

$$0 = \frac{\partial \phi}{\partial c} = \sum_{k=1}^{m} 2(e^{cx_k} - y_k)e^{cx_k}\,x_k$$

This equation can be rearranged slightly, but it is certainly nonlinear in c. One could contemplate solving it by Newton's method or the secant method. On the other hand, the problem of minimizing $\phi(c)$ could be attacked directly. Since there can be multiple roots in the normal equation and local minima in ϕ itself, a direct minimization of ϕ would be safer. This type of difficulty is typical of nonlinear least squares problems. Consequently, other methods of curve fitting are often preferred if the unknown parameters do not occur linearly in the problem.

Alternatively, this particular example can be "linearized" by a change of variables $z = \ln y$ and by considering

$$z = cx$$

The problem of minimizing the function

$$\phi(c) = \sum_{k=1}^{m} (cx_k - z_k)^2 \qquad z_k = \ln y_k$$

is easy and leads to

$$c = \frac{\sum_{k=1}^{m} z_k x_k}{\sum_{k=1}^{m} x_k}$$

This value of c is not the solution of the original problem but may be satisfactory in some applications.

The final example contains elements of linear and nonlinear theory. Suppose that an (x_k, y_k) table is given with m entries and that a functional relationship like

$$y = a \sin bx$$

is suspected. Can the least squares principle be used to obtain the appropriate values of the parameters a and b?

Notice that parameter b enters the function in a nonlinear way—a source of some difficulty, as will be seen. According to the principle of least squares, the parameters should be so chosen that the expression

$$\sum_{k=1}^{m} (a \sin bx_k - y_k)^2$$

is a minimum. The minimum is sought by differentiating this expression with respect to a and b and setting these partial derivatives equal to zero. The results are

$$\sum_{k=1}^{m} 2(a \sin bx_k - y_k) \sin bx_k = 0$$

$$\sum_{k=1}^{m} 2(a \sin bx_k - y_k) a x_k \cos bx_k = 0$$

If b were known, a could be obtained from either equation. The correct value of b is the one for which these corresponding two a values are identical. So each of the preceding equations should be solved for a and the results set equal to each other. This process leads to the equation

$$\frac{\sum_{k=1}^{m} y_k \sin bx_k}{\sum_{k=1}^{m} (\sin bx_k)^2} = \frac{\sum_{k=1}^{m} x_k y_k \cos bx_k}{\sum_{k=1}^{m} x_k \sin bx_k \cos bx_k}$$

which can now be solved for parameter b, using, for example, the bisection or the secant method. Then either side of this equation can be evaluated as the value of a.

Problems 12.3

1. Analyze the least squares problem of fitting data by a function of the form $y = x^c$.
2. Show that the Hilbert matrix (Problem 1, Sec. 6.2) arises in the normal equations when we minimize

$$\int_0^1 \left(\sum_{j=0}^{n} c_j x^j - f(x) \right)^2 dx$$

3. Find a function of the form $y = e^{cx}$ that best fits this table.

x	0	1
y	$\frac{1}{2}$	1

4. (Continuation) Repeat for the table below.

x	0	1
y	a	b

5. (Continuation) Repeat under the supposition that b is negative.
6. Show that the normal equation for the problem of fitting $y = e^{cx}$ to points $(1, -12)$, $(2, 7.5)$ has two real roots: $c = \ln 2$ or $c = 0$. Which value is correct for the fitting problem?
7. Consider the inconsistent system (1). Suppose that each equation has associated with it a positive number w_i indicating its relative importance or reliability. How should Eqs. (2) and (3) be modified to reflect this?
8. Determine the best approximate solution of the inconsistent system of linear equations

$$\begin{cases} 2x + 3y = 1 \\ x - 4y = -9 \\ 2x - y = -1 \end{cases}$$

in the least squares sense.
9. Find the constant c for which cx is the best approximation in the sense of least squares to the function $\sin x$ on interval $[0, \pi/2]$. Do the same for e^x on $[0, 1]$.
10. Analyze the problem of fitting a function $y = (c - x)^{-1}$ to a table of m points.
11. Show that the normal equations for the least squares solution of $\mathbf{Ax} = \mathbf{b}$ can be written $(\mathbf{A}^T\mathbf{A})\mathbf{x} = \mathbf{A}^T\mathbf{b}$.
12. Derive the normal equations given by system (5).

Computer Problems 12.3

1. Using the method suggested in the text, fit the data

x	0.1	0.2	0.3	0.4	0.5	0.6	0.7	0.8
y	0.6	1.1	1.6	1.8	2.0	1.9	1.7	1.3

by a function $y = a \sin bx$.
2. (Prony's Method, $n = 1$) To fit a table of the form

x	1	2	$\cdots$	m
y	y_1	y_2	$\cdots$	y_m

by the function $y = ab^x$, we can proceed as follows. If y is actually ab^x, then $y_k = ab^k$ and $y_{k+1} = by_k$ for $k = 1, \ldots, m - 1$. So we determine b by "solving" this system of equations, using the least squares method. Having found b, we find a by "solving" the equations $y_k = ab^k$ in the least squares sense. Write a program to carry out this procedure and test it on an artificial example.
3. Modify the procedure of Problem 2 to handle any case of equally spaced points.

4. A quick way of fitting a function of the form

$$f(x) \approx (a + bx)/(1 + cx)$$

is to apply the least squares method to the problem

$$f(x)(1 + cx) \approx a + bx$$

Use this technique to fit the world population data given here.

Year	Population (in Millions)
1000	340
1650	545
1800	907
1900	1610
1950	2509
1970	3650

Verify that the world population will become infinite between 5 and 6 P.M. on Monday, August 30, in the year 2010.

5. Write a subroutine that takes as input an $m \times n$ matrix $\mathbf{A}$ and an $m \times 1$ vector $\mathbf{b}$ and returns the least-squares solution of the system $\mathbf{Ax} = \mathbf{b}$.

Supplementary Problems (Chapter 12)

1. A table of values (x_k, y_k), $k = 1, \ldots, m$, is obtained from an experiment. When plotted on "semilogarithmic" graph paper, the points are found to lie nearly on a straight line, implying that $y \approx e^{ax+b}$. Suggest a simple procedure for obtaining parameters a and b.
2. In fitting a table of values to a function of the form $a + bx^{-1} + cx^{-2}$, we try to make each point lie on the curve. This leads to the equation $a + bx_k^{-1} + cx_k^{-2} = y_k$ $(1 \leqq k \leqq m)$. An equivalent equation is $ax_k^2 + bx_k + c = y_kx_k^2$ $(1 \leqq k \leqq m)$. Are the least squares problems for these systems of equations equivalent?
3. A table of points (x_k, y_k) is plotted and appears to lie on a hyperbola with form $y = (a + bx)^{-1}$. How can the *linear* theory of least squares be used to obtain good estimates of a and b?
4. An experiment involves two independent variables x and y and one dependent variable z. How can a function $z = a + bx + cy$ be fitted to the table of points (x_k, y_k, z_k)? Give the normal equations.
5. Find the quadratic polynomial that best fits the following data, in the sense of least squares.

x	-2	-1	0	1	2
y	2	1	1	1	2

6. Find the best function (in the least squares sense) of form $f(x) = a \sin \pi x + b \cos \pi x$ that fits these data points.

x	-1	$-\frac{1}{2}$	0	$\frac{1}{2}$	1
y	-1	0	1	2	1

7. Consider $f(x) = e^{2x}$ over $[0, \pi]$. We wish to approximate the function by a trigonometric polynomial of the form $p(x) = a + b \cos(x) + c \sin(x)$. Determine the linear system to be solved for determining the best least squares fit of p to f.

8. Show that in every least squares problem the normal equations have a symmetric coefficient matrix.
9. What straight line best fits the data

x	1	2	3	4
y	0	1	1	2

in the least squares sense?
10. In analytic geometry we learn that the distance from a point (x_0, y_0) to a line represented by the equations $ax + by = c$ is $(ax_0 + by_0 - c)(a^2 + b^2)^{-1/2}$. Determine a straight line that fits a table of data points (x_i, y_i), $1 \leqq i \leqq n$, in such a way that the sum of the squares of the distances from the points to the line is minimized.
11. Show that if a straight line is fitted to a table (x_i, y_i) by the method of least squares, then the line will pass through the point (x^*, y^*), where x^* and y^* are the average of the x's and y's, respectively.
12. Viscosity V of a liquid is known to vary with temperature according to a quadratic law $V = a + bT + cT^2$. Find the best values of a, b, c from this table.

T	1	2	3	4	5	6	7
V	2.31	2.01	1.80	1.66	1.55	1.47	1.41

13

Systems of Ordinary Differential Equations

A simple model to account for the way in which two different animal species sometimes react is the "predator-prey" model. If $u(t)$ is the number of individuals in the predator species and $v(t)$ the number of individuals in the prey species, then under suitable simplifying assumptions and with appropriate constants a, b, c, d

$$\begin{cases} \dfrac{du}{dt} = au(v + b) \\ \dfrac{dv}{dt} = cv(u + d) \end{cases}$$

This is a pair of nonlinear ordinary differential equations governing the populations of the two species (as functions of time t). In this chapter numerical procedures are developed for solving such problems.

13.1 Methods for First-Order Systems

Mar 9

In Chapter 8 ordinary differential equations were considered in the simplest context. That is, we restricted our attention to a single differential equation of the first order with an accompanying auxiliary condition. Scientific and technological problems often lead to more complicated situations, however. The next degree of complication occurs with *systems* of several first-order equations.

For example, the sun and the nine planets form a system of "particles" moving under the jurisdiction of Newton's law of gravitation. The position vectors of the planets constitute a system of 27 functions, and the Newtonian laws of motion can be written, then, as a system of 54 first-order ordinary differential equations. In principle, the past and future positions of the planets can be obtained by solving these equations numerically.

Taking an example of more modest scope, we consider two equations with two auxiliary conditions. Let $x = x(t)$ and $y = y(t)$ be two functions subject to the system

$$\begin{cases} x'(t) = x(t) - y(t) + 2t - t^2 - t^3 \\ y'(t) = x(t) + y(t) - 4t^2 + t^3 \end{cases} \tag{1}$$

with initial conditions

$$\begin{cases} x(0) = 1 \\ y(0) = 0 \end{cases}$$

This is an example of an initial value problem involving a system of two first-order differential equations. The reader is invited to verify that the analytic solution is

$$\begin{cases} x(t) = e^t \cos(t) + t^2 \\ y(t) = e^t \sin(t) - t^3 \end{cases}$$

If we denote by $\mathbf{X}$ a vector whose two components are x and y, then our system has the form

$$\begin{bmatrix} x' \\ y' \end{bmatrix} = \begin{bmatrix} x - y + 2t - t^2 - t^3 \\ x + y - 4t^2 + t^3 \end{bmatrix}$$

or using vector notation

$$\begin{cases} \mathbf{X}' = \mathbf{F}(t, \mathbf{X}) \\ \mathbf{X}(a) \text{ specified} \end{cases}$$

where $\mathbf{X} = (x, y)^T$, $\mathbf{X}' = (x', y')^T$, $a = 0$, and $\mathbf{F}$ is the vector whose two components are given by the right-hand sides in Eq. (1). Since $\mathbf{F}$ depends on t and $\mathbf{X}$, we write $\mathbf{F}(t, \mathbf{X})$.

Note that in the example given it is not possible to solve either of the two differential equations by itself, for the first equation governing x' involves the unknown function y, whereas the second equation governing y' involves the unknown function x. In this situation, we say that the two differential equations are *coupled*.

Let us look at another example that is superficially similar to the first but is actually simpler:

$$\begin{cases} x'(t) = x(t) + 2t - t^2 - t^3 \\ y'(t) = y(t) - 4t^2 + t^3 \end{cases} \tag{2}$$

with

$$\begin{cases} x(0) = 1 \\ y(0) = 0 \end{cases}$$

These two equations are *not* coupled and can be solved separately as two unrelated initial value problems (using, for instance, the methods of Chapter 8).

Naturally, our concern here is with systems that are coupled, although methods that solve coupled systems also solve those that are not. The procedures discussed in Chapter 8 extend to systems whether coupled or uncoupled.

We illustrate the Taylor series method for the system (1) and begin by differentiating the equations constituting it.

$$\begin{cases} x' = x - y + 2t - t^2 - t^3 \\ y' = x + y - 4t^2 + t^3 \end{cases}$$

$$\begin{cases} x'' = x' - y' + 2 - 2t - 3t^2 \\ y'' = x' + y' - 8t + 3t^2 \end{cases}$$

$$\begin{cases} x''' = x'' - y'' - 2 - 6t \\ y''' = x'' + y'' - 8 + 6t \end{cases}$$

$$\begin{cases} x^{(iv)} = x''' - y''' - 6 \\ y^{(iv)} = x''' + y''' + 6 \end{cases}$$

etc.

A program to proceed from $x(t)$ to $x(t + h)$ and from $y(t)$ to $y(t + h)$ is easily written by using a few terms of the Taylor series

$$x(t + h) = x + hx' + \frac{h^2}{2}x'' + \frac{h^3}{6}x''' + \frac{h^4}{24}x^{(iv)} + \cdots$$

$$y(t + h) = y + hy' + \frac{h^2}{2}y'' + \frac{h^3}{6}y''' + \frac{h^4}{24}y^{(iv)} + \cdots$$

together with equations for the various derivatives. Here $x = x(t)$, $y = y(t)$, $x' = x'(t)$, $y'' = y''(t)$, and so on. A program that generates and prints a numerical solution from 0 to 1 in steps of 10^{-2} is as follows. Terms up to h^4 have been used in the Taylor series.

```
      DATA  T/0.0/,X/1.0/,Y/0.0/,H/0.01/
      PRINT 3,T,X,Y
      DO 2 K = 1,100
      X1 = X - Y + T*(2.0 - T*(1.0 + T))
      Y1 = X + Y + T*T*(-4.0 + T)
      X2 = X1 - Y1 + 2.0 - T*(2.0 + 3.0*T)
      Y2 = X1 + Y1 + T*(-8.0 + 3.0*T)
      X3 = X2 - Y2 - 2.0 - 6.0*T
      Y3 = X2 + Y2 - 8.0 + 6.0*T
      X4 = X3 - Y3 - 6.0
      Y4 = X3 + Y3 + 6.0
      X = X + H*(X1 + H*(X2/2.0 + H*(X3/6.0 + H*X4/24.0)))
      Y = Y + H*(Y1 + H*(Y2/2.0 + H*(Y3/6.0 + H*Y4/24.0)))
      T = FLOAT(K)*H
    2 PRINT 3,T,X,Y
    3 FORMAT(5X,3(E20.13,5X))
      END
```

Before describing how other methods of Chapter 8 can be used for systems of equations, we introduce a slight simplification. When we wrote the system of differential equations in vector form $\mathbf{X}' = \mathbf{F}(t, \mathbf{X})$, we assumed that the variable t was explicitly separated from the other variables x and y and treated differently. It is not necessary to do so. Indeed, we can introduce a new variable x_1 that is t in disguise and add a new differential equation $x_1' = 1$. A new initial condition must also be provided, $x_1(a) = a$. In this way, we increase the number of differential equations by one and obtain a system written in the simpler vector form

$$\begin{cases} \mathbf{X}' = \mathbf{F}(\mathbf{X}) \\ \mathbf{X}(a) \text{ specified} \end{cases}$$

Consider the system of two equations given by Eq. (1). We write it as a system with three variables by letting $x_1 = t$, $x_2 = x$, $x_3 = y$. Thus we have

$$\begin{bmatrix} x_1' \\ x_2' \\ x_3' \end{bmatrix} = \begin{bmatrix} 1 \\ x_2 - x_3 + 2x_1 - x_1^2 - x_1^3 \\ x_2 + x_3 - 4x_1^2 + x_1^3 \end{bmatrix} \tag{3}$$

The auxiliary condition for the vector $\mathbf{X}$ is $\mathbf{X}(0) = (0, 1, 0)^T$.

As a result of the preceding remarks, we sacrifice no generality in considering a system of n first-order differential equations written

$$\begin{cases} x_1'(t) = f_1(x_1, x_2, \ldots, x_n) \\ x_2'(t) = f_2(x_1, x_2, \ldots, x_n) \\ \quad\vdots \\ x_n'(t) = f_n(x_1, x_2, \ldots, x_n) \end{cases}$$

or

$$\mathbf{X}' = \mathbf{F}(\mathbf{X}) \tag{4}$$

where

$$\mathbf{X} = (x_1, x_2, \ldots, x_n)^T$$
$$\mathbf{X}' = (x_1', x_2', \ldots, x_n')^T$$
$$\mathbf{F} = (f_1, f_2, \ldots, f_n)^T$$

plus a prescribed auxiliary condition on the $\mathbf{X}$ vector at $t = a$.

Runge–Kutta methods of Chapter 8 also extend to systems of differential equations. The fourth-order Runge–Kutta method for system (4) uses these formulas.

$$\mathbf{X}(t + h) = \mathbf{X}(t) + \frac{h}{6}(\mathbf{F}_1 + 2\mathbf{F}_2 + 2\mathbf{F}_3 + \mathbf{F}_4)$$

$$\begin{cases} \mathbf{F}_1 = \mathbf{F}(\mathbf{X}) \\ \mathbf{F}_2 = \mathbf{F}(\mathbf{X} + \frac{1}{2}h\mathbf{F}_1) \\ \mathbf{F}_3 = \mathbf{F}(\mathbf{X} + \frac{1}{2}h\mathbf{F}_2) \\ \mathbf{F}_4 = \mathbf{F}(\mathbf{X} + h\mathbf{F}_3) \end{cases}$$

All quantities here, except t and h, are vectors with n components.

A subroutine for carrying out the Runge–Kutta procedure is given next. It is assumed that the system to be solved is in the form of Eq. (4) and that the number of equations, N, is not greater than 10. The user furnishes the initial value of T, the initial value of X, step size H, and the number of steps to be taken, NSTEP. Furthermore, a subprogram XPSYS (N, X, F) is needed that evaluates the right-hand side of Eq. (4) for a given N-component array X and stores the result in array F. ("XPSYS" is chosen as an abbreviation of "X-prime for a system.")

```
      SUBROUTINE RK4SYS(T,X,H,NSTEP)
      DIMENSION  X(10),Y(10),F1(10),F2(10),F3(10),F4(10)
      CALL XPSYS(N,X,F1)
      PRINT 7,T,(X(I),I=1,N)
      H2 = 0.5*H
      START = T
      DO 6 K = 1,NSTEP
      CALL XPSYS(N,X,F1)
      DO 2 I = 1,N
    2 Y(I) = X(I) + H2*F1(I)
      CALL XPSYS(N,Y,F2)
      DO 3 I = 1,N
    3 Y(I) = X(I) + H2*F2(I)
      CALL XPSYS(N,Y,F3)
      DO 4 I = 1,N
    4 Y(I) = X(I) + H*F3(I)
      CALL XPSYS(N,Y,F4)
      DO 5 I = 1,N
    5 X(I) = X(I) + H*(F1(I) + 2.0*(F2(I) + F3(I)) + F4(I))/6.0
      T = START + H*FLOAT(K)
    6 PRINT 7,T,(X(I),I = 1,N)
    7 FORMAT(2X,4HT,X:,E10.3,5(2X,E20.13))
      RETURN
      END
```

To illustrate the use of this subroutine, we again use system (1) for our example. Of course, it must be rewritten in the form of Eq. (3). A suitable main program and a subroutine for computing the right side of Eq. (3) is given on the next page.

A numerical experiment to compare the results of the Taylor series method and the Runge–Kutta method with the analytic solution of system (1) is suggested in computer Problem 1. At the point $t = 1.0$ the results are as follows.

Taylor Series	*Runge–Kutta*	*Analytic Solution*
$x = 2.46869\ 39397$	2.46869 39407	2.46869 39399
$y = 1.28735\ 52884$	1.28735 52875	1.28735 52872

```
      DIMENSION  X(3)
      DATA  M/100/, H/0.01/, T/0.0/, X/0.0,1.0,0.0/
      PRINT 2,T,(X(I),I = 1,3)
      CALL RK4SYS(T,X,H,M)
    2 FORMAT(5X,F10.5,3(5X,E20.13))
      END

      SUBROUTINE XPSYS(N,X,F)
      DIMENSION  X(3),F(3)
      N = 3
      F(1) = 1.0
      F(2) = X(2) - X(3) + X(1)*(2.0 - X(1)*(1.0 + X(1)))
      F(3) = X(2) + X(3) - X(1)*X(1)*(4.0 - X(1))
      RETURN
      END
```

Computer Problems 13.1

1. Solve the system of differential equations (1) by using the two methods given in the text and compare the results with the analytic solution.
2. Solve the initial value problem

$$\begin{cases} x' = t + x^2 - y \\ y' = t^2 - x + y^2 \\ x(0) = 3, \qquad y(0) = 2 \end{cases}$$

by means of the Taylor series method, using $h = 1/128$, on the interval $[0, 0.38]$. Include terms up to and including three derivatives in x and y. How accurate are the computed function values?
3. Recode the Taylor series method for system (1), using statement functions with the same parameter list for each—namely, (T,X,Y). Rerun the program to test the recoded version.
4. Write a driver program for subroutine RK4SYS that solves the ODE system given by Eq. (2). Use $h = -10^{-2}$ and print out the values of x_1, x_2, x_3, together with the true solution on the interval $[-1, 0]$. Verify that the true solution is $x(t) = e^t + 6 + 6t + 4t^2 + t^3$ and $y(t) = e^t - t^3 + t^2 + 2t + 2$.
5. Using the Runge–Kutta procedure, solve the following initial value problem on the interval $0 \leqq t \leqq 2\pi$. Plot the resulting curves $(x_1(t), x_2(t))$ and $(x_3(t), x_4(t))$. They should be circles.

$$\begin{cases} \mathbf{X}' = \begin{bmatrix} x_3 \\ x_4 \\ -x_1(x_1^2 + x_2^2)^{-3} \\ -x_2(x_1^2 + x_2^2)^{-3} \end{bmatrix} \\ \mathbf{X}(0) = (1, 0, 0, 1)^T \end{cases}$$

6. Recode subroutine RK4SYS so that printing of output occurs only at each Jth step, where J is a parameter in the calling sequence. Test this version of RK4SYS.
7. Write a program, using the Taylor series method of order 5, to solve the system

$$\begin{cases} x' = tx - y^2 + 3t \\ y' = x^2 - ty - t^2 \\ x(5) = 2, \qquad y(5) = 3 \end{cases}$$

on the interval [5, 6], using $h = 10^{-3}$. Print values of x and y at steps of 0.1.

8. Print a table of sin t and cos t on the interval $[0, \pi/2]$ by numerically solving the system

$$\begin{cases} x' = y \\ y' = -x \\ x(0) = 0, \qquad y(0) = 1 \end{cases}$$

13.2 Higher-Order Equations and Systems

Consider next the initial value problem for ordinary differential equations of order higher than one. A differential equation of order n is normally accompanied by n auxiliary conditions. This many initial conditions are needed to specify the solution of the differential equation precisely. Take, for example, a particular second-order initial value problem.

$$\begin{cases} x''(t) = -3\cos^2(t) + 2 \\ x(0) = 0, \qquad x'(0) = 0 \end{cases} \tag{1}$$

Without the auxiliary conditions the general analytic solution is $x(t) = \frac{1}{4}t^2 + \frac{3}{8}\cos(2t) + c_1 t + c_2$, where c_1 and c_2 are arbitrary constants. In order to select one specific solution, c_1 and c_2 must be fixed, and two initial conditions enable this to be done. In fact, $x(0) = 0$ forces $c_2 = -\frac{3}{8}$ and $x'(0) = 0$ forces $c_1 = 0$.

In general, higher-order equations can be much more complicated than this simple example, for system (1) has the special property that the function on the right side of the differential equation does not involve x. The most general form of an ordinary differential equation with initial conditions that we consider is

$$\begin{cases} x^{(n)} = f(t, x, x', x'', \ldots, x^{(n-1)}) \\ x(a), x'(a), x''(a), \ldots, x^{(n-1)}(a) \text{ all specified} \end{cases} \tag{2}$$

which can be solved numerically by turning it into a system of *first-order* differential equations. To do so, we define new variables $x_1, x_2, x_3, \ldots, x_{n+1}$ as follows.

$$x_1 = t, \quad x_2 = x, \quad x_3 = x', \quad x_4 = x'', \quad x_5 = x''', \ldots, x_n = x^{(n-2)}, \quad x_{n+1} = x^{(n-1)}$$

Consequently, the original initial value problem (2) is equivalent to

$$\begin{cases} x_1' = 1 \\ x_2' = x_3 \\ x_3' = x_4 \\ x_4' = x_5 \\ \quad \vdots \end{cases}$$

$$\begin{cases} \quad \vdots \\ x_n' = x_{n+1} \\ x_{n+1}' = f(x_1, x_2, x_3, \ldots, x_{n+1}) \\ x_1(a), x_2(a), x_3(a), \ldots, x_{n+1}(a) \text{ all specified} \end{cases}$$

or in vector notation

$$\begin{cases} \mathbf{X}' = \mathbf{F}(\mathbf{X}) \\ \mathbf{X}(a) \text{ specified} \end{cases}$$

where $\mathbf{X} = (x_1, x_2, x_3, \ldots, x_{n+1})^T$, $\mathbf{X}' = (x_1', x_2', x_3', \ldots, x_{n+1}')^T$, and $\mathbf{F} = (1, x_3, x_4, x_5, \ldots, x_{n+1}, f)^T$.

It is recommended,whenever a problem must be transformed by introducing new variables, that a ''dictionary'' be given to show the relationship between the new and the old variables. At the same time this information, together with the differential equations and the initial values, can be displayed in a chart. Such systematic bookkeeping can be helpful in a complicated situation. To illustrate, let us transform the inital value problem

$$\begin{cases} x''' = \cos x + \sin x' - e^{x''} + t^2 \\ x(0) = 3, \quad x'(0) = 7, \quad x''(0) = 13 \end{cases}$$

into a form suitable for solution by the Runge–Kutta procedure. We notice that t is present on the right-hand side and so the equations $x_1 = t$ and $x_1' = 1$ are necessary. A chart summarizing the transformed problem is as follows.

Old Variable	*New Variable*	*Initial Value*	*Differential Equation*
t	x_1	0	$x_1' = 1$
x	x_2	3	$x_2' = x_3$
x'	x_3	7	$x_3' = x_4$
x''	x_4	13	$x_4' = \cos x_2 + \sin x_3 - e^{x_4} + x_1^2$

So the corresponding first-order system is

$$\mathbf{X}' = \begin{bmatrix} 1 \\ x_3 \\ x_4 \\ \cos x_2 + \sin x_3 - e^{x_4} + x_1^2 \end{bmatrix}$$

and $\mathbf{X} = (0, 3, 7, 13)^T$ at $x_1 = 0$.

By systematically introducing new variables, a system of differential equations of various orders can be transformed into a larger system of first-order equations. For instance, the system

$$\begin{cases} x'' = x - y - (3x')^2 + (y')^3 + 6y'' + 2t \\ y''' = y'' - x' + e^x - t \\ x(1) = 2, \quad x'(1) = -4, \quad y(1) = -2, \quad y'(1) = 7, \quad y''(1) = 6 \end{cases}$$

can be solved by the Runge–Kutta procedure if we first transform it according to the following chart:

Old Variable	*New Variable*	*Initial Value*	*Differential Equation*
t	x_1	1	$x_1' = 1$
x	x_2	2	$x_2' = x_3$
x'	x_3	-4	$x_3' = x_2 - x_4 - 9x_3^2 + x_5^3 + 6x_6 + 2x_1$
y	x_4	-2	$x_4' = x_5$
y'	x_5	7	$x_5' = x_6$
y''	x_6	6	$x_6' = x_6 - x_3 + e^{x_2} - x_1$

Problems 13.2

1. Rewrite each as a system of first-order differential equations with t not appearing on the right-hand side.

   (a) $$\begin{cases} x' = x^2 + \log(y) + t^2 \\ y' = e^y - \cos(x) + \sin(tx) - (xy)^7 \\ x(0) = 1, \quad y(0) = 3 \end{cases}$$

   (b) $$\begin{cases} x^{(iv)} = (x''')^2 + \cos(x'\,x'') - \sin(tx) + \log(x/t) \\ x(0) = 1, \quad x'(0) = 3, \quad x''(0) = 4, \quad x'''(0) = 5 \end{cases}$$

2. Consider

   $$\begin{cases} x' = y \\ y' = x \end{cases} \quad \text{with} \quad \begin{cases} x(0) = -1 \\ y(0) = 0 \end{cases}$$

   Write down the equations, without derivatives, to be used in the Taylor series method of order 5.

3. Turn this differential equation into a system of first-order equations suitable for applying the Runge–Kutta method.

   $$\begin{cases} x''' = 2x' + \log(x'') + \cos(x) \\ x(0) = 1, \quad x'(0) = -3, \quad x''(0) = 5 \end{cases}$$

4. Assuming that a program is available for solving initial value problems of the form $\mathbf{X}' = \mathbf{F}(\mathbf{X})$, where $\mathbf{X} = (x_1, x_2, \ldots, x_n)^T$, how can it be used to solve the differential equation $x''' = t + x + 2x' + 3x''$ with initial conditions $x(1) = 3$, $x'(1) = -7$, $x''(1) = 4$? How would this problem be solved if the initial conditions were $x(1) = 3$, $x'(1) = -7$, and $x'''(1) = 0$?

5. How would you solve this differential problem numerically?

   $$\begin{cases} x_1' = x_1^2 + e^t - t^2 \\ x_2' = x_2 - \cos t \\ x_1(0) = 0, \quad x_2(1) = 0 \end{cases}$$

6. How would you solve this differential problem numerically?

   $$\begin{cases} x_1'' = x_1' + x_1^2 - \sin t \\ x_2'' = x_2 - (x_2')^{1/2} + t \\ x_1(0) = 1, \quad x_2(1) = 3, \quad x_1'(0) = 0, \quad x_2'(1) = -2 \end{cases}$$

7. How would you solve the initial value problem

$$\begin{cases} x_1'(t) = x_1(t)e^t + \sin t - t^2 \\ x_2'(t) = [x_2(t)]^2 - e^t + x_2(t) \\ x_1(1) = 2, \qquad x_2(1) = 4 \end{cases}$$

if a program was available to solve an initial value problem of the form $x' = f(t, x)$ involving a single unknown function $x = x(t)$?

8. Consider

$$\begin{cases} x'' = x' - x \\ x(0) = 0, \qquad x'(0) = 1 \end{cases}$$

Determine the associated first-order system and its auxiliary condition vector.

9. The problem

$$\begin{cases} x''(t) = x + y - 2x' + 3y' + \log t, & x(0) = 1, \;\; x'(0) = 2 \\ y''(t) = 2x - 3y + 5x' + ty' - \sin t, & y(0) = 3, \;\; y'(0) = 4 \end{cases}$$

is to be put into the form of a system of five first-order equations. Give the resulting system and the appropriate initial values.

10. Write subroutine XPSYS that can be used with the fourth-order Runge–Kutta routine RK4SYS for this differential equation.

$$\begin{cases} x''' = 10e^{x''} - x''' \sin(x'x) - (xt)^{10} \\ x(2) = 6.5, \;\; x'(2) = 4.1, \;\; x''(2) = 3.2 \end{cases}$$

11. If we are going to solve the initial value problem

$$\begin{cases} x''' = x' - tx'' + x + \ln t \\ x(1) = x'(1) = x''(1) = 1 \end{cases}$$

using Runge–Kutta formulas, how must the problem be transformed?

Computer Problems 13.2

1. Use RK4SYS to solve each of the following for $0 \leqq t \leqq 1$. Use $h = 2^{-k}$ with $k = 5, 6, 7$ and compare results.

   (a) $$\begin{cases} x'' = 2(e^{2t} - x^2)^{1/2} \\ x(0) = 0, \qquad x'(0) = 1 \end{cases}$$

   (b) $$\begin{cases} x'' = x^2 - y + e^t \\ y'' = x - y^2 - e^t \\ x(0) = 0, \qquad x'(0) = 0 \\ y(0) = 1, \qquad y'(0) = -2 \end{cases}$$

2. Write a program for using the Taylor series method of order 3 to solve the system

$$\begin{cases} x' = tx + y' - t^2 \\ y' = ty + 3t \\ z' = tz - y' + 6t^3 \\ x(0) = 1, \quad y(0) = 2, \quad z(0) = 3 \end{cases}$$

on the interval $[0, 0.75]$, using $h = .01$.

3. Solve the Airy differential equation

$$\begin{cases} x'' = tx \\ x(0) = 0.35502\ 80538\ 87817 \\ x'(0) = -0.25881\ 94037\ 92807 \end{cases}$$

on the interval $[0, 4.5]$, using the Runge–Kutta method. The value $x(4.5) = 0.00033\ 02503$ is correct.

4. Modify subroutine RK45 of Sec. 8.3 so that it can solve a system of at most ten differential equations. Test it on problem 3.
5. Solve

$$\begin{cases} x'' + x' + x^2 - 2t = 0 \\ x(0) = 0, \qquad x'(0) = 0.1 \end{cases}$$

on $[0, 3]$ by any convenient method. If a plotter is available, graph the solution.

6. Solve

$$\begin{cases} x'' = 2x' - 5x \\ x(0) = 0, \qquad x'(0) = 0.4 \end{cases}$$

on the interval $[-2, 0]$.

13.3 Adams–Moulton Method

The procedures explained so far have solved the initial value problem

$$\begin{cases} \mathbf{X}' = \mathbf{F}(\mathbf{X}) \\ \mathbf{X}(a) \text{ specified} \end{cases} \tag{1}$$

by means of *single-step* numerical methods. In other words, if the solution $\mathbf{X}(t)$ is known at a particular point t, then $\mathbf{X}(t + h)$ can be computed without knowing the solution at points earlier than t. The Runge–Kutta and Taylor series methods compute $\mathbf{X}(t + h)$ in terms of $\mathbf{X}(t)$ and various values of $\mathbf{F}$.

More efficient methods can be devised if several of the values $\mathbf{X}(t - h), \mathbf{X}(t - 2h), \ldots$ are utilized in computing $\mathbf{X}(t + h)$. Such methods are called ***multistep*** methods. They have the obvious drawback that at the beginning of the numerical solution no prior values of $\mathbf{X}$ are available. So it is usual to start a numerical solution with a single-step method, such as the Runge–Kutta procedure, and transfer to a multistep procedure for efficiency as soon as enough starting values have been computed.

One example of a multistep formula is known by the name of *Adams–Bashforth*. (See Supplementary Computing Problem 1 of Chapter 8.) It is

$$\mathbf{X}(t+h) = \mathbf{X}(t) + \frac{h}{24}\,[55\mathbf{F}(\mathbf{X}(t)) - 59\mathbf{F}(\mathbf{X}(t-h)) + 37\mathbf{F}(\mathbf{X}(t-2h)) - 9\mathbf{F}(\mathbf{X}(t-3h))] \tag{2}$$

If the solution $\mathbf{X}$ has been computed at the four points t, $t - h$, $t - 2h$, and $t - 3h$, then formula (2) can be used to compute $\mathbf{X}(t + h)$. If this is done systematically, then only *one* evaluation of $\mathbf{F}$ is required for each step. This represents a considerable saving over the fourth-order Runge–Kutta procedure; the latter requires *four* evaluations of $\mathbf{F}$ per step.

In practice, formula (2) is never used by itself. Instead it is used as a *predictor* and then another formula is used as a *corrector*. The corrector usually employed with (2) is the *Adams–Moulton* formula

$$\mathbf{X}(t+h) = \mathbf{X}(t) + \frac{h}{24}\,[9\mathbf{F}(\mathbf{X}^{(p)}) + 19\mathbf{F}(\mathbf{X}(t)) - 5\mathbf{F}(\mathbf{X}(t-h)) + \mathbf{F}(\mathbf{X}(t-2h))] \tag{3}$$

Here $\mathbf{X}^{(p)}$ is the tentative value of $\mathbf{X}(t + h)$ computed from formula (2). Thus Eq. (2) predicts a tentative value of $\mathbf{X}(t + h)$, and Eq. (3) computes this $\mathbf{X}$ value more accurately. The combination of the two formulas results in a *predictor–corrector* scheme.

With initial values of $\mathbf{X}$ specified at a, three steps of a Runge–Kutta method can be performed to determine enough $\mathbf{X}$ values so that the Adams–Bashforth–Moulton procedure can continue. The fourth-order Adams–Bashforth and Adams–Moulton formulas, started with the fourth-order Runge–Kutta method, is referred to as the *Adams–Moulton method*. Predictor and corrector formulas of the same order are used so that only one application of the corrector formula is needed.

Storage of the approximate solution at previous steps in the Adams–Moulton method is usually handled either by storing in an array of dimension larger than the total number of steps to be taken or by physically shifting data after each step (discarding the oldest data and storing the newest in its place). If an adaptive process is used, the total number of steps to be taken cannot be determined beforehand. Physical shifting of data can be eliminated by cycling the indices of a storage array of fixed dimension. For the Adams–Moulton method, the x_i data for $\mathbf{X}(t)$ are stored in a two-dimensional array $z_{i,m}$ in locations $m = 1, 2, 3, 4, 5, 1, 2, \ldots$ for $t = a, a + h, a + 2h, a + 3h, a + 4h, a + 5h, a + 6h, \ldots$, respectively. Following is a sketch of the first several t values with corresponding m values and abbreviations for the formulas used.

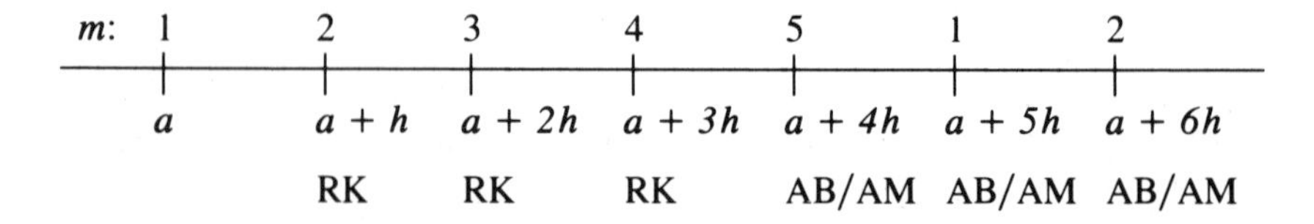

An error analysis is conducted after each step of the Adams–Moulton method. If $x_i^{(p)}$ is the numerical approximation of the ith equation in system (1) at $t + h$ obtained by predictor formula (2) and x_i is that from corrector formula (3), then it can be shown that the single-step error at $t + h$ is given approximately by

$$\varepsilon_i = \frac{19}{270} \frac{|x_i - x_i^{(p)}|}{|x_i|}$$

So we compute EST $= \max_{1 \leq i \leq n} |\varepsilon_i|$.

The Adams–Moulton method for a system and the computation of the single-step error is accomplished in the following subroutine.

```
      SUBROUTINE AMSYS(N, M, T, X, H, F, EST)
      DIMENSION  X(10,5), XP(10), F(10,5), CP(4), CC(4), SUM(10)
      DATA  CP/55.0,-59.0,37.0,-9.0/
      DATA  CC/ 9.0,  19.0,-5.0,  1.0/
      MP1 = 1 + MOD(M,5)
      CALL XPSYS(N, X(1,M), F(1,M))
      DO 2 I = 1,N
    2 SUM(I) = 0.0
      DO 3 K = 1,4
      J = 1 + MOD(M-K+5,5)
      DO 3 I = 1,N
    3 SUM(I) = SUM(I) + H*CP(K)*F(I,J)/24.0
      DO 4 I = 1,N
    4 XP(I) = X(I,M) + SUM(I)
      CALL XPSYS(N, XP, F(1,MP1))
      DO 5 I = 1,N
    5 SUM(I) = 0.0
      DO 6 K = 1,4
      J = 1 + MOD(MP1-K+5,5)
      DO 6 I = 1,N
    6 SUM(I) = SUM(I) + H*CC(K)*F(I,J)/24.0
      DO 7 I = 1,N
    7 X(I,MP1) = X(I,M) + SUM(I)
      T = T + H
      M = MP1
      DLTMAX = 0.0
      DO 8 I = 1,N
      DLT = ABS( X(I,M) - XP(I) )
      IF(DLT .LT. DLTMAX) GO TO 8
      DLTMAX = DLT
      IDLT = I
    8 CONTINUE
      EST = 19.0*DLTMAX/( 270.0*ABS(X(IDLT,M)) )
      RETURN
      END
```

Here the function evaluations are stored cyclically in $f_{i,m}$ for use by formulas (2) and (3). A companion Runge–Kutta subroutine is needed that involves a slight modification of routine RK4SYS from Sec. 13.1.

```
      SUBROUTINE RKSYSM(N,M,T,X,H,F)
      DIMENSION  X(10,5),XP(10),F(10,5),F2(10),F3(10),F4(10)
      MP1 = 1 + MOD(M,5)
      H2 = 0.5*H
      CALL XPSYS(N,X(1,M),F(1,M))
      DO 2 I = 1,N
    2 XP(I) = X(I,M) + H2*F(I,M)
      CALL XPSYS(N,XP,F2)
      DO 3 I = 1,N
    3 XP(I) = X(I,M) + H2*F2(I)
      CALL XPSYS(N,XP,F3)
      DO 4 I = 1,N
    4 XP(I) = X(I,M) + H*F3(I)
      CALL XPSYS(N,XP,F4)
      DO 5 I = 1,N
    5 X(I,MP1) = X(I,M) + H*(F(I,M) + 2.0*(F2(I) + F3(I)) + F4(I))/6.0
      T = T + H
      M = MP1
      RETURN
      END
```

A control routine that calls the Runge–Kutta subroutine three times is needed, and then the remaining steps are computed by the Adams–Moulton predictor–corrector scheme. Such a subroutine for doing NSTEP steps with a fixed step size H follows.

```
      SUBROUTINE AMRK(T,X,H,NSTEP)
      DIMENSION X(10),Z(10,5),F(10,5)
      M = 1
      CALL XPSYS(N,X,F(1,M))
      PRINT 6,T,H
      PRINT 7,(X(I),I=1,N)
      DO 2 I=1,N
    2 Z(I,M) = X(I)
      DO 3 K = 1,3
      CALL RKSYSM(N,M,T,Z,H,F)
      PRINT 6,T,H
    3 PRINT 7,(Z(I,M),I = 1,N)
      DO 4 K = 4,NSTEP
      CALL AMSYS(N,M,T,Z,H,F,EST)
      PRINT 6,T,H,EST
    4 PRINT 7,(Z(I,M),I = 1,N)
      DO 5 I=1,N
    5 X(I) = Z(I,M)
      RETURN
    6 FORMAT(2X,8HT,H,EST:,3(E10.3,1X))
    7 FORMAT(8X,2HX:,5(5X,E20.13))
      END
```

To use the Adams–Moulton code, we supply the subroutine XPSYS, which defines the system of ordinary differential equations as in Sec. 13.1 and write a driver program with a call to AMRK. The complete program would then consist of the following five parts: (a) main program, (b) subroutine XPSYS, (c) subroutine AMRK, (d) subroutine RKSYSM,

Since an estimate of the error is available from the Adams–Moulton method, it is natural to replace subroutine AMRK with an adaptive procedure — that is, one that changes the step size. A procedure similar to the one used in Sec. 8.3 over interval $[a, b]$ is as follows. If error analysis determines that halving of the step size is necessary on the first step of the Adams–Moulton formula after the Runge–Kutta method has been used to calculate the initial values, then the step size is halved. A retreat is made to the starting value (four steps), and new initial values are obtained with the new step size. This process is repeated until the error analysis allows at least one forward step by the Adams–Moulton formula. If error analysis indicates that halving is required at some point within interval $[a, b]$, then the step size is halved. A retreat must be made to an appropriate previous value so that when the values needed by the Adams–Moulton formula are calculated from the Runge–Kutta method with the new step size, the point at which the error was too large is computed by the Adams–Moulton formula, not the Runge–Kutta formula. To simplify this process, always back up four steps whether halving or doubling the step size. (See Computer Problem 3.)

Computer Problems 13.3

1. Test the subroutines of this section on the system given in Problem 7 of Sec. 13.2.
2. The single-step error is closely controlled by using fourth-order formulas; however, the roundoff error in performing $\mathbf{X}(t) + h \sum_i c_i \mathbf{F}(t + ih)$ can be large. It is logical to preserve $\mathbf{X}(t)$ in double-precision and perform the addition by what is known as *partial double-precision* arithmetic. A rounded value of $\mathbf{X}(t)$ in single precision is used to perform the multiplication involved in $h \sum_i b_i \mathbf{F}(t + ih)$. Then the addition of $\mathbf{X}(t)$ to this result is done in double precision. Recode the Adams–Moulton method so that partial double-precision arithmetic is used. Compare this code to that in the text for a system with a known solution. How do they compare with regard to roundoff error at each step?
3. Write and test an adaptive process similar to RK45AD in Sec. 8.3 called

   ```
   SUBROUTINE AMRKAD(T, X, H, TB, ITMAX, EMIN, EMAX, HMIN, HMAX, IFLAG)
   ```

   This routine should carry out the adaptive procedure outlined in this section and be used in place of the AMRK subroutine.
4. How can subroutine AMSYS be improved? Recode and test.

14

Boundary Value Problems for Ordinary Differential Equations

In the design of pivots and bearings the mechanical engineer encounters the following problem. The cross section of a pivot is determined by a curve $y = y(x)$ that must pass through two fixed points (0, 1) and (1, a), as in Fig. 14.1. Moreover, for optimal performance (principally low friction), the unknown function must minimize the value of a certain integral

$$\int_0^1 [y(y')^2 + b(x)y^2]\, dx$$

in which $b(x)$ is a known function. From this it is possible to obtain a second-order differential equation (the so-called Euler equation) for y. It and the initial and terminal values are

$$\begin{cases} -(y')^2 + 2b(x)y - 2yy'' = 0 \\ y(0) = 1, \qquad y(1) = a \end{cases}$$

This is a *two-point boundary value problem,* and methods for solving it numerically are discussed in this chapter.

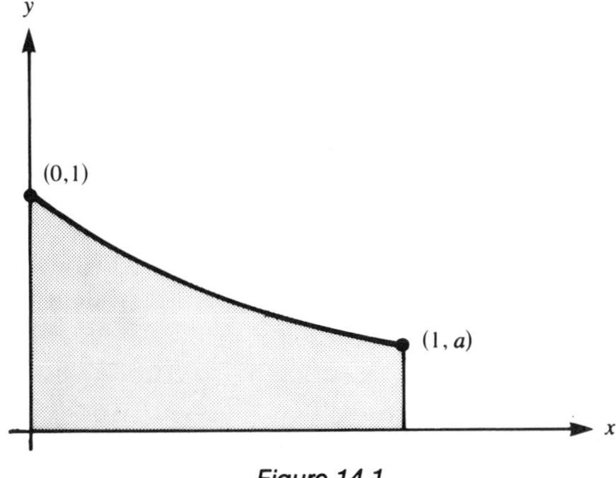

Figure 14.1

We now consider another type of numerical problem involving ordinary differential equations. A *boundary value problem* is exemplified by a second-order ordinary differential equation whose solution function is prescribed at the endpoints of the interval of interest. An instance of such a problem is

$$\begin{cases} x'' = -x \\ x(0) = 1, \qquad x\left(\dfrac{\pi}{2}\right) = -3 \end{cases} \tag{1}$$

Here we have a differential equation whose general solution involves two arbitrary parameters. In order to specify a particular solution, two conditions must be given. If this were an initial value problem, x and x' would be specified at some initial value. In problem (1), however, we are given two points of the form $(t, x(t))$—namely, $(0, 1)$ and $(\pi/2, -3)$—through which the solution curve passes. The general solution of the differential equation is $x(t) = c_1 \sin(t) + c_2 \cos(t)$, and the two conditions (known as *boundary values*) enable us to determine that $c_1 = -3$ and $c_2 = 1$.

Now suppose that we have a similar problem in which we are unable to determine the general solution as above. We take as our model the problem

$$\begin{cases} x''(t) = f(t, x(t), x'(t)) \\ x(a) = \alpha, \qquad x(b) = \beta \end{cases} \tag{2}$$

A step-by-step numerical solution of problem (2) by the methods of Chapter 13 requires two initial conditions. But in problem (2) only one condition is present at $t = a$. This fact makes a problem like (2) considerably more difficult than an initial value problem. Several ways to attack it are considered in this chapter.

14.1 The Shooting Method

One way to proceed in solving Eq.(2) is to "guess" $x'(a)$, then carry out the solution of the resulting initial value problem as far as b, and pray that the computed solution agrees with β—that is, that $x(b) = \beta$. If it does not (which is quite likely), we can go back and change our guess for $x'(a)$. Repeating this procedure until we hit the "target" β may be a good method if we can learn something from the various trials. There are systematic ways of utilizing this information, and the resulting method is known by the nickname *shooting*.

We observe that the final value of the solution of our initial value problem—that is, $x(b)$—depends upon the guess that was made for $x'(a)$. Everything else remains fixed in this problem. Thus the differential equation $x'' = f(t, x, x')$ and the first initial value, $x(a) = \alpha$, do not change. If we assign a real value z to the missing initial condition, $x'(a) = z$, then the initial value problem can be solved numerically. The value of x at b is now a function of z, which we denote by $\phi(z)$. In other words, for each choice of z, we obtain a new value for $x(b)$, and ϕ is the name of the function with this behavior. We know very little about $\phi(z)$, but we can compute or evaluate it. It is, however, an *expensive* function to evaluate, for each value of $\phi(z)$ is obtained only after solving an initial value problem.

To summarize, a function $\phi(z)$ is computed as follows. Solve the initial value problem

$$\begin{cases} x'' = f(t, x(t), x'(t)) \\ x(a) = \alpha, \qquad x'(a) = z \end{cases}$$

on the interval $[a, b]$. Let $\phi(z) = x(b)$.

Our objective is to adjust z until we find a value for which $\phi(z) = \beta$. One way to do so is to use linear interpolation between $\phi(z_1)$ and $\phi(z_2)$, where z_1 and z_2 are two guesses for the initial condition $x'(a)$. That is, given two values of ϕ, we pretend that ϕ is a linear function and determine an appropriate value of z based on this hypothesis. A sketch of the values of z versus $\phi(z)$ might look like Fig 14.2.

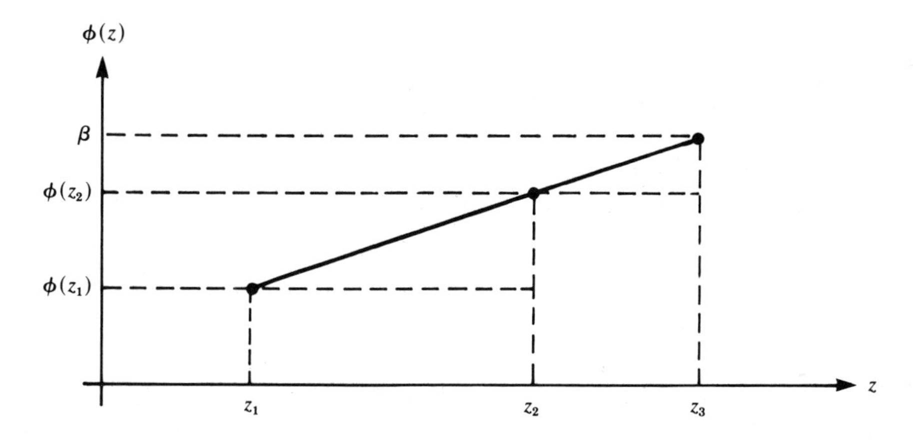

Figure 14.2

To obtain an estimating formula for the next value z_3, we compute $\phi(z_1)$ and $\phi(z_2)$ based on values z_1 and z_2, respectively. By considering similar triangles, we have

$$\frac{z_3 - z_2}{\beta - \phi(z_2)} = \frac{z_2 - z_1}{\phi(z_2) - \phi(z_1)}$$

from which

$$z_3 = z_2 + \frac{(\beta - \phi(z_2))(z_2 - z_1)}{\phi(z_2) - \phi(z_1)}$$

In general, we have the sequence

$$z_{n+1} = z_n + \frac{(\beta - \phi(z_n))(z_n - z_{n-1})}{\phi(z_n) - \phi(z_{n-1})} \qquad (n \geqq 2) \tag{1}$$

all based on two starting values z_1 and z_2.

This procedure for solving the two-point boundary value problem

$$\begin{cases} x'' = f(t, x, x') \\ x(a) = \alpha, \qquad x(b) = \beta \end{cases} \tag{2}$$

is then as follows. Solve the initial value problem

$$\begin{cases} x'' = f(t, x, x') \\ x(a) = \alpha, \qquad x'(a) = z \end{cases} \tag{3}$$

from $t = a$ to $t = b$, letting the value of the solution at b be denoted by $\phi(z)$. Do this twice with two different values of z, say z_1 and z_2, and compute $\phi(z_1)$ and $\phi(z_2)$. Now calculate a new z, called z_3, by formula (1). Then compute $\phi(z_3)$ by again solving (3). Obtain z_4 from z_2 and z_3 in the same way and so on. Monitor $|\phi(z_{n+1}) - \beta|$ to see whether progress is being made. When it is satisfactorily small, stop. Note that the numerically obtained values $x(t_i)$, for $a \leqq t_i \leqq b$, must be saved until better ones are obtained [i.e., one whose terminal value $x(b)$ is closer to β than the present one] because the objective in solving (2) is to obtain values of x for values of t between a and b.

The shooting method may be very time consuming if each solution of the associated initial value problem involves a small value for h. Consequently, we use a relatively large value of h until $|\phi(z_{n+1}) - \beta|$ is small and then reduce h to obtain the required accuracy.

Many modifications and refinements are possible. For instance, when $\phi(z_{n+1})$ is near β, one can use higher-order interpolation formulas to estimate successive values of z_i. Suppose, for example, that instead of utilizing two values $\phi(z_1)$ and $\phi(z_2)$ to obtain z_3, we were to utilize four values, $\phi(z_1)$, $\phi(z_2)$, $\phi(z_3)$, $\phi(z_4)$, to estimate z_5. We could set up a cubic interpolating polynomial p_3 for the data

z_1	z_2	z_3	z_4
$\phi(z_1)$	$\phi(z_2)$	$\phi(z_3)$	$\phi(z_4)$

(4)

and solve $p_3(z_5) = \beta$ for z_5. Since p_3 is a cubic, this would entail some additional work. A better way is to set up a polynomial $\hat{p}_3$ to interpolate the data

$\phi(z_1)$	$\phi(z_2)$	$\phi(z_3)$	$\phi(z_4)$
z_1	z_2	z_3	z_4

(5)

and then use $\hat{p}_3(\beta)$ as the estimate for z_5. This procedure is known as *inverse interpolation*. (See Supplementary Problem 28, Chapter 5.)

Further remarks on the shooting method will be made in the next section after the discussion of an alternative procedure.

Problems 14.1

1. Verify that $x = (2t + 1)e^t$ is the solution to each of the following problems.

$$\begin{cases} x'' = x + 4e^t \\ x(0) = 1, \qquad x(\tfrac{1}{2}) = 2e^{1/2} \end{cases} \qquad \begin{cases} x'' = x' + x - (2t - 1)e^t \\ x(1) = 3e, \qquad x(2) = 5e^2 \end{cases}$$

2. Verify that $x = c_1 e^t + c_2 e^{-t}$ solves the boundary value problem

$$\begin{cases} x'' = x \\ x(0) = 1, \qquad x(1) = 2 \end{cases}$$

if appropriate values of c_1 and c_2 are chosen.

3. Solve these boundary value problems by adjusting the general solution of the differential equation.
(a) $x'' = x, \quad x(0) = 0, \quad x(\pi) = 1$
(b) $x'' = t^2, \quad x(0) = 1, \quad x(1) = -1$

4. Determine all pairs (α, β) for which the problem

$$x'' = -x, \qquad x(0) = \alpha, \quad x\left(\frac{\pi}{2}\right) = \beta$$

has a solution. Repeat for $x(0) = \alpha$, $x(\pi) = \beta$.

5. Verify the following algorithm for the inverse interpolation technique suggested in the text. Here we have set $\phi_i = \phi(z_i)$.

$$u = \frac{z_2 - z_1}{\phi_2 - \phi_1}$$

$$s = \frac{z_3 - z_2}{\phi_3 - \phi_2}$$

$$e = \frac{z_4 - z_3}{\phi_4 - \phi_3}$$

$$v = \frac{s - u}{\phi_3 - \phi_1}$$

$$r = \frac{e - s}{\phi_4 - \phi_2}$$

$$w = \frac{r - v}{\phi_4 - \phi_1}$$

$$z_5 = z_1 + (\beta - \phi_1)\{u + (\beta - \phi_2)[v + w(\beta - \phi_3)]\}$$

6. Determine the function ϕ explicitly in the case of this two-point boundary value problem.

$$\begin{cases} x'' = -x \\ x(0) = 1, \qquad x\left(\frac{\pi}{2}\right) = 3 \end{cases}$$

7. (Continuation) Do the same for $x'' = -(x')^2/x$, $x(1) = 3$, $x(2) = 5$. Using your result, solve the boundary value problem. *Hint*: The general solution of the differential equation is $x(t) = c_1\sqrt{c_2 + t}$.

8. Determine the function ϕ explicitly in the case of this two-point boundary value problem.

$$\begin{cases} x'' = x \\ x(-1) = e, \qquad x(1) = \frac{1}{2}e \end{cases}$$

9. Boundary value problems may involve differential equations of degree higher than 2. For example,

$$\begin{cases} x''' = f(t, x, x', x'') \\ x(a) = \alpha, \quad x'(a) = \gamma, \quad x(b) = \beta \end{cases}$$

List the ways it can be solved using the shooting method.

10. Solve analytically this three-point boundary value problem.

$$\begin{cases} x''' = -e^t + 4(t+1)^{-3} \\ x(0) = -1, \qquad x(1) = 3 - e + 2\ln 2, \qquad x(2) = 6 - e^2 + 2\ln 3 \end{cases}$$

Computer Problems 14.1

1. The classical nonlinear two-point boundary value problem

$$\begin{cases} x'' = e^x \\ x(0) = \alpha, \qquad x(1) = \beta \end{cases}$$

has the closed-form solution

$$x = \ln c_1 - 2\ln\{\cos[(\tfrac{1}{2}c_1)^{1/2}t + c_2]\}$$

where c_1 and c_2 are the solutions of

$$\begin{cases} \alpha = \ln c_1 - 2\ln\cos c_2 \\ \beta = \ln c_1 - 2\ln\{\cos[(\tfrac{1}{2}c_1)^{1/2} + c_2]\} \end{cases}$$

The corresponding discrete problem, as discussed in the next section, involves a system of nonlinear equations with no closed-form solution. Use the shooting method to solve this problem when $\alpha = \beta = \ln 8\pi^2$. Start with $z_1 = -1$ and $z_2 = 2$. Also, determine c_1 and c_2 so that a comparison with the true solution can be made.

14.2 A Discretization Method

We turn now to a completely different approach to solving the two-point boundary value problem, one based on a direct discretization of the differential equation. The problem that we want to solve is

$$\begin{cases} x'' = f(t, x, x') \\ x(a) = \alpha, \qquad x(b) = \beta \end{cases} \tag{1}$$

Select a set of equally spaced points $t_0, t_1, \ldots, t_{n+1}$ on the interval $[a, b]$ by putting $h = (b-a)/(n+1)$ and $t_i = a + ih$. Next, approximate the derivatives, using the standard central difference formulas from Sec. 5.3.

$$\begin{aligned} x'(t) &\approx \frac{1}{2h}[x(t+h) - x(t-h)] \\ x''(t) &\approx \frac{1}{h^2}[x(t+h) - 2x(t) + x(t-h)] \end{aligned} \tag{2}$$

The approximate value of $x(t_i)$ is denoted by x_i. Hence the problem becomes

$$\begin{cases} x_0 = \alpha \\ \dfrac{1}{h^2}(x_{i-1} - 2x_i + x_{i+1}) = f\left(t_i, x_i, \dfrac{1}{2h}(x_{i+1} - x_{i-1})\right) \qquad (1 \leqq i \leqq n) \\ x_{n+1} = \beta \end{cases} \tag{3}$$

This is usually a nonlinear system of equations in the n unknowns $x_1, \ldots, x_n$, since f will generally involve the x_i in a nonlinear way. The solution of such a system is seldom easy.

In some cases, system (3) *is* linear. This situation occurs exactly when f in Eq. (1) has the form

$$f(t, x, x') = u(t) + v(t)x + w(t)x' \tag{4}$$

In this special case, the principal equation in system (3) looks like this:

$$\frac{1}{h^2}(x_{i-1} - 2x_i + x_{i+1}) = u(t_i) + v(t_i)x_i + w(t_i)\left[\frac{1}{2h}(x_{i+1} - x_{i-1})\right]$$

or, equivalently,

$$-\left(1 + \frac{h}{2}w_i\right)x_{i-1} + (2 + h^2 v_i)x_i - \left(1 - \frac{h}{2}w_i\right)x_{i+1} = -h^2 u_i \tag{5}$$

where $u_i = u(t_i)$, $v_i = v(t_i)$, and $w_i = w(t_i)$. Now let

$$\begin{cases} a_i = -\left(1 + \dfrac{h}{2}w_i\right) \qquad (1 \leqq i \leqq n) \\ d_i = (2 + h^2 v_i) \\ c_i = -\left(1 - \dfrac{h}{2}w_i\right) \\ b_i = -h^2 u_i \end{cases}$$

where $h = (b - a)/(n + 1)$ and $t_i = a + ih$. Then the principal equation becomes

$$a_i x_{i-1} + d_i x_i + c_i x_{i+1} = b_i$$

The equations corresponding to $i = 1$ and $i = n$ are different, since we know x_0 and x_{n+1}. The system, therefore, can be written

$$\begin{cases} d_1x_1 + c_1x_2 = b_1 - a_1\alpha \\ a_ix_{i-1} + d_ix_i + c_ix_{i+1} = b_i \qquad (2 \leqq i \leqq n-1) \\ a_nx_{n-1} + d_nx_n = b_n - c_n\beta \end{cases} \tag{6}$$

In matrix form system (6) looks like this.

$$\begin{bmatrix} d_1 & c_1 & & & & \\ a_2 & d_2 & c_2 & & 0 & \\ & a_3 & d_3 & c_3 & & \\ & & \ddots & \ddots & \ddots & \\ & 0 & & a_{n-1} & d_{n-1} & c_{n-1} \\ & & & & a_n & d_n \end{bmatrix} \begin{bmatrix} x_1 \\ x_2 \\ x_3 \\ \vdots \\ x_{n-1} \\ x_n \end{bmatrix} = \begin{bmatrix} b_1 - a_1\alpha \\ b_2 \\ b_3 \\ \vdots \\ b_{n-1} \\ b_n - c_n\beta \end{bmatrix}$$

Since this system is tridiagonal, we can solve it with the special Gaussian procedure for tridiagonal systems developed in Sec. 6.3. (That procedure does not include pivoting, however, and may fail in cases where subroutine GAUSS would succeed. See Problem 5.)

The ideas just explained are now used to write a program for a specific test case. The problem is of the form (1) with f a linear function as in Eq. (4).

$$\begin{cases} x'' = e^t + x' - 3\sin(t) - x \\ x(1) = 1.09737491, \qquad x(2) = 8.63749661 \end{cases} \tag{7}$$

The solution, known in advance to be $x(t) = e^t - 3\cos(t)$, can be used to check the computer solution. We employ the discretization technique described earlier and subroutine TRI for solving the resulting linear system.

First, we decide to use 100 points, including points $a = 1$ and $b = 2$. Thus, $n = 98$, $h = 1/99$, and $t_i = 1 + ih$ for $0 \leqq i \leqq 99$. Then $t_0 = 1$, $x_0 = x(t_0) \approx 1.09737491$, $t_{99} = 2$, and $x_{99} = x(t_{99}) \approx 8.63749661$. The unknowns in our problem are the remaining values of x_i—namely, $x_1, x_2, \ldots, x_{98}$. We discretize the derivatives by the central difference formulas (2) and obtain a linear system of type (3). Our principal equation is of the form (5) and is

$$-\left(1 + \frac{h}{2}\right)x_{i-1} + (2 - h^2)x_i - \left(1 - \frac{h}{2}\right)x_{i+1} = -h^2[e^{t_i} - 3\sin(t_i)]$$

for $2 \leqq i \leqq 98$. The boundary values given are correct only to the number of digits shown.

We generalize the code so that with only a few changes it can accommodate any two-point boundary value problem of type (1) with right-hand side of form (4).

We have just seen that this discretization method (also called a *finite-difference method*) is rather simple in the case of the linear two-point boundary value problem

$$\begin{cases} x'' = u(t) + v(t)x + w(t)x' \\ x(a) = \alpha, \qquad x(b) = \beta \end{cases} \tag{8}$$

The shooting method is also especially simple in this case. Recall that the shooting method requires us to solve an initial value problem

$$\begin{cases} x'' = u(t) + v(t)x + w(t)x' \\ x(a) = \alpha, \qquad x'(a) = z \end{cases} \tag{9}$$

```
      DIMENSION  A(98),B(98),C(98),D(98)
      DATA  N/98/, TA/1.0/, TB/2.0/, ALPHA/1.09737491/, BETA/8.63749661/
      U(X) = EXP(X) - 3.0*SIN(X)
      V(X) = -1.0
      W(X) =  1.0
      H = (TB - TA)/FLOAT(N+1)
      HSQ = H*H
      H2 = 0.5*H
      DO 2 I = 1,N
      T = TA + H*FLOAT(I)
      A(I) = -(1.0 + H2*W(T))
      D(I) =  (2.0 + HSQ*V(T))
      C(I) = -(1.0 - H2*W(T))
    2 B(I) = -HSQ*U(T)
      B(1) = B(1) - A(1)*ALPHA
      B(N) = B(N) - C(N)*BETA
      CALL TRI(N,A(2),D,C,B)
      E = EXP(TA) - 3.0*COS(TA) - ALPHA
      PRINT 4,TA,ALPHA,E
      DO 3 I = 9,N,9
      T = TA + H*FLOAT(I)
      E = EXP(T) - 3.0*COS(T) - B(I)
    3 PRINT 4,T,B(I),E
      E = EXP(TB) - 3.0*COS(TB) - BETA
      PRINT 4,TB,BETA,E
    4 FORMAT(5X,F10.5,2(5X,E20.13))
      END
```

Computed Results

T value	*Numerical Solution*	*Error*
1.00000	1.0973749100000E+00	8.5461238086282E−10
1.09091	1.5919420991874E+00	−3.6591942631503E−07
1.18182	2.1225680421291E+00	−6.9431588656244E−07
1.27273	2.6895533356390E+00	−9.6362370527459E−07
1.36364	3.2933439678210E+00	−1.1563301001161E−06
1.45455	3.9345700498940E+00	−1.2590325866313E−06
1.54545	4.6140870640167E+00	−1.2633944379559E−06
1.63636	5.3330196706161E+00	−1.1671422726067E−06
1.72727	6.0928081341454E+00	−9.7509939678275E−07
1.81818	6.8952574446102E+00	−7.0025032528065E−07
1.90909	7.7425892337858E+00	−3.6482839504970E−07
2.00000	8.6374966100000E+00	−1.4279635252024E−09

and look upon the terminal value $x(b)$ as a function of z. We call that function ϕ and seek a value of z for which $\phi(z) = \beta$. For the linear problem (9), ϕ is a *linear* function of z, and so the figure sketched in Sec 14.1 is actually realistic. Consequently, we need only solve (9) with two values of z in order to determine the function precisely. In order to establish these facts, let us do a little more analysis.

Suppose that we have solved (9) twice with particular values z_1 and z_2. Let the solutions so obtained be denoted by $x_1(t)$ and $x_2(t)$. Then we claim that the function

$$g(t) = \lambda x_1(t) + (1 - \lambda)x_2(t) \tag{10}$$

has properties

$$\begin{cases} g'' = u + vg + wg' \\ g(a) = \alpha \end{cases}$$

which are left to the reader to verify. (The value of λ in this analysis is completely arbitrary.)

The function g nearly solves the two-point boundary value problem (8), and g contains a parameter λ at our disposal. Imposing the condition $g(b) = \beta$, we obtain

$$\lambda x_1(b) + (1 - \lambda)x_2(b) = \beta$$

from which

$$\lambda = \frac{\beta - x_2(b)}{x_1(b) - x_2(b)}$$

Perhaps the simplest way to implement these ideas is to solve two initial value problems

$$\begin{cases} x'' = u(t) + v(t)x + w(t)x' \\ x(a) = \alpha, \qquad x'(a) = 0 \end{cases}$$

and

$$\begin{cases} y'' = u(t) + v(t)y + w(t)y' \\ y(a) = \alpha, \qquad y'(a) = 1 \end{cases}$$

Then the solution to the original two-point boundary value problem (8) is

$$\lambda x(t) + (1 - \lambda)y(t) \quad \text{with} \quad \lambda = \frac{\beta - y(b)}{x(b) - y(b)} \tag{11}$$

In the computer realization of this procedure we must save the entire solution curves x and y. They are stored in vector arrays X and Y.

As an example of the shooting method, consider the problem of Eq. (7). We solve the two problems

$$\begin{cases} x'' = e^t - 3\sin(t) + x' - x \\ x(1) = 1.09737491, \\ x'(1) = 0 \end{cases} \qquad \begin{cases} y'' = e^t - 3\sin(t) + y' - y \\ y(1) = 1.09737491, \\ y'(1) = 1 \end{cases} \tag{12}$$

by using the fourth-order Runge–Kutta method. To do so, we introduce variables $x_1 = t$, $x_2 = x$, and $x_3 = x'$. Then the first initial value problem is

$$\begin{bmatrix} x_1' \\ x_2' \\ x_3' \end{bmatrix} = \begin{bmatrix} 1 \\ x_3 \\ e^{x_1} - 3\sin(x_1) + x_3 - x_2 \end{bmatrix}, \qquad \begin{bmatrix} x_1(1) \\ x_2(1) \\ x_3(1) \end{bmatrix} = \begin{bmatrix} 1 \\ 1.09737491 \\ 0 \end{bmatrix}$$

Now let $y_1 = t$, $y_2 = y$, and $y_3 = y'$. The second initial value problem that we must solve is similar except that we modify the initial vector

$$\begin{bmatrix} y_1' \\ y_2' \\ y_3' \end{bmatrix} = \begin{bmatrix} 1 \\ y_3 \\ e^{y_1} - 3\sin(y_1) + y_3 - y_2 \end{bmatrix}, \qquad \begin{bmatrix} y_1(1) \\ y_2(1) \\ y_3(1) \end{bmatrix} = \begin{bmatrix} 1 \\ 1.09737491 \\ 1 \end{bmatrix}$$

It is more efficient to solve these two problems together as a single system. Introducing $x_4 = y$ and $x_5 = y'$ into the first system, we have

$$\begin{bmatrix} x_1' \\ x_2' \\ x_3' \\ x_4' \\ x_5' \end{bmatrix} = \begin{bmatrix} 1 \\ x_3 \\ e^{x_1} - 3\sin(x_1) + x_3 - x_2 \\ x_5 \\ e^{x_1} - 3\sin(x_1) + x_5 - x_4 \end{bmatrix}, \qquad \begin{bmatrix} x_1(1) \\ x_2(1) \\ x_3(1) \\ x_4(1) \\ x_5(1) \end{bmatrix} = \begin{bmatrix} 1 \\ 1.09737491 \\ 0 \\ 1.09737491 \\ 1 \end{bmatrix}$$

Clearly the $x_2(t)$ and $x_4(t)$ components of the solution vector at each t satisfy the first and the second problems, respectively. Consequently, the solution is

$$\lambda x_2(t_i) + (1 - \lambda)x_4(t_i) \qquad (1 \leqq i \leqq n)$$

where $\lambda = [8.63749661 - x_4(2)]/[x_2(2) - x_4(2)]$. We shall use 100 points as before so that $n = 98$.

```
      DIMENSION  X(5),X2(99),X4(99)
      DATA  NP1/99/,TA/1.0/,TB/2.0/,ALPHA/1.09737491/,BETA/8.63749661/
      DATA  X/1.0,1.09737491,0.0,1.09737491,1.0/
      H = (TB - TA)/FLOAT(NP1)
      T = TA
      DO 2 I = 1,NP1
      CALL RK4SYS(T,X,H,1,0)
      X2(I) = X(2)
      X4(I) = X(4)
    2 T = TA + H*FLOAT(I)
      P = (BETA - X4(NP1))/(X2(NP1) - X4(NP1))
      Q = 1.0 - P
      DO 3 I = 1,NP1
    3 X2(I) = P*X2(I) + Q*X4(I)
      E = EXP(TA) - 3.0*COS(TA) - ALPHA
      PRINT 5,TA,ALPHA,E
      DO 4 I = 9,NP1,9
      T = TA + H*FLOAT(I)
      E = EXP(T) - 3.0*COS(T) - X2(I)
    4 PRINT 5,T,X2(I),E
    5 FORMAT(5X,F10.5,2(5X,E20.13))
      END
```

```
SUBROUTINE XPSYS(N,X,F)
DIMENSION  X(5),F(5)
N = 5
F(1) = 1.0
F(2) = X(3)
F(3) = EXP(X(1)) - 3.0*SIN(X(1)) + X(3) - X(2)
F(4) = X(5)
F(5) = EXP(X(1)) - 3.0*SIN(X(1)) + X(5) - X(4)
RETURN
END
```

Here the `SUBROUTINE RK4SYS(T,X,H,NSTEP,J)` has been called. This routine has been modified according to Computer Problem 6 of Sec. 13.1 so that only the Jth step of the Runge–Kutta method is printed. The computer results are shown.

T value	*Numerical Solution*	*Error*
1.00000	1.0973749100000E+00	8.5461238086282E−10
1.09091	1.5919417325402E+00	7.2773786996549E−10
1.18182	2.1225673472277E+00	5.8544458170218E−10
1.27273	2.6895523715877E+00	4.2760461838043E−10
1.36364	3.2933428112372E+00	2.5373481093993E−10
1.45455	3.9345687907979E+00	6.3550942286383E−11
1.54545	4.6140858007655E+00	−1.4318857211038E−10
1.63636	5.3330185038404E+00	−3.6658320823335E−10
1.72727	6.0928071596526E+00	−6.0666138779197E−10
1.81818	6.8952567452235E+00	−8.6367890617112E−10
1.90909	7.7425888700949E+00	−1.1374652331142E−09
2.00000	8.6374966100000E+00	−1.4279066817835E−09

Notice that the errors are smaller than those obtained in the discretization method for the same problem.

Problems 14.2

1. If standard finite-difference approximations to derivatives are used to solve a two-point boundary value problem involving the differential equation $x'' = t + 2x - x'$, what is the typical equation in the resulting linear system of equations?
2. Consider the two-point boundary value problem

$$\begin{cases} x'' = -x \\ x(0) = 0, \qquad x(1) = 1 \end{cases}$$

Set up and solve the tridiagonal system arising from the finite-difference method when $h = \frac{1}{4}$. Explain any differences from the analytic solution $x(\frac{1}{4}) \approx 0.29401$, $x(\frac{1}{2}) \approx 0.56975$, $x(\frac{3}{4}) \approx 0.81006$.
3. Verify that Eq. (11) is the solution of boundary value problem (8).
4. Consider the two-point boundary value problem

$$\begin{cases} x'' = x^2 - t + tx \\ x(0) = 1, \qquad x(1) = 3 \end{cases}$$

Suppose that we have solved two initial value problems

$$\begin{cases} u'' = u^2 - t + tu \\ u(0) = 1, \quad u'(0) = 1 \end{cases} \qquad \begin{cases} v'' = v^2 - t + tv \\ v(0) = 1, \quad v'(0) = 2 \end{cases}$$

numerically and have found as terminal values $u(1) = 2$, $v(1) = 3.5$. What is a reasonable initial value problem to try *next* in attempting to solve the original two-point value problem?

5. Consider the tridiagonal system (6). Show that if $v_i > 0$, then some choice of h exists for which the matrix is diagonally dominant.
6. Establish the properties claimed for the function g in Eq. (10).
7. Show that for the simple problem

$$\begin{cases} x'' = -x \\ x(a) = \alpha, \quad x(b) = \beta \end{cases}$$

the tridiagonal system to be solved can be written

$$\begin{cases} (2 - h^2)x_1 - x_2 = \alpha \\ -x_{i-1} + (2 - h^2)x_i - x_{i+1} = 0 \qquad (2 \leqq i \leqq n - 1) \\ -x_{n-1} + (2 - h^2)x_n = \beta \end{cases}$$

Computer Problems 14.2

1. Explain the main steps in setting up a program to solve this two-point boundary value problem by the finite-difference method.

$$\begin{cases} x'' = x \sin t + x' \cos t - e^t \\ x(0) = 0, \quad x(1) = 1 \end{cases}$$

Show any preliminary work that must be done before programming. Exploit the linearity of the differential equation. Program and compare the results when different values of N are used, say $N = 10$, 100, and 1000.

2. Solve the following two-point boundary value problem numerically. For comparisons, the exact solutions are given.

(a) $x'' = \dfrac{(1 - t)x + 1}{(1 + t)^2}, \qquad x(0) = 1, \quad x(1) = 0.5$

Solution: $x = \dfrac{1}{1 + t}$

(b) $x'' = \dfrac{1}{3}[(2 - t)e^{2x} + (1 + t)^{-1}], \qquad x(0) = 0, \quad x(1) = -\log 2$

Solution: $x = -\log(1 + t)$

3. Solve the boundary value problem

$$\begin{cases} x'' = -x + tx' - 2t \cos t + t \\ x(0) = 0, \quad x(\pi) = \pi \end{cases}$$

by discretization. Compare to the exact solution, which is $x(t) = t + 2 \sin t$.

Supplementary Problems (Chapter 14)

1. Solve

$$\begin{cases} x'' = -x \\ x(0) = 2, \quad x(\pi) = 3 \end{cases}$$

analytically and analyze any difficulties.

2. Show that the following two problems are equivalent in the sense that a solution of one is easily obtained from a solution of the other.

(a) $$\begin{cases} y'' = f(t, y) \\ y(0) = \alpha, \quad y(1) = \beta \end{cases}$$

(b) $$\begin{cases} z'' = f(t, z + \alpha - \alpha t + \beta t) \\ z(0) = 0, \quad z(1) = 0 \end{cases}$$

3. Discuss in general terms the numerical solution of the following two-point boundary value problems. Recommend specific methods for each, being sure to take advantage of any special structure.

(a) $$\begin{cases} x'' = \sin t + (e^t\sqrt{t^2 + 1})x + (\cos t)x' \\ x(0) = 0, \quad x(1) = 5 \end{cases}$$

(b) $$\begin{cases} x_1' = x_1^2 + (t - 3)x_1 + \sin t \\ x_2' = x_2^3 + \sqrt{t^2 + 1} + (\cos t)x_1 \\ x_1(0) = 1, \quad x_2(2) = 3 \end{cases}$$

4. Write down the system of equations $\mathbf{Ax} = \mathbf{b}$ that results from using the usual second-order central difference approximation to solve

$$\begin{cases} x'' = (1 + t)x \\ x(0) = 0, \quad x(1) = 1 \end{cases}$$

5. Let u be a solution of the initial value problem

$$\begin{cases} u'' = e^t u + t^2 u' \\ u(1) = 0, \quad u'(1) = 1 \end{cases}$$

How do we solve the following two-point boundary value problem by utilizing u?

$$\begin{cases} x'' = e^t x + t^2 x' \\ x(1) = 0, \quad x(2) = 7 \end{cases}$$

15

Partial Differential Equations

In the theory of elasticity it is shown that the stress in a cylindrical beam under torsion can be derived from a function $u(x, y)$ that satisfies the Poisson equation

$$\frac{\partial^2 u}{\partial x^2} + \frac{\partial^2 u}{\partial y^2} + 2 = 0$$

In the case of a beam whose cross section is the square defined by $|x| \leqq 1$, $|y| \leqq 1$, the function u must satisfy Poisson's equation *inside* the square and must be zero at each point on the *perimeter* of the square. Using the methods of this chapter, a table of approximate values of u can be constructed.

Many physical phenomena can be modeled mathematically by differential equations. When the function being studied involves two or more independent variables, the differential equation will usually be a *partial* differential equation. Since functions of several variables are intrinsically more complicated than those of one variable, partial differential equations can lead to the most challenging of numerical problems. In fact, their numerical solution is one type of scientific calculation in which the resources of the biggest and fastest computing systems easily become taxed. We shall see later why this is so.

Some important partial differential equations and the physical phenomena that they govern are listed here.

1. The *wave equation* in three spatial variables (x, y, z) and time t is

$$\frac{\partial^2 u}{\partial x^2} + \frac{\partial^2 u}{\partial y^2} + \frac{\partial^2 u}{\partial z^2} - \frac{\partial^2 u}{\partial t^2} = 0$$

   The function u represents the displacement at time t of a particle whose position at rest is (x, y, z). With appropriate boundary conditions, this equation governs vibrations of a three-dimensional elastic body.

2. The *heat equation* is

$$\frac{\partial^2 u}{\partial x^2} + \frac{\partial^2 u}{\partial y^2} + \frac{\partial^2 u}{\partial z^2} - \frac{\partial u}{\partial t} = 0$$

The function u represents the temperature at time t at the point whose coordinates are (x, y, z).

3. *Laplace's equation* is

$$\frac{\partial^2 u}{\partial x^2} + \frac{\partial^2 u}{\partial y^2} + \frac{\partial^2 u}{\partial z^2} = 0$$

It governs the steady-state distribution of heat in a body or the steady-state distribution of electrical charge in a body. Laplace's equation also governs gravitational, electric, and magnetic potentials, and velocity potentials in irrotational flows of incompressible fluids. The form of Laplace's equation given above applies to rectangular coordinates. In cylindrical and spherical coordinates it takes these respective forms:

$$\frac{\partial^2 u}{\partial r^2} + \frac{1}{r}\frac{\partial u}{\partial r} + \frac{1}{r^2}\frac{\partial^2 u}{\partial \phi^2} + \frac{\partial^2 u}{\partial z^2} = 0$$

$$\frac{1}{r}\frac{\partial^2}{\partial r^2}(ru) + \frac{1}{r^2 \sin\theta}\frac{\partial}{\partial \theta}\left(\sin\theta \frac{\partial u}{\partial \theta}\right) + \frac{1}{r^2 \sin^2\theta}\frac{\partial^2 u}{\partial \phi^2} = 0$$

4. The *biharmonic equation* is

$$\frac{\partial^4 u}{\partial x^4} + 2\frac{\partial^4 u}{\partial x^2 \partial y^2} + \frac{\partial^4 u}{\partial y^4} = 0$$

It occurs in the study of elastic stress, and from its solution the shearing and normal stresses can be derived for an elastic body.

5. The *Navier–Stokes equations* are

$$\frac{\partial u}{\partial t} + u\frac{\partial u}{\partial x} + v\frac{\partial u}{\partial y} + \frac{\partial p}{\partial x} = \frac{\partial^2 u}{\partial x^2} + \frac{\partial^2 u}{\partial y^2}$$

$$\frac{\partial v}{\partial t} + u\frac{\partial v}{\partial x} + v\frac{\partial v}{\partial y} + \frac{\partial p}{\partial y} = \frac{\partial^2 v}{\partial x^2} + \frac{\partial^2 v}{\partial y^2}$$

Here u and v are components of the velocity vector in a fluid flow. The function p is the pressure, and the fluid is assumed to be incompressible but viscous.

Additional examples from quantum mechanics, electromagnetism, hydrodynamics, elasticity, and so on could also be given. But the five examples shown already exhibit a great diversity. Example 5, in particular, illustrates a very complicated problem: a pair of nonlinear simultaneous partial differential equations.

To specify a unique solution to a partial differential equation, additional conditions must be imposed upon the solution function. Typically these conditions occur in the form of boundary values that are prescribed on all or part of the perimeter of the region in which the solution is sought. The nature of the boundary and the boundary values are usually the determining factors in setting up an appropriate numerical scheme for obtaining the approximate solution.

15.1 Parabolic Problems

In this section we consider a model problem of modest scope to introduce some of the essential ideas. For technical reasons, the problem is said to be of *parabolic type*. In it we have the heat equation in one spatial variable accompanied by boundary conditions appropriate to a certain physical phenomenon:

$$\begin{cases} \dfrac{\partial^2}{\partial x^2} u(x, t) = \dfrac{\partial}{\partial t} u(x, t) \\ u(0, t) = u(1, t) = 0, \qquad u(x, 0) = \sin \pi x \end{cases} \tag{1}$$

These equations govern the temperature $u(x, t)$ in a thin rod of length 1 when the ends are held at temperature 0, under the assumption that the initial temperature in the rod is given by the function $\sin \pi x$ (see Fig. 15.1).

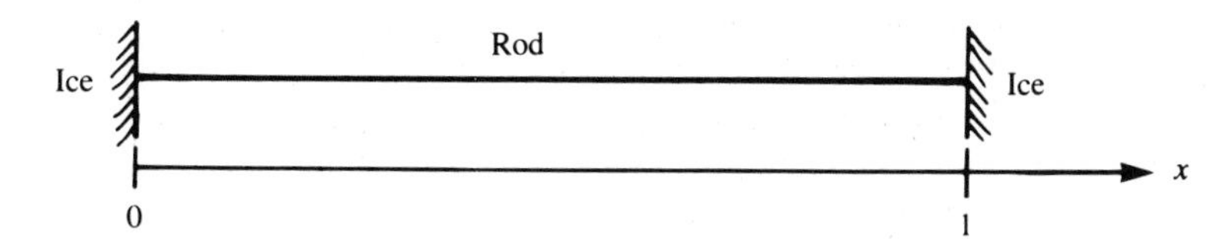

Figure 15.1

In the xt plane, the region in which the solution is sought is described by inequalities $0 \leqq x \leqq 1$ and $t \geqq 0$. On the boundary of this region (shaded in Fig. 15.2) the values of u have been prescribed.

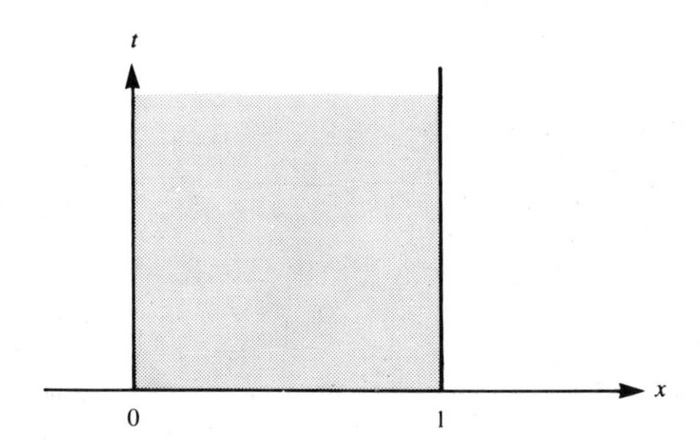

Figure 15.2

A principal approach to the numerical solution of such a problem is the *finite-difference method*. It proceeds by replacing the derivatives in the equation by finite differences. Two formulas from Sec. 5.3 are useful in this context:

$$f'(x) \approx \frac{1}{h} [f(x + h) - f(x)]$$

$$f''(x) \approx \frac{1}{h^2} [f(x + h) - 2f(x) + f(x - h)]$$

If used in the differential equation (1), with possibly different step lengths h and k, the result is

$$\frac{1}{h^2} [u(x + h, t) - 2u(x, t) + u(x - h, t)] = \frac{1}{k} [u(x, t + k) - u(x, t)] \tag{2}$$

This equation is now interpreted as a means of advancing the solution step by step in the t variable. That is, if $u(x, t)$ is known for $0 \leqq x \leqq 1$ and $0 \leqq t \leqq t_0$, then Eq. (2) allows us to evaluate the solution for $t = t_0 + k$.

Equation (2) can be rewritten in the form

$$u(x, t + k) = su(x + h, t) + (1 - 2s)u(x, t) + su(x - h, t) \tag{3}$$

where $s = k/h^2$. A sketch showing the location of the four points involved in this equation is given in Fig. 15.3. Since the solution is known on the boundary of the region, it is possible to compute an approximate solution inside the region by systematically using Eq. (3). It is, of course, an *approximate* solution because Eq. (2) is only a finite-difference analog of Eq. (1).

In order to obtain an approximate solution on a computer, we select values for h and k and use Eq. (3). An analysis of this procedure, which is outside the scope of this text, shows that for "stability" of the computation the coefficient $1 - 2s$ in Eq. (3) should be nonnegative. (If this condition is not met, errors made at one step will probably be magnified at subsequent steps, ultimately spoiling the solution.)

For utmost simplicity, we select $h = 0.1$ and $k = 0.005$. This choice makes the coefficient $1 - 2s$ equal to zero. Coefficient s is now 0.5. Our program first prints $u(ih, 0)$ for $0 \leqq i \leqq 10$, since they are known boundary values. Then it computes and prints $u(ih, k)$ for $0 \leqq i \leqq 10$, using Eq. (3) and boundary values $u(0, t) = u(1, t) = 0$. This procedure is continued until t reaches the value 0.1. The singly subscripted arrays U and V are used to store the values of the approximate solution at t and $t + k$, respectively. Since the analytic solution of the problem is $u(x, t) = e^{-\pi^2 t} \sin \pi x$ (see Problem 3), the error can be printed out at each step.

```
      DIMENSION  U(11),V(11)
      DATA  U(1),V(1),U(11),V(11)/4*0.0/
      DATA  M/11/, H/0.1/, HK/0.005/
      PI = 4.0*ATAN(1.0)
      PI2 = PI*PI
      DO 2 I = 2,M-1
    2 U(I) = SIN(PI*H*FLOAT(I-1))
      PRINT 6, (U(I),I = 1,M)
      DO 5 J = 1,20
      DO 3 I = 2,M-1
    3 V(I) = 0.5*(U(I-1) + U(I+1))
      PRINT 6, (V(I),I = 1,M)
      T = HK*FLOAT(J)
      DO 4 I = 2,M-1
    4 U(I) = EXP(-PI2*T)*SIN(PI*H*FLOAT(I-1)) - V(I)
      PRINT 6, (U(I),I = 1,M)
      DO 5 I = 2,M-1
    5 U(I) = V(I)
    6 FORMAT(//(5(5X,E20.13)))
      END
```

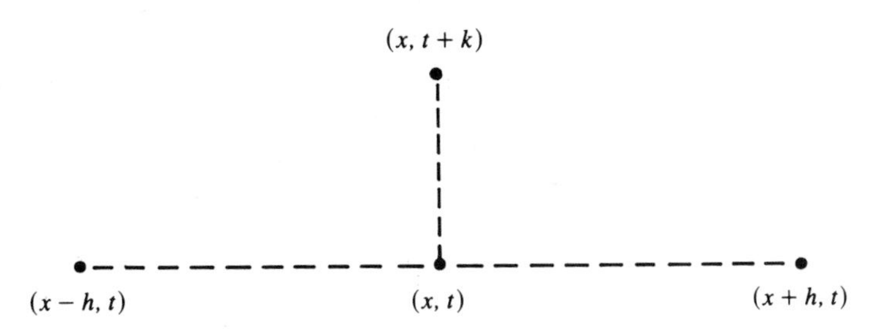

Figure 15.3

The procedure just described is an example of an *explicit method*. The approximate values of $u(x, t + k)$ are calculated explicitly in terms of $u(x, t)$. Not only is this situation atypical, but even in this problem the procedure is rather slow, since considerations of stability force us to select $k \leqq h^2/2$. Since h must be rather small in order to represent the derivative accurately by the finite-difference formula, the corresponding k must be extremely small. Values $h = 0.1$ and $k = 0.005$ are representative, as are $h = 0.01$ and $k = 0.00005$. With such small values of k, an inordinate amount of computation is necessary to make much progress in the t variable.

An alternative procedure of implicit type goes by the name of its inventors, Crank and Nicolson, and is based on a simple variant of Eq. (2).

$$\frac{1}{h^2}[u(x + h, t) - 2u(x, t) + u(x - h, t)] = \frac{1}{k}[u(x, t) - u(x, t - k)] \tag{4}$$

If a numerical solution at grid points $x = ih$, $t = jk$ has been obtained up to a certain level in the t variable, Eq. (4) governs the values of u on the next t level. Therefore Eq. (4) should be rewritten

$$-u(x - h, t) + ru(x, t) - u(x + h, t) = su(x, t - k) \tag{5}$$

in which $r = 2 + s$ and $s = h^2/k$. The locations of the four points in this equation are shown in Fig. 15.4.

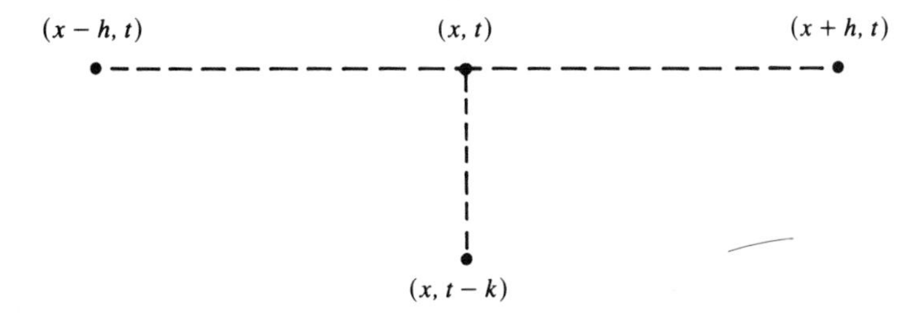

Figure 15.4

On the t level, u is unknown; but on the $(t - k)$ level, u is known. So we can introduce unknowns $u_i = u(ih, t)$ and known quantities $b_i = su(ih, t - k)$ and write Eq. (5) in matrix form.

$$\begin{bmatrix} r & -1 & & & & \\ -1 & r & -1 & & 0 & \\ & -1 & r & -1 & & \\ & & -1 & r & -1 & \\ & 0 & & \ddots & \ddots & \ddots \\ & & & & -1 & r \end{bmatrix} \begin{bmatrix} u_1 \\ u_2 \\ u_3 \\ u_4 \\ \vdots \\ u_{n-1} \end{bmatrix} = \begin{bmatrix} b_1 \\ b_2 \\ b_3 \\ b_4 \\ \vdots \\ b_{n-1} \end{bmatrix} \quad (6)$$

The simplifying assumption that $u(0, t) = u(1, t) = 0$ has been used here. Also, $h = 1/n$. The system of equations is tridiagonal and diagonally dominant, since $|r| = 2 + h^2/k > 2$. Hence it can be solved by the efficient methods of Sec. 6.3.

An elementary argument shows that the method contemplated here is stable. We shall see that if the initial values $u(x, 0)$ lie in an interval $[\alpha, \beta]$, then values subsequently calculated by using Eq. (5) will also lie in $[\alpha, \beta]$, thereby ruling out any unstable growth. Since the solution is built up line by line in a uniform way, we need only verify that the values on the first computed line, $u(x, k)$, lie in $[\alpha, \beta]$. Let j be the index of the largest u_i that occurs on this line $t = k$. Then

$$-u_{j-1} + ru_j - u_{j+1} = b_j$$

Since u_j is the largest of the u's, $u_{j-1} \leqq u_j$ and $u_{j+1} \leqq u_j$. Thus

$$ru_j = b_j + u_{j-1} + u_{j+1} \leqq b_j + 2u_j$$

Since $r = 2 + s$ and $b_j = su(jh, 0)$, the previous inequality leads at once to $u_j \leqq u(jh, 0) \leqq \beta$. Since u_j is the largest of the u_i, we have $u_i \leqq \beta$ for all i. Similarly, $u_i \geqq \alpha$ for all i, establishing our assertion.

A program to carry out the *Crank–Nicolson method* on the model program is given next. In it $h = 0.1$, $k = h^2$, and the solution is continued until $t = 0.1$. The value of r is 3

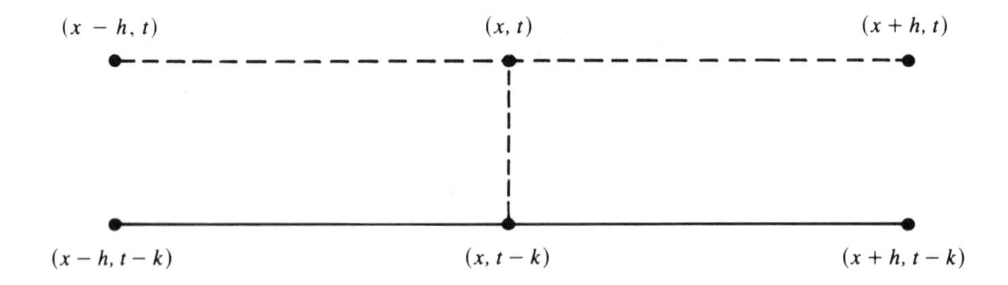

Figure 15.5

and $s = 1$. It is easier to compute and print only the values of u at interior points on each horizontal line. At boundary points we have $u(0, t) = u(1, t) = 0$. The program calls subroutine TRI from Sec. 6.3

```
      DIMENSION  C(9),D(9),U(9),V(9)
      DATA  N/9/, H/0.1/, HK/0.005/
      PI = 4.0*ATAN(1.0)
      PI2 = PI*PI
      S = H*H/HK
      R = 2.0 + S
      DO 2 I = 1,N
      D(I) = R
      C(I) = -1.0
    2 U(I) = SIN(PI*H*FLOAT(I))
      PRINT 6, (U(I),I = 1,N)
      DO 5 J = 1,20
      DO 3 I = 1,N
      D(I) = R
    3 V(I) = S*U(I)
      CALL TRI(N,C,D,C,V)
      PRINT 6, (V(I),I = 1,N)
      T = HK*FLOAT(J)
      DO 4 I = 1,N
    4 U(I) = EXP(-PI2*T)*SIN(PI*H*FLOAT(I)) - V(I)
      PRINT 6, (U(I),I = 1,N)
      DO 5 I = 1,N
    5 U(I) = V(I)
    6 FORMAT(//(5(5X,E20.13)))
      END
```

Another version of the Crank–Nicolson method is obtained as follows. The central differences at $(x, t - \frac{1}{2}k)$ in Eq. (4) produce

$$\frac{1}{h^2}\left[u\left(x + h, t - \frac{k}{2}\right) - 2u\left(x, t - \frac{k}{2}\right) + u\left(x - h, t - \frac{k}{2}\right)\right]$$

$$= \frac{1}{k}\,[u(x, t) - u(x, t - k)]$$

Since the u values are known only at integer multiples of k, terms like $u(x, t - \frac{1}{2}k)$ are replaced by the average of u values at adjacent grid points; that is,

$$u\left(x, t - \frac{1}{2}k\right) \approx \frac{1}{2}\,[u(x, t) + u(x, t - k)]$$

So we have

$$\frac{1}{2h^2}[u(x+h,t) - 2u(x,t) + u(x-h,t) + u(x+h,t-k) - 2u(x,t-k) + u(x-h,t-k)] = \frac{1}{k}[u(x,t) - u(x,t-k)]$$

The computational form of this equation is

$$-u(x-h,t) + 2(1+s)u(x,t) - u(x+h,t) = u(x-h,t-k) + 2(s-1)u(x,t-k) + u(x+h,t-k) \qquad (7)$$

where $s = h^2/k$. The six points in this equation are shown in Fig. 15.5 on p. 320. This leads to a tridiagonal system of form (6) with $r = 2(1+s)$ and

$$b_i = u((i-1)h, t-k) + 2(s-1)u(ih, t-k) + u((i+1)h, t-k).$$

Notice the special form of (6) when $k = h^2$.

Problems 15.1

1. A second-order linear differential equation with two variables will have the form

$$A\frac{\partial^2 u}{\partial x^2} + B\frac{\partial^2 u}{\partial x\,\partial y} + C\frac{\partial^2 u}{\partial y^2} + \cdots = 0$$

Here A, B, and C are functions of x and y, and the terms not written are of lower order. The equation is said to be elliptic, parabolic, or hyperbolic at a point (x, y), depending on whether $B^2 - 4AC$ is negative, zero, or positive, respectively. Classify each equation in this manner.
(a) $u_{xx} + u_{yy} + u_x + \sin xu_y - u = x^2 + y^2$
(b) $u_{xx} - u_{yy} + 2u_x + 2u_y + e^x u = x - y$
(c) $u_{xx} = u_y + u - u_x + y$
(d) $u_{xy} = u - u_x - u_y$
(e) $3u_{xx} + u_{xy} + u_{yy} = e^{xy}$
(f) $e^x u_{xx} + \cos yu_{xy} - u_{yy} = 0$
(g) $u_{xx} + 2u_{xy} + u_{yy} = 0$
(h) $xu_{xx} + yu_{xy} + u_{yy} = 0$
2. Write the two-dimensional form of Laplace's equation in polar coordinates.
3. Show that the function

$$u(x,t) = \sum_{n=1}^{N} c_n e^{-(n\pi)^2 t} \sin n\pi x$$

is a solution of the heat conduction problem $u_{xx} = u_t$ and satisfies the boundary condition

$$u(0,t) = u(1,t) = 0,\ u(x,0) = \sum_{n=1}^{N} c_n \sin n\pi x \quad \text{for all } N \geqq 1.$$

4. Refer to the model problem solved numerically in this section and show that if there is no round-off, the approximate solution values obtained by using Eq. (3) lie in the interval $[0, 1]$. (Assume $1 \geqq 2k/h^2$.)
5. Find a solution of Eq. (3) having the form $u(x,t) = a^t \sin \pi x$, where a is a constant.

6. When using Eq. (5), how must the linear system (6) be modified for $u(0, t) = c_0$ and $u(1, t) = c_n$ with $c_0 \neq 0$, $c_n \neq 0$? When using Eq. (7)?
7. Describe in detail how problem (1) with boundary conditions $u(0, t) = q(t)$, $u(1, t) = g(t)$ and $u(x, 0) = f(x)$ can be solved numerically using (6). Here q, g and f are known functions.
8. What finite-difference equation would be a suitable replacement for the equation $\partial^2 u/\partial x^2 = \partial u/\partial t + \partial u/\partial x$ in numerical work?
9. Consider the partial differential equation $\partial u/\partial x + \partial u/\partial t = 0$ with $u = u(x, t)$ in the region $[0, 1] \times [0, \infty]$, subject to the boundary conditions $u(0, t) = 0$ and $u(x, 0)$ specified. For fixed t, we discretize only the first term, using $(u_{i+1} - u_{i-1})/(2h)$ for $i = 1, 2, \ldots, n - 1$ and $(u_n - u_{n-1})/h$, where $h = 1/n$. Here $u_i = u(x_i, t)$ and $x_i = ih$ with fixed t. By so doing, the original problem can be considered a first-order initial value problem

$$\frac{d\mathbf{y}}{dx} + \frac{1}{2h}\mathbf{A}\mathbf{y} = 0$$

where

$$\mathbf{y} = (u_1, u_2, \ldots, u_n)^T, \quad \frac{d\mathbf{y}}{dx} = (u_1', u_2', \ldots, u_n')^T, \quad u_i' = \left(\frac{\partial u}{\partial t}\right)_i$$

Determine the $n \times n$ matrix $\mathbf{A}$.
10. Refer to the discussion of stability of the Crank–Nicolson procedure, and establish the inequality $u_i \geqq \alpha$.

Computer Problems 15.1

1. Solve the same heat conduction problem as in the text except use $h = 2^{-4}$, $10k = 2^{-10}$, and $u(x, 0) = x(1 - x)$. Carry out the solution until $t = 0.0125$.
2. Modify the Crank–Nicolson code in the text so that it uses the second scheme (7). Compare the two programs on the same problems with the same spacing.

15.2 Hyperbolic Problems

The "wave equation" with one space variable

$$\frac{\partial^2 u}{\partial x^2} = \frac{\partial^2 u}{\partial t^2} \tag{1}$$

governs the vibration of a string (transverse vibration in a plane) or vibration in a rod (longitudinal vibration). It is an example of a second-order linear differential equation of hyperbolic type. If (1) is used to model the vibrating string, then $u(x, t)$ represents the deflection at time t of a point on the string whose coordinate is x when the string is at rest.

To pose a definite model problem, we suppose the points on the string have coordinates x in interval $0 \leqq x \leqq 1$. (See Fig. 15.6.). Let us suppose that at time $t = 0$ the deflections satisfy equations $u(x, 0) = f(x)$ and $u_t(x, 0) = 0$. Assume also that the ends of the string remain fixed. Then $u(0, t) = u(1, t) = 0$. A fully defined boundary value problem then is

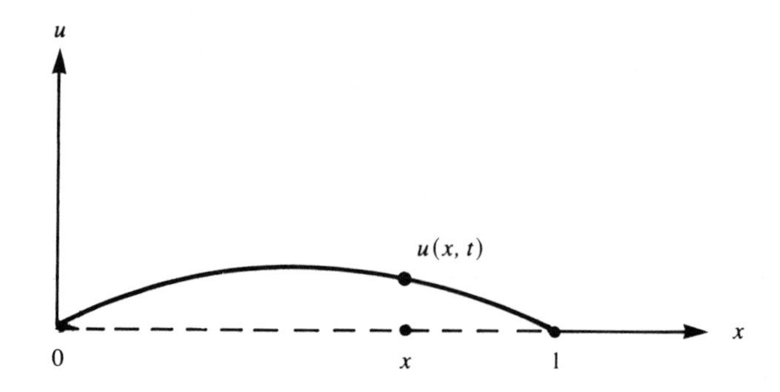

Figure 15.6

$$\begin{cases} u_{xx} - u_{tt} = 0 \\ u(x, 0) = f(x) \\ u_t(x, 0) = 0 \\ u(0, t) = u(1, t) = 0 \end{cases} \tag{2}$$

The region in the tx plane where a solution is sought is the semi-infinite strip defined by inequalities $0 \leqq x \leqq 1$, $t \geqq 0$. As in the heat conduction problem of Sec. 15.1, the values of the unknown function are prescribed on the boundary of the region shown (Fig. 15.7).

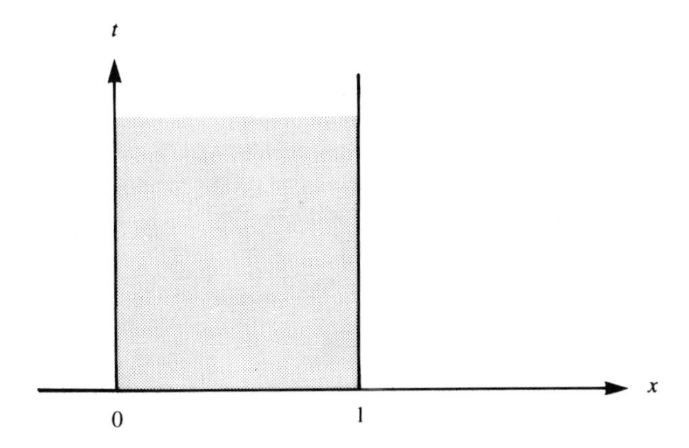

Figure 15.7

The model problem in (2) is so simple that it can be immediately solved. Indeed, the solution is

$$u(x, t) = \frac{1}{2} [f(x + t) + f(x - t)] \tag{3}$$

provided that f possesses two derivatives and has been extended to the whole real line by defining $f(-x) = -f(x)$ and $f(x + 2) = f(x)$. To verify that (3) is a solution, we compute derivatives, using the chain rule.

$$u_x = \frac{1}{2}[f'(x+t) + f'(x-t)] \qquad u_t = \frac{1}{2}[f'(x+t) - f'(x-t)]$$

$$u_{xx} = \frac{1}{2}[f''(x+t) + f''(x-t)] \qquad u_{tt} = \frac{1}{2}[f''(x+t) - f''(x-t)]$$

Obviously $u_{xx} = u_{tt}$; also, $u(x, 0) = f(x)$. Furthermore, we have $u_t(x, 0) = \frac{1}{2}[f'(x) - f'(x)] = 0$. In checking endpoint conditions, we use the formulas by which f was extended.

$$u(0, t) = \frac{1}{2}[f(t) + f(-t)] = 0$$

$$\begin{aligned} u(1, t) &= \frac{1}{2}[f(1+t) + f(1-t)] \\ &= \frac{1}{2}[f(1+t) - f(t-1)] \\ &= \frac{1}{2}[f(1+t) - f(t-1+2)] = 0 \end{aligned}$$

The extension of f from its original domain to the entire real line makes it an *odd periodic* function of period 2. "Odd" means that $f(x) = -f(-x)$, and the periodicity is expressed by $f(x+2) = f(x)$ for all x. To compute $u(x, t)$, we need only know f at two points on the x axis, $x + t$ and $x - t$, as in Fig. 15.8.

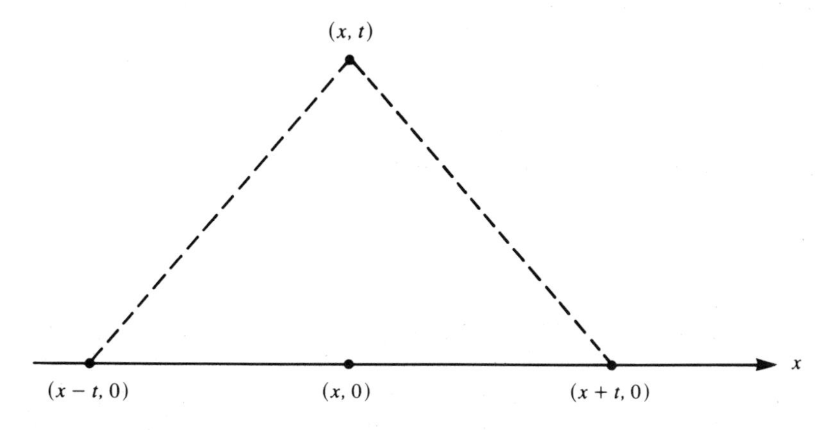

Figure 15.8

The model problem is used next to illustrate again the principle of numerical solution. Choosing step sizes h and k for x and t, respectively, and using the familiar approximations for derivatives, we have from Eq. (1)

$$\frac{1}{h^2}[u(x+h,t)-2u(x,t)+u(x-h,t)]$$
$$=\frac{1}{k^2}[u(x,t+k)-2u(x,t)+u(x,t-k)]$$

which can be rearranged as

$$u(x,t+k)$$
$$=\rho u(x+h,t)+2(1-\rho)u(x,t)+\rho u(x-h,t)-u(x,t-k) \quad (4)$$

Here ρ denotes k^2/h^2. Figure 15.9 shows the point $(x, t+k)$ and the nearby points that enter into Eq. (4).

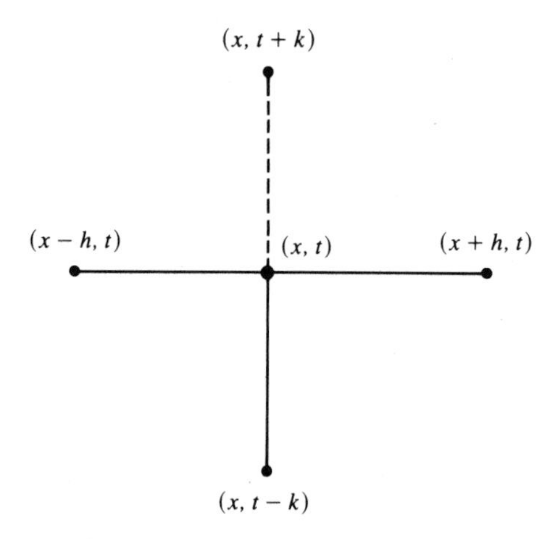

Figure 15.9

The boundary conditions in problem (2) can be written

$$\begin{cases} u(x,0)=f(x) \\ \frac{1}{k}[u(x,k)-u(x,0)]=0 \\ u(0,t)=u(1,t)=0 \end{cases} \quad (5)$$

The problem defined by Eqs. (4) and (5) can be solved by beginning at the line $t = 0$ where u is known and then progressing one line at a time with $t = k$, $t = 2k$, $t = 3k$, ... Note that, because of (5), our approximate solution satifies

$$u(x,k)=u(x,0)=f(x) \quad (6)$$

The use of the $\mathcal{O}(h)$ approximation for u_t leads to low accuracy in the computed solution to problem (2). Suppose that there was a row of grid points $(x, -k)$. Letting $t = 0$ in Eq. (4), we have

$$u(x,k)=\rho u(x+h,0)+2(1-\rho)u(x,0)+\rho u(x-h,0)-u(x,-k)$$

Now the central difference approximation

$$\frac{1}{2k}[u(x, k) - u(x, -k)] = 0$$

for $u_t(x, 0) = 0$ can be used to eliminate the fictitious grid point $(x, -k)$. So instead of (6), we set

$$u(x, k) = \frac{r}{2}[f(x + h) + f(x - h)] + (1 - r)f(x) \tag{7}$$

since $u(x, 0) = f(x)$. Values of $u(x, nk)$, $n \geqq 2$, can now be computed from Eq. (4).

A Fortran program to carry out this numerical process is given next. For simplicity, three one-dimensional arrays, U, V, W, are used. U represents the solution being computed on the new t line, whereas V and W represent solutions on the preceding two t lines. Step size k is denoted by HK in the program.

```
      DIMENSION  U(11),V(11),W(11)
      DATA  N/11/,  M/20/,  H/0.1/,  HK/0.05/
      DATA  U(1),V(1),W(1),U(11),V(11),W(11)/6*0.0/
      RHO  =  (HK/H)**2
      PHO  = 2.0*(1.0  -  RHO)
      DO 2 I = 2,N-1
      X = H*FLOAT(I-1)
      W(I)  = F(X)
    2 V(I)  = 0.5*( RHO*(F(X-H)  + F(X+H))  + PHO*F(X)  )
      DO 5 K = 2,M
      DO 3 I = 2,N-1
    3 U(I)  = RHO*(V(I+1)  + V(I-1))  + PHO*V(I)  -  W(I)
      PRINT 6,K, (U(I),I = 1,N)
      DO 4 I = 2,N-1
      W(I)  = V(I)
      V(I)  = U(I)
      T = HK*FLOAT(K)
      X = H*FLOAT(I-1)
    4 U(I)  = TRUE(X,T)  -  V(I)
    5 PRINT 6,K, (U(I),I = 1,N)
    6 FORMAT(//5X,I5,//(4(5X,E20.13)))
      END

      FUNCTION F(X)
      DATA  PI/3.14159 26535 8979/
      F = SIN(PI*X)
      RETURN
      END

      FUNCTION TRUE(X,T)
      DATA  PI/3.14159 26535 8979/
      TRUE = SIN(PI*X)*COS(PI*T)
      RETURN
      END
```

This program requires an accompanying function subprogram to compute values of $f(x)$. It is assumed that the x interval is $[0, 1]$, but by changing h or n, the interval can be $[0, b]$—that is, $(n - 1)h = b$. The numerical solution is printed on the t lines corresponding to $2k, 3k, \ldots, (m + 1)k$.

More advanced treatments show that the ratios $\rho = k^2/h^2$ must not exceed one if the solution of the finite-difference equations is to converge to a solution of the differential problem as $k \to 0$ and $h \to 0$. Furthermore, if $\rho > 1$, roundoff errors occurring at one stage of the computation would probably be magnified at later stages and thereby ruin the numerical solution.

Problems 15.2

1. What is the solution of the boundary value problem
$$u_{xx} = u_{tt}, \quad u(x, 0) = x(1 - x), \quad u(0, t) = u(1, t)$$
at the point where $x = 3.3$ and $t = 4$?
2. Show that the function $u(x, t) = f(x + at) + g(x - at)$ satisfies the wave equation $a^2u_{xx} = u_{tt}$.
3. (Continuation) Using the idea in Problem 2, solve this boundary value problem.
$$u_{xx} = u_{tt}, \quad u(x, 0) = F(x), \quad u_t(x, 0) = G(x), \quad u(0, t) = u(1, t) = 0$$
4. Show that the boundary value problem $u_{xx} = u_{tt}$, $u(x, 0) = 2f(x)$, $u_t(x, 0) = 2g(x)$ has the solution $u(x, t) = f(x + t) + f(x - t) + G(x + t) - G(x - t)$, where G is an antiderivative of g. Assume that $-\infty < x < \infty$ and $t \geqq 0$.
5. (Continuation) Solve Problem 4 on a finite x interval, say $0 \leqq x \leqq 1$, adding boundary condition $u(0, t) = u(1, t) = 0$. In this case, f and g are defined only for $0 \leqq x \leqq 1$.

Computer Problems 15.2

1. Given a subprogram for calculating $f(x)$ when $0 \leqq x \leqq 1$, write a subprogram for calculating the extended f that obeys the equations $f(-x) = -f(x)$, $f(x + 2) = f(x)$.
2. (Continuation) Write a program to compute the solution $u(x, t)$, at any given point (x, t), for the boundary value problem of Eq. (2).
3. Compare the accuracy of the computed solution, using first (6) and then (7) on the same problem.
4. Use the program in the text to solve boundary value problem (2) with
$$f(x) = \frac{1}{4}\left(\frac{1}{2} - \left|x - \frac{1}{2}\right|\right), \quad h = \frac{1}{16}, \quad k = \frac{1}{32}$$
5. Modify the program in the text to solve boundary value problem (2) when $u_t(x, 0) = g(x)$. *Hint:* Equations (5) and (7) will be slightly different (a fact that affects only the initial DO loop in the program).
6. (Continuation) Use the program that you wrote for Problem 5 to solve the following boundary value problem.
$$\begin{cases} u_{xx} = u_{tt} \qquad (0 \leqq x \leqq 1,\ t \geqq 0) \\ u(x, 0) = \sin \pi x \\ u_t(x, 0) = \dfrac{1}{4} \sin 2\pi x \\ u(0, t) = u(1, t) = 0 \end{cases}$$

7. Modify the program in the text to solve the following boundary value problem.

$$\begin{cases} u_{xx} = u_{tt} & (-1 \leqq x \leqq 1,\ t \geqq 0) \\ u(x, 0) = |x| - 1 \\ u_t(x, 0) = 0 \\ u(-1, t) = u(+1, t) = 0 \end{cases}$$

8. Modify the program in the text to avoid storage of the V and W arrays.
9. Simplify the program in the text for the special case $\rho = 1$. Compare the numerical solution at the same grid points for a problem when $\rho = 1$ and $\rho \neq 1$.

15.3 Elliptic Problems

One of the most important partial differential equations in mathematical physics and engineering is *Laplace's equation,* which has the following form in three variables.

$$\nabla^2 u \equiv \frac{\partial^2 u}{\partial x^2} + \frac{\partial^2 u}{\partial y^2} + \frac{\partial^2 u}{\partial z^2} = 0$$

Closely related to it is *Poisson's equation.*

$$\nabla^2 u = \frac{\partial^2 u}{\partial x^2} + \frac{\partial^2 u}{\partial y^2} + \frac{\partial^2 u}{\partial z^2} = g(x, y, z)$$

These are examples of *elliptic* equations. (Refer to Problem 1 of Sec. 15.1 for the classification of equations.) The boundary conditions associated with elliptic equations generally differ from those for parabolic and hyperbolic equations. A model problem is considered here to illustrate the numerical procedures often used.

Suppose that a function $u = u(x, y)$ of two variables is the solution to a certain physical problem. This function is unknown but has some properties that, theoretically, determine it uniquely. We assume that on a given region R in the xy plane

$$\begin{cases} \dfrac{\partial^2 u}{\partial x^2} + \dfrac{\partial^2 u}{\partial y^2} + fu = g \\ u(x, y) \text{ known on the boundary of } R. \end{cases} \tag{1}$$

Here $f = f(x, y)$ and $g = g(x, y)$ are given continuous functions defined in R. The boundary values could be given by a third function: $u(x, y) = q(x, y)$ on the perimeter of R. When f is a constant, this partial differential equation is called the *Helmholtz equation*.

As before, we contemplate an approximate solution of such a problem by the finite-difference method. The first step is to select approximate formulas for the derivatives in our problem. In the present situation, we use the standard formula

$$f''(x) \approx \frac{1}{h^2}\,[f(x + h) - 2f(x) + f(x - h)] \tag{2}$$

derived in Sec. 5.3. If used on a function of two variables, we obtain the ''five-point formula''

$$\nabla^2 u \approx \frac{1}{h^2}[u(x+h, y) + u(x-h, y) + u(x, y+h) + u(x, y-h) - 4u(x, y)] \tag{3}$$

This formula involves the five points displayed in Fig. 15.10. The local error inherent in the five-point formula is

$$\frac{-h^2}{12}\left[\frac{\partial^4 u}{\partial x^4}(\xi, y) + \frac{\partial^4 u}{\partial y^4}(x, \eta)\right] \tag{4}$$

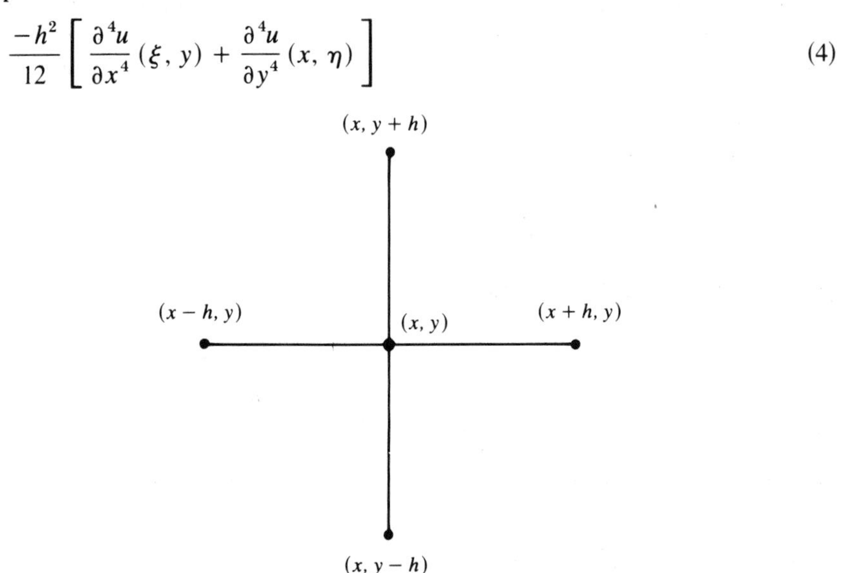

Figure 15.10

and, for this reason, formula (3) is said to provide an approximation of order h^2. In other words, if grids are used with smaller and smaller spacing, $h \to 0$, then the error committed in replacing $\nabla^2 u$ by its finite-difference approximation goes to zero as rapidly as h^2. Equation (3) is called the *five-point formula* because it involves values of u at (x, y) and at the four nearest grid points.

It should be emphasized that when the differential equation in (1) is replaced by the finite-difference analog, we have changed the problem. Even if the analogous finite-difference problem is solved with complete precision, the solution is that of a problem that only *simulates* the original one. This simulation of one problem by another becomes better and better as h is made to decrease to zero, but computing cost will inevitably increase.

We should also note that more accurate representations of the derivatives can be used. For example, the often-used *nine-point formula* is

$$\nabla^2 u \approx \frac{1}{6h^2}[4u(x+h, y) + 4u(x-h, y) + 4u(x, y+h) + 4u(x, y-h) + u(x+h, y+h) + u(x-h, y+h) + u(x+h, y-h) + u(x-h, y-h) - 20u(x, y)] \tag{5}$$

This formula is of order h^6.

If the mesh spacing is not regular (say, h_1, h_2, h_3, h_4 are the left, bottom, right, top spacing, respectively), then it is not difficult to show that at (x, y)

$$\begin{aligned}\nabla^2 u \approx\ & \frac{1}{\frac{1}{2}h_1 h_3(h_1 + h_3)}[h_1 u(x + h_3, y) + h_3 u(x - h_1, y)] \\ & + \frac{1}{\frac{1}{2}h_2 h_4(h_2 + h_4)}[h_2 u(x, y + h_4) + h_4 u(x, y - h_2)] \\ & - 2\left(\frac{1}{h_1 h_3} + \frac{1}{h_2 h_4}\right)u(x, y) \qquad (6)\end{aligned}$$

which is only of order h when $h_i = \alpha_i h$ for $0 < \alpha_i < 1$. This formula is usually used near boundary points, as in Fig. 15.11. If the mesh is small, however, the boundary points can be moved over slightly to avoid the use of (6). This perturbation of R (in most cases, for small h) produces an error no greater than that introduced by using the irregular scheme (6).

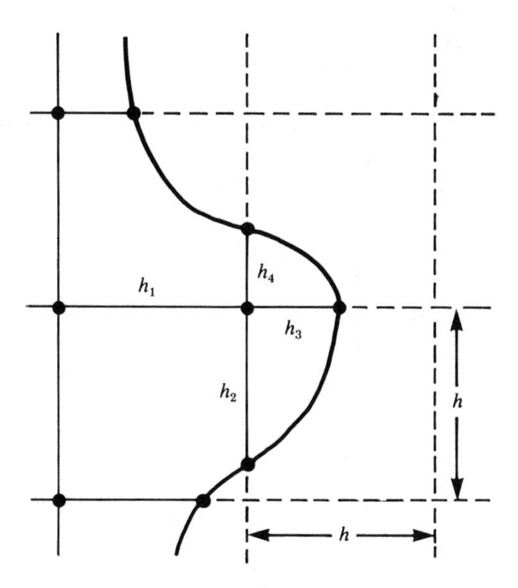

Figure 15.11

Returning to the model problem (1), we cover the region R by mesh points

$$x_i = ih, \qquad y_j = jh \quad (i, j \geqq 0) \qquad (7)$$

It is convenient next to introduce an abbreviated notation

$$u_{ij} = u(x_i, y_j), \quad f_{ij} = f(x_i, y_j), \quad g_{ij} = g(x_i, y_j) \qquad (8)$$

With it the five-point formula takes on a simple form at the point (x_i, y_j).

$$(\nabla^2 u)_{ij} \approx \frac{1}{h^2}(u_{i+1,j} + u_{i-1,j} + u_{i,j+1} + u_{i,j-1} - 4u_{ij}) \qquad (9)$$

If this approximation is made in the differential equation (1), the result is (the reader should verify it)

$$-u_{i+1,j} - u_{i-1,j} - u_{i,j+1} - u_{i,j-1} + (4 - h^2 f_{ij})u_{ij} = -h^2 g_{ij} \tag{10}$$

The coefficients of this equation can be illustrated by a "five-point star" in which each point corresponds to the coefficient of the u in the grid (Fig. 15.12).

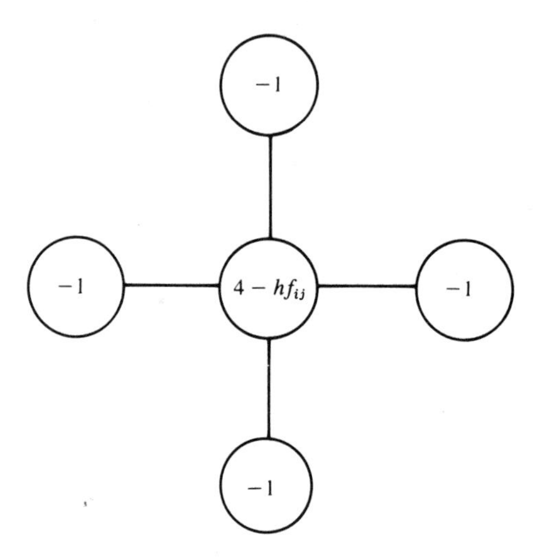

Figure 15.12

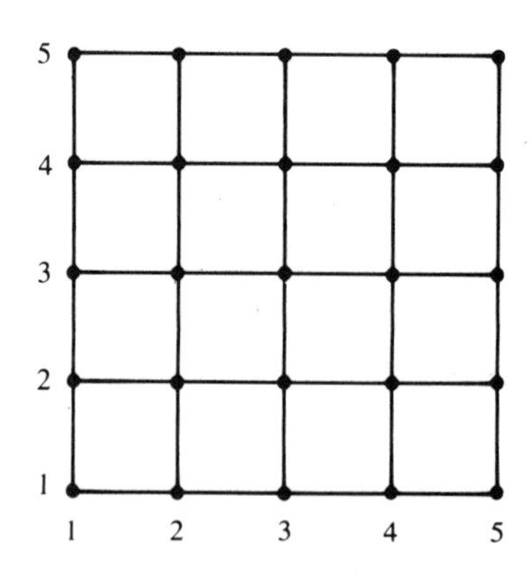

Figure 15.13

To be specific, we assume that the region R is a square and the grid has spacing $h = \frac{1}{4}$ (see Fig. 15.13). We obtain a single linear equation of the form (10) for each of the nine interior grid points. These nine equations are as follows.

$$-u_{12} - u_{32} - u_{23} - u_{21} + (4 - h^2 f_{22})u_{22} = -h^2 g_{22}$$

$$-u_{22} - u_{42} - u_{33} - u_{31} + (4 - h^2 f_{32})u_{32} = -h^2 g_{32}$$

$$-u_{52} - u_{32} - u_{43} - u_{41} + (4 - h^2 f_{42})u_{42} = -h^2 g_{42}$$

$$-u_{13} - u_{33} - u_{24} - u_{22} + (4 - h^2 f_{23})u_{23} = -h^2 g_{23}$$
$$-u_{43} - u_{23} - u_{34} - u_{32} + (4 - h^2 f_{33})u_{33} = -h^2 g_{33}$$
$$-u_{53} - u_{33} - u_{44} - u_{42} + (4 - h^2 f_{43})u_{43} = -h^2 g_{43}$$
$$-u_{34} - u_{14} - u_{25} - u_{23} + (4 - h^2 f_{24})u_{24} = -h^2 g_{24}$$
$$-u_{44} - u_{24} - u_{35} - u_{33} + (4 - h^2 f_{34})u_{34} = -h^2 g_{34}$$
$$-u_{54} - u_{34} - u_{45} - u_{43} + (4 - h^2 f_{44})u_{44} = -h^2 g_{44}$$

This system of equations could be solved through Gaussian elimination, but let us examine them more closely. There are 45 coefficients. Since u is known at the boundary points, we move these 12 terms to the right-hand side, leaving only 33 nonzero entries out of 81 in our 9×9 system. The standard Gaussian elimination causes a great deal of "fill-in"—that is, zero entries being replaced by nonzero values—in the forward elimination phase. So we seek a method that retains the "sparse" structure of this system. To illustrate how sparse this system of equations is, we write it in matrix notation.

$$\mathbf{A}\mathbf{u} = \mathbf{b} \tag{11}$$

Suppose that we order the unknowns from left to right and up.

$$\mathbf{u} = [u_{22}, u_{32}, u_{42}, u_{23}, u_{33}, u_{43}, u_{24}, u_{34}, u_{44}]^T \tag{12}$$

This is called the *natural ordering*. Now the coefficient matrix is

$$\mathbf{A} = \begin{bmatrix} 4-h^2f_{22} & -1 & 0 & -1 & 0 & 0 & 0 & 0 & 0 \\ -1 & 4-h^2f_{32} & -1 & 0 & -1 & 0 & 0 & 0 & 0 \\ 0 & -1 & 4-h^2f_{42} & 0 & 0 & -1 & 0 & 0 & 0 \\ -1 & 0 & 0 & 4-h^2f_{23} & -1 & 0 & -1 & 0 & 0 \\ 0 & -1 & 0 & -1 & 4-h^2f_{33} & -1 & 0 & -1 & 0 \\ 0 & 0 & -1 & 0 & -1 & 4-h^2f_{43} & 0 & 0 & -1 \\ 0 & 0 & 0 & -1 & 0 & 0 & 4-h^2f_{24} & -1 & 0 \\ 0 & 0 & 0 & 0 & -1 & 0 & -1 & 4-h^2f_{34} & -1 \\ 0 & 0 & 0 & 0 & 0 & -1 & 0 & -1 & 4-h^2f_{44} \end{bmatrix}$$

and the right-hand side is

$$\mathbf{b} = \begin{bmatrix} -h^2g_{22} + u_{12} + u_{21} \\ -h^2g_{32} + u_{31} \\ -h^2g_{42} + u_{52} + u_{41} \\ -h^2g_{23} + u_{13} \\ -h^2g_{33} \\ -h^2g_{43} + u_{53} \\ -h^2g_{24} + u_{14} + u_{25} \\ -h^2g_{34} + u_{35} \\ -h^2g_{44} + u_{54} + u_{45} \end{bmatrix}$$

Since each equation is similar in form, iterative methods are often used to solve such sparse systems. Solving for the diagonal unknown, we have from Eq. (10)

$$u_{ij} = \frac{1}{4 - h^2 f_{ij}} (u_{i+1,j} + u_{i-1,j} + u_{i,j+1} + u_{i,j-1} - h^2 g_{ij})$$

If we use a trial value for each grid point, this relation generates new values. Moreover, with the natural ordering, the new values can be used in this equation as soon as they become available. This procedure is called the *Gauss–Seidel iteration*. We call $u^{(n)}$ the current values of the unknowns and $u^{(n+1)}$ the value in the next iteration. Thus the *Gauss–Seidel method* is given by

$$u_{ij}^{(n+1)} = \frac{1}{4 - h^2 f_{ij}} (u_{i+1,j}^{(n)} + u_{i-1,j}^{(n+1)} + u_{i,j+1}^{(n)} + u_{i,j-1}^{(n+1)} - h^2 g_{ij})$$

The code for this method on a rectangle is as follows.

```
      SUBROUTINE SEIDEL(AX, AY, NX, NY, H, ITMAX, U)
      DIMENSION  U(NX, NY)
      H2 = H*H
      DO 3 K = 1, ITMAX
      DO 2 J = 2, NY-1
      DO 2 I = 2, NX-1
      X = AX + H*FLOAT(I-1)
      Y = AY + H*FLOAT(J-1)
      V = U(I+1, J) + U(I-1, J) + U(I, J+1) + U(I, J-1)
    2 U(I, J) = (V - H2*G(X, Y))/(4.0 - H2*F(X, Y))
    3 PRINT 4, K, ((U(I, J), I = 1, NX), J = 1, NY)
      RETURN
    4 FORMAT(//4X, I5, //(5(5X, E20.13)))
      END
```

In using this subroutine, one must decide on the number of iterative steps to be computed, ITMAX. The coordinates of the lower left-hand corner of the rectangle, (AX, AY), and the step size H are specified. The number of x grid points is NX, and the number of y grid points is NY.

Let us illustrate this procedure on the boundary value problem

$$\begin{cases} \nabla^2 u + 2u = g \text{ inside } R \\ u = 0 \text{ on the boundary of } R \end{cases} \tag{13}$$

where $g(x, y) = (xy + 1)(xy - x - y) + x^2 + y^2$ and R is the unit square. This problem has the known solution $u = \frac{1}{2}xy(x - 1)(y - 1)$. A driver program for the Gauss–Seidel subroutine, starting with $u = xy$ and taking 30 iterations, is given next. Notice that only 25 words of storage are needed for the array in solving the linear system iteratively.

```
      DIMENSION  U(5,5)
      DATA  AX,AY/2*0.0/. BX,BY/2*1.0/,NX,NY/2*5/. ITMAX/30/
      H = (BX - AX)/FLOAT(NX - 1)
      DO 2 J = 1,NY
      Y = AY + H*FLOAT(J-1)
      U(1,J) = BNDY(AX,Y)
    2 U(NX,J) = BNDY(BX,Y)
      DO 3 I = 1,NX
      X = AX + H*FLOAT(I-1)
      U(I,1)  = BNDY(X,AY)
    3 U(I,NY) = BNDY(X,BY)
      DO 4 J = 2,NY-1
      DO 4 I = 2,NX-1
      X = AX + H*FLOAT(I-1)
      Y = AY + H*FLOAT(J-1)
    4 U(I,J) = USTART(X,Y)
      PRINT 6,((U(I,J),I = 1,NX),J = 1,NY)
      CALL SEIDEL(AX,AY,NX,NY,H,ITMAX,U)
      DO 5 J = 2,NY-1
      DO 5 I = 2,NX-1
      X = AX + H*FLOAT(I-1)
      Y = AY + H*FLOAT(J-1)
    5 U(I,J) = TRUE(X,Y) - U(I,J)
      PRINT 6,((U(I,J),I = 1,NX),J = 1,NY)
    6 FORMAT(//(5(5X,E20.13)))
      END
```

The accompanying function subprograms for this model problem are given next.

```
      FUNCTION F(X,Y)
      F = 2.0
      RETURN
      END

      FUNCTION G(X,Y)
      G = X*X + Y*Y + (X*Y + 1.0)*(X*Y - X - Y)
      RETURN
      END

      FUNCTION BNDY(X,Y)
      BNDY = 0.0
      RETURN
      END

      FUNCTION USTART(X,Y)
      USTART = X*Y
      RETURN
      END
```

```
FUNCTION TRUE(X,Y)
TRUE = 0.5*X*(X - 1.0)*Y*(Y - 1.0)
RETURN
END
```

The initial values at the nine interior grid points are

```
6.2500000000000E-02  1.2500000000000E-01  1.8750000000000E-01
1.2500000000000E-01  2.5000000000000E-01  3.7500000000000E-01
1.8750000000000E-01  3.7500000000000E-01  5.6250000000000E-01
```

and after 30 iterations the values are

```
1.7578128649170E-02  2.3437503766885E-02  1.7578126944199E-02
2.3437503766885E-02  3.1250003888397E-02  2.3437502006915E-02
1.7578126944199E-02  2.3437502006915E-02  1.7578126035827E-02
```

which are in error by only

```
-3.6491697441576E-09  -3.7668849151018E-09  -1.9441985799062E-09
-3.7668849151018E-09  -3.8883971598125E-09  -2.0069146344781E-09
-1.9441985799062E-09  -2.0069146344781E-09  -1.0358268687227E-09
```

Problems 15.3

1. Establish the formula for the error in
   (a) the "five-point formula," Eq. (3).
   (b) the "nine-point formula," Eq. (5).
2. Establish the irregular five-point formula (6) and its error term.
3. Write the matrices that occur in Eq. (11) when the unknowns are ordered according to the vector $\mathbf{u} = [u_{22}, u_{42}, u_{33}, u_{24}, u_{44}, u_{32}, u_{23}, u_{43}, u_{34}]$. This is known as *red-black* or *checkerboard ordering*.
4. Verify that the solution of Eq. (13) is as given in the text.
5. Consider the problem of solving the partial differential equation

$$20u_{xx} - 30u_{yy} + \frac{5}{x+y}u_x + \frac{1}{y}u_y = 69$$

   in a region R with u prescribed on the boundary. Derive a five-point finite-difference equation of order $\mathcal{O}(h^2)$ that corresponds to this equation at some interior point (x_i, y_j).
6. Solve this boundary value problem to estimate $u(\frac{1}{2}, \frac{1}{2})$ and $u(0, \frac{1}{2})$.

$$\begin{cases} \nabla^2 u = 0, & (x, y) \in R \\ u = x, & (x, y) \in \partial R \end{cases}$$

   The region R with boundary ∂R is shown in Fig. 15.14. (The arc is circular.) Use $h = \frac{1}{2}$. *Cultural note:* This problem (and many others in this text) can be stated in physical terms also. For example, in Problem 6 we are finding the steady-state temperature in a beam of cross section R if the surface of the beam is held at temperature $u(x, y) = x$.

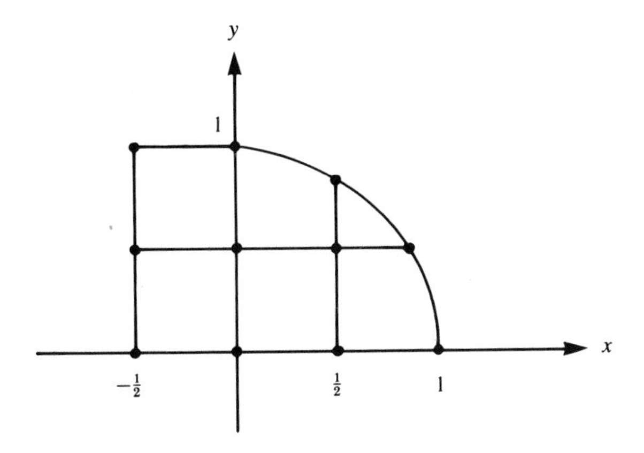

Figure 15.14

7. Consider the boundary value problem

$$\begin{cases} \nabla^2 u = 9(x^2 + y^2), & (x, y) \in R \\ u = x - y, & (x, y) \in \partial R_1 \end{cases}$$

for the region in the unit square with $h = \frac{1}{3}$ in Fig. 15.15. Here ∂R is the boundary of R, $\partial R_2 = \{(x, y) \in \partial R : \frac{2}{3} \leqq x < 1, \frac{2}{3} \leqq y < 1\}$, $\partial R_1 = \partial R - \partial R_2$. At the unknown mesh points, determine the system of linear equations that yields an approximate value for $u(x, y)$. Write the system in the form $\mathbf{Au} = \mathbf{b}$.
8. Determine the linear system to be solved if the nine-point formula (5) is used as the approximation in the problem of Eq. (1). Notice the pattern in the coefficient matrix with both the five- and nine-point formulas when unknowns in each row are grouped together. (Draw dotted lines through $\mathbf{A}$ to form 3×3 submatrices.)

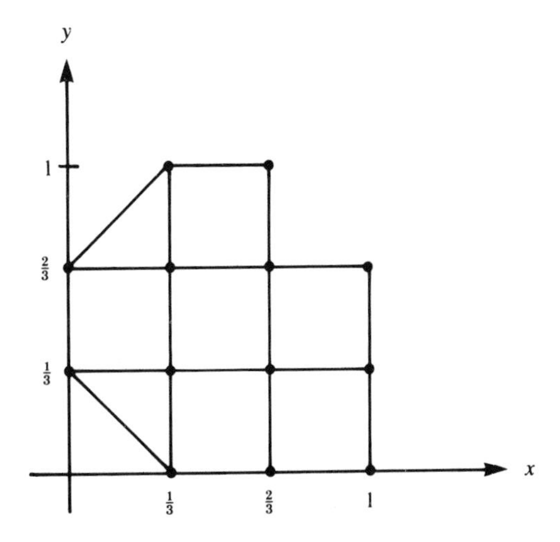

Figure 15.15

Computer Problems 15.3

1. Print the system of linear equations for (13) with $h = \frac{1}{4}$ and $\frac{1}{8}$. Solve these systems, using subroutines GAUSS and SOLVE of Chapter 6.
2. Try the Gauss–Seidel subroutine on the problem

$$\begin{cases} \nabla^2 u = 2e^{x+y}, & (x, y) \in R \\ u = e^{x+y}, & (x, y) \in \partial R \end{cases}$$

R is the rectangle shown in Fig. 15.16. Starting values and mesh sizes are in the following table. Compare to the exact solution after ITMAX iterations.

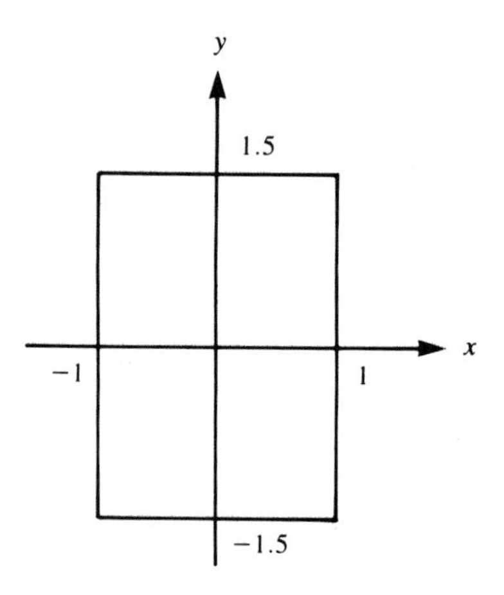

Figure 15.16

Starting Values	h	ITMAX
$u = xy$	0.1	15
$u = 0$	0.2	20
$u = (1 + x)(1 + y)$	0.25	40
$u = \left(1 + x + \frac{x^2}{2}\right)\left(1 + y + \frac{y^2}{2}\right)$	0.05	100
$u = 1 + xy$	0.25	200

3. Modify the Gauss–Seidel subroutine to handle the red-black ordering. Redo the previous problem with this ordering. Does the ordering make any difference? (See Problem 3.)
4. Rewrite the Gauss–Seidel code so that it can handle any ordering; that is, introduce an ordering array L. Try several different orderings — natural, red-black, spiral, diagonal.
5. Consider the heat transfer problem on the irregular region shown in Fig. 15.17. The mathematical statement of this problem is as follows.

$$\begin{cases} \dfrac{\partial^2 u}{\partial x^2} + \dfrac{\partial^2 u}{\partial y^2} = 0 & \text{inside} \\ u = 0 & \text{top} \\ u = 100 & \text{bottom} \\ \dfrac{\partial u}{\partial x} = 0 & \text{sides} \end{cases}$$

Establish that the insulated boundaries act as mirrors so that we can assume the temperature is the same as at an adjacent interior grid point. Determine the associated linear system and solve for the temperature u_i, $1 \leqq i \leqq 10$.

6. Modify the SEIDEL routine so that it uses the nine-point formula (5). Re-solve Problem 13 and compare results.

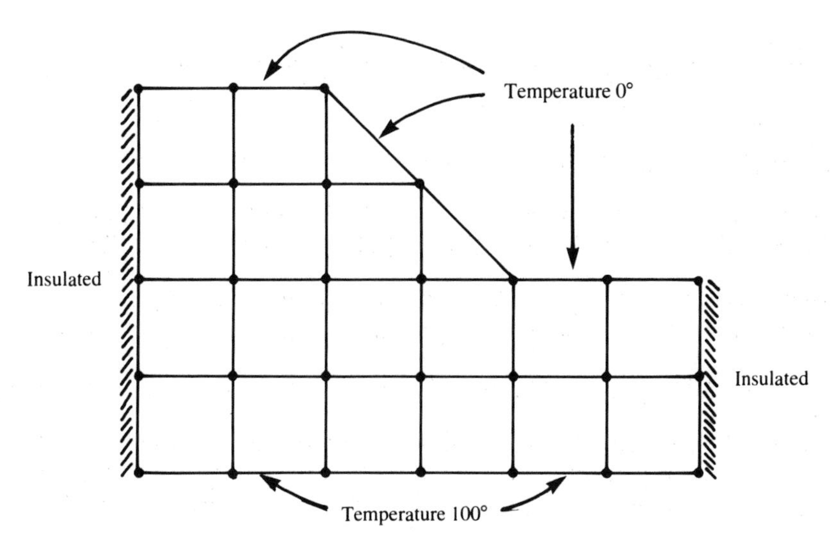

Figure 15.17

Appendix on Fortran Programming

A.1 Brief Review of Fortran

The purpose of this section is to reacquaint the student with the essentials of the Fortran language in a brief informal outline. For precise rules, a Fortran reference manual should be consulted.

Each line of Fortran code is typed in specified columns as follows.

Columns	*Contents*
1	"C" for a comment line
1–5	statement numbers
6	character for continuation line
7–72	Fortran statements
73–80	sequence numbers

Blanks may be used anywhere in the statement portion to improve readability. Also, entire blank lines may be inserted to separate portions of the code.

Constants. The value of a constant is fixed throughout the execution of a program. An integer constant is a whole number written without commas or a decimal point. A real constant is a rational number written with a decimal point. It may also be represented in scientific notation where the real number is followed by the letter E and a positive or negative integer power of 10 without a decimal point. Hence 0.005 could be written 5.0E−3 for example.

Variables. A variable may assume many values within a program and is a quantity that is referred to by name. Variable names may be from 1 to 7 characters long (letters or digits only), the first being a letter, not a digit. Variables beginning with I, J, K, L, M, N are presumed to be integers unless declared otherwise in a declaration statement. Conversely, variables beginning with other letters are presumed to be real unless declared otherwise.

Declaration Statements. The types of variables can be established with declaration statements. For example, typical declaration statements are

```
INTEGER X,Y,Z
REAL IA,J,NUM
DOUBLE PRECISION U,V,W
LOGICAL A,B,C
COMPLEX Z
```

Space for arrays may be allocated with a DIMENSION statement such as

```
DIMENSION M(10),X(56,75),A(20,10,5)
```

DATA statements are used to initialize variables and arrays. Some simple forms are

```
DATA  EE/1.5154/,G/0.5772/,NF/362880/
DATA  EE,G,NF/1.5154,0.5772,362880/
DATA  A/2.3,6.7,8.9,6.3,1.1,6.9,5.4/
```

Here A is an array that has been dimensioned. If the value of a variable in a DATA statement is changed during the execution of a subprogram, reentering the subprogram does not reinitialize the variable. Labeled COMMON statements in subprograms allow variables to be referenced in each routine containing them. Variable names in a COMMON list cannot also be in the parameter list of a subprogram. Moreover, a variable dimension is not possible in COMMON statements. An example of a labeled COMMON statement is:

```
COMMON /BLKA/A,X(10),Z
```

When the parameter list of a subprogram contains another subprogram name, the latter must appear in an EXTERNAL statement in every routine from which the former subprogram is called or referenced. (See Chapter 3 for an example of the usage of the EXTERNAL statement.)

Arrays. An array consists of a set of consecutive storage locations in memory. Each element in an array is referred to by means of a subscripted variable. In referring to an element of a two-dimensional array, the first subscript is the row number and the second is the column number. However, the elements of an array are *stored* in memory locations corresponding to consecutive columns. An array may have one, two, or three subscripts representing one-, two-, or three-dimensional arrays.

Operations. The five basic arithmetic operations are represented by the symbols

+	addition
−	subtraction
*	multiplication
/	division
**	exponentiation

Two symbols may not appear together.

Expressions. An expression is formed by combining variable names and/or constants with operational symbols. Parentheses may be used to clarify the meaning of an expression. When parentheses are not used, the expression is evaluated according to the following hierarchy: first, all exponentiation; secondly, all multiplication and division; and then all addition and subtraction. In a line with no parentheses the operations are performed according to the hierarchy from left to right. Any expression may be raised to a positive or negative integer power; however, only real expressions may be raised to a real power. Moreover, a negative real number cannot be raised to a real power nor zero to the zero power. Mixing of variables of real and integer type is risky and should be avoided.

When two integers are divided, truncation may result in an unexpected integer value.

Assignment Statement. An arithmetic assignment statement has the form

<variable> = <expression>

When a line of code of this form is reached, the value of the expression is computed and that value is assigned to the variable. A simple rule to remember is that the value of a variable is not destroyed when it is fetched from memory, but the old value is destroyed when a new value is assigned to the variable.

Logical `IF` *Statement.* The form of a logical `IF` statement is

`IF` (<logical expression>) <statement>

If the <logical expression> is true, then the <statement> is executed; otherwise the next line of code after the `IF` statement is executed. Simple relational operators that can be used in the <logical expression> are

`.LT.`	less than
`.LE.`	less than or equal to
`.GT.`	greater than
`.GE.`	greater than or equal to
`.EQ.`	equal to
`.NE.`	not equal to

and logical operators are

`.AND.`	and
`.OR.`	or (one or both)
`.NOT.`	not

For example,

```
IF(X .LE. 0.5) N = 5
IF(X*X + Y*Y .GE. 4.2 .AND. X .NE. Y) RETURN
```

Arithmetic `IF` *Statement.* The form of the arithmetic `IF` statement is

`IF` (<expression>) <*n*1>,<*n*2>, <*n*3>

where <*n*1>, <*n*2>, <*n*3> are statement numbers. Transfer is made to the statement with label <*n*1>, <*n*2>, or <*n*3> if the value of the <expression> is negative, zero, or positive, respectively.

`GO TO` *Statement.* `GO TO` causes an unconditional transfer of control to the indicated statement number and is of the form

`GO TO` <*n*>

where <*n*> is a statement number in the routine.

Computed `GO TO` *Statement.* The form of the computed `GO TO` is

`GO TO` (<*n*1>, <*n*2>, . . . , <*nk*>), <integer variable>

Transfer is made to the first statement number <*n*1> if the value of the <integer variable> is 1, to the second <*n*2> if it is 2, . . . , to <*nk*> if it is *k*.

`DO` *Statements.* The form of the `DO` statement is

`DO` <*n*> <integer variable> = <*i*1>, <*i*2>

or

`DO` <*n*> <integer variable> = <*i*1>, <*i*2>, <*i*3>

where <*n*> is a statement number and <*i*1>, <*i*2>, <*i*3> are integer constants or integer variables. The execution is as follows. The <integer variable> is called the `DO` loop index and is set equal to <*i*1>; a check is made to determine whether this value of the index is greater than <*i*2>; if so, execution is transferred to the first statement after <*n*>; if not, all statements after the `DO` statement, including the one with statement number <*n*>, are executed; the value of the index is increased by 1 in the first form and by <*i*3> in the second; a check is made. . . . `DO` loops may be nested—that is, a `DO` loop may have one or more `DO` loops within its range. The last statement in the range of a `DO` loop may not be another `DO` statement, `GO TO`, or `IF` statement. The value of the `DO` loop index should not be changed within the range of the loop. Transfer into the range of a `DO` loop from without is not allowed.

Input/Output. Some Fortran compilers allow format-free `I/O` for variables that are being read or printed. However, greater flexibility is available through use of a `FORMAT` statement. The forms of the read and print statements are

`READ` <*n*>, <variable 1>, . . . , <variable *k*>

`PRINT` <*n*>, <variable 1>, . . . , <variable *k*>

where <*n*> is the statement number of the format statement. An example of a `FORMAT` statement is

<*n*> `FORMAT(5X, I8, F10.5, E20.13)`

where the `I` specification is used for integer values, the `F` for fixed decimal point real values, and the `E` for variable decimal point real values. In printing, the `X` causes four blanks to appear since the first character of information in each line is used for carriage control of the printer. Some of the control characters are

<blank>	single spacing
`0`	double spacing
`1`	skip to top of next page

Adding either `1H_`, `1H0`, or `1H1` to the format statement is one way to use them.

Statement Functions. Statement functions are defined before the first executable statement and have the form

<function name> (<argument 1>, . . . , <argument *k*>) = <expression>

where the <function name> is a valid variable name with type specified by the first letter or a type declaration statement. The statement function is used to replace relatively simple computations that recur throughout a routine. The arguments in the definition have no relation to any variable of the same name that may appear elsewhere.

Function Subprograms. The form of a function subprogram is

```
FUNCTION <name> (<argument 1>, . . . ,<argument k>)
{ statements for function subprogram
RETURN
END
```

The type of the function is determined by the type of its name. The subprogram must contain at least one `RETURN` statement. The <name> of the function appears to the left of an equals sign at least once in the subprogram and a value is assigned to the function <name> at that point. The variable names in a subprogram are completely independent of all variable names in the main program or in other subprograms.

Subroutines. These are similar to function subprograms. However, no value or type is associated with its <name>. The form of a `SUBROUTINE` is

```
SUBROUTINE <name> (<argument 1>, . . . ,<argument k>)
{ statements for subroutine
RETURN
END
```

The <name> of the subroutine does *not* appear to the left of an equals sign, and values are passed via the arguments. A statement of the form

```
CALL <name> (<argument 1>, . . . ,<argument k>)
```

is used to execute the subroutine.

Mathematical Functions. Common mathematical functions, such as on the back endsheet in Table A.1, are available. These definitions are used:

$$x_1 \bmod (x_2) = x_1 - \left[\frac{x_1}{x_2}\right] x_2$$

$$\text{positive difference}\,(x_1, x_2) = \begin{cases} x_1 - x_2, & x_1 > x_2 \\ 0, & x_1 \leqq x_2 \end{cases}$$

$$\text{nearest whole number}\,(x) = \begin{cases} [x + 0.5], & x \geqq 0 \\ [x - 0.5], & x < 0 \end{cases}$$

$$\text{truncation}\,(x) = [x]$$

where $[z]$ is the integer whose magnitude is the largest integer that does not exceed the magnitude of z and whose sign is the same as the sign of z.

A.2 Conversion of Fortran 77 Codes to Fortran 66

Much of the Fortran code in this book is written in "extended" Fortran (or Fortran 77) as established by the American National Standards Institute and approved as the American National Standard Fortran. The standard specifies the form and establishes the interpretation of programs expressed in Fortran language. The purpose of the standard is to promote portability of Fortran programs written for a variety of data processing systems.

Since Fortran 77 compilers may not be available everywhere this book is used, we now list the changes that may have to be made to the codes so that they will compile on older Fortran 66 compilers. Almost all our code is written according to Fortran 66 standards with a few minor exceptions. The more elegant Fortran 77 code is used primarily when *not* using it would require additional lines of code with no added informational value. Many attractive features of Fortran 77, such as negative DO-loop increments and IF–THEN–ELSE statements, have not been used for changing the routines back to Fortran 66 code would require major rewriting by the users.

Program Statement. Some compilers require the main program to begin with a program statement that includes the name of the program, say PROG, and the names of files that the program may use. For example, the program statement may have the form

```
PROGRAM PROG (INPUT, OUTPUT, TAPE5=INPUT, TAPE6=OUTPUT)
```

The reason for tape five and six is that they are used in the form of the read and write statements explained below. For some computers, statements of the form

```
OPEN(UNIT=5, DEVICE='DSK', ACCESS='SEQIN', FILE='IN.DAT')
OPEN(UNIT=6, DEVICE='DSK', ACCESS='SEQOUT', FILE='OUT.LPT')
```

are used in place of a program statement. Here the input is on file IN.DAT and the output on file OUT.LPT.

READ *and* WRITE *Statements.* Read and print statements should be changed to a form similar to

```
      READ(5,2) X,L
      WRITE(6,2) X,L
    2 FORMAT(5X,E20.13,I10)
```

DO *Loops.* Simple computation in the indices of a DO loop should be done before the loop. Thus,

```
      DO 2 I = 1,N-1
```

becomes

```
      NM1 = N-1
      DO 2 I = 1,NM1
```

Array Indices. An array reference, such as A(L(I),K), should be changed as follows.

```
LI = L(I)
R = A(LI,K)
```

Alphanumeric Variables. For portability between different computers, the names of variables and subroutines should be limited to no more than six alphanumeric characters.

STOP *Statements.* The main program should end with a STOP statement.

Since all these changes are straightforward and may not be necessary on many computers, we have not written the code so that it works on all Fortran 66 compilers. The Fortran 77 code presented herein is easy to understand and to modify if necessary.

Bibliography

Abramowitz, M., and I. A. Stegun (Eds.) [1964], *Handbook of Mathematical Functions with Formulas, Graphs, and Mathematical Tables,* National Bureau of Standards; reprint [1965], Dover, New York.

Acton, F. S. [1959], *Analysis of Straight-Line Data,* Wiley, New York; reprint [1966], Dover, New York.

American National Standard Programming Language FORTRAN [1978], American National Standards Institute, New York.

Barrodale, I., F. D. K. Roberts, and B. L. Ehle [1971], *Elementary Computer Applications,* Wiley, New York.

Barrodale, I., and F. D. K. Roberts [1974], "Solution of an overdetermined system of equations in the ℓ_1 norm," *CACM,* **17,** 319–320.

Barrodale, I., and C. Phillips [1975], "Solution of an overdetermined system of linear equations in the Chebyshev norm," *ACM Tran. On Math. Software,* **1,** 264–270.

Bartels, R. H. [1971], "A stabilization of the simplex method," *Numerische Mathematik,* **16,** 414–434.

de Boor, C. [1978], *A Practical Guide to Splines,* Springer-Verlag, New York.

de Boor, C. [1971], "CADRE: An algorithm for numerical quadrature," in *Mathematical Software* (J. R. Rice, Ed.), Academic Press, New York, pp. 417–449.

Brent, R. P. [1976], "Fast multiple precision evaluation of elementary functions," *JACM,* **23,** 242–251.

Chaitlin, G. J. [1975], "Randomness and mathematical proof," *Scientific American 232,* May, 47–52.

Collatz, L. [1966], *The Numerical Treatment of Differential Equations*, 3rd ed., Springer-Verlag, Berlin.

Conte, S. D., and C. de Boor [1972], *Elementary Numerical Analysis,* McGraw-Hill, New York.

Cooper, L., and D. Steinberg [1974], *Methods and Applications of Linear Programming,* Saunders, Philadelphia.

Dahlquist, G., and A. Björck [1974], *Numerical Methods,* Prentice-Hall, Englewood Cliffs, N. J.

Davis, P. J., and P. Rabinowitz [1975], *Methods of Numerical Integration,* Academic Press, New York.

Dorn, W. S., and D. D. McCracken [1972], *Numerical Methods with FORTRAN IV Case Studies,* Wiley, New York.

Evans, G. W., G. F. Wallace, and G. L. Sutherland [1967], *Simulation Using Digital Computers,* Prentice-Hall, Englewood Cliffs, N.J.

Fehlberg, E. [1969], "Klassische Runge-Kutta formeln fünfter und siebenter ordnung mit schrittweitenkontrolle," *Computing,* **4,** 93–106.

Forsythe, G. E., and W. R. Wasow [1960], *Finite Difference Methods for Partial Differential Equations,* Wiley, New York.

Forsythe, G. E. [1965], "Generation and use of orthogonal polynomials for data-fitting with a digital computer," *SIAM J,* **5,** 74–88.

Forsythe, G. E., and C. B. Moler [1967], *Computer Solution of Linear Algebraic Systems,* Prentice-Hall, Englewood Cliffs, N.J.

Forsythe, G. E., M. A. Malcolm, and C. B. Moler [1977], *Computer Methods for Mathematical Computations,* Prentice-Hall, Englewood Cliffs, N.J.

Fröberg, C.-E. [1969], *Introduction to Numerical Analysis,* Addison-Wesley, Reading, MA.

Gear, C. W. [1971], *Numerical Initial Value Problems in Ordinary Differential Equations,* Prentice-Hall, Englewood Cliffs, N.J.

Gerald, C. F. [1978], *Applied Numerical Analysis*, Addison-Wesley, Reading, MA.

Hammersley, J. M., and D. C. Handscomb [1964], *Monte Carlo Methods,* Methuen, London.

Henrici, P. [1962], *Discrete Variable Methods in Ordinary Differential Equations,* Wiley, New York.

Hildebrand, F. B. [1974], *Introduction to Numerical Analysis,* McGraw-Hill, New York.

Householder, A. S. [1970], *The Numerical Treatment of a Single Nonlinear Equation,* McGraw-Hill, New York.

Hull, T. E., and A. R. Dobell [1962], "Random number generators," *SIAM J. Appl. Math.,* **4,** 230–254.

Hull, T. E., W. H. Enright, B. M. Fellen, and A. E. Sedgwick [1972], "Comparing numerical methods for ordinary differential equations," *SIAM J Numer. Anal.,* **9,** 603–637.

Issacson, E., and H. B. Keller [1966], *Analysis of Numerical Methods,* Wiley, New York.

Jennings, A. [1977], *Matrix Computation for Engineers and Scientists,* Wiley, New York.

Kernighan, B. W., and P. J. Plauger [1974], *The Elements of Programming Style,* McGraw-Hill, New York.

Kincaid, D. R., and D. M. Young [1979], "Survey of iterative methods," in *Encyclopedia of Computer Science and Technology* (J. Belzer, A. G. Holzman, and A. Kent, Eds.), Dekker, New York.

Lambert, J. D. [1973], *Computational Methods in Ordinary Differential Equations,* Wiley, New York.

Lapidus, L., and J. H. Seinfeld [1971], *Numerical Solution of Ordinary Differential Equations,* Academic Press, New York.

Lawson, C. L., and R. J. Hanson [1974], *Solving Least-Squares Problems,* Prentice-Hall, Englewood Cliffs, N.J.

Marsaglia, G. [1968], "Random numbers fall mainly in the planes," *Proc. Nat. Acad. Sci.,* **61,** 25–28.

Nerinckx, D., and A. Haegemans [1976], "A comparison of nonlinear equation solvers," *J. Computational and Applied Math.,* **2,** 145–148.

Niederreiter, H. [1978], "Quasi-Monte Carlo Methods," *Bull. Amer. Math. Soc.,* **84,** 957–1041.

Nievergelt, J., J. G. Farrar, and E. M. Reingold [1974], *Computer Approaches to Mathematical Problems,* Prentice-Hall, Englewood Cliffs, N.J.

Noble, B., and J. W. Daniel [1977], *Applied Linear Algebra,* Prentice-Hall, Englewood Cliffs, N.J.

Orchard-Hays, W. [1968], *Advanced Linear Programming Computing Techniques,* McGraw-Hill, New York.

Ortega, J. M., and W. C. Rheinboldt [1970], *Iterative Solution of Nonlinear Equations in Several Variables,* Academic Press, New York.

Phillips, G. M., and P. J. Taylor [1973], *Theory and Applications of Numerical Analysis,* Academic Press, New York.

Rabinowitz, P. [1968], "Applications of linear programming to numerical analysis," *SIAM Review,* **10,** 121–159.

Rabinowitz, P. [1970], *Numerical Methods for Nonlinear Algebraic Equations,* Gordon & Breach, London.

Ralston, A. [1965], *A First Course in Numerical Analysis,* McGraw-Hill, New York.

Rice, J. R., and J. S. White [1964], "Norms for smoothing and estimation," *SIAM Review,* **6,** 243–256.

Rice, J. R. [1971], "SQUARS: An algorithm for least squares approximation," in *Mathematical Software* (J. R. Rice, Ed.), Academic Press, New York.

Rivlin, T. J. [1974], *The Chebyshev Polynomials,* Wiley, New York.

Salamin, E. [1976], "Computation of π using arithmetic-geometric mean," *Math. Comp.,* **30,** 565–570.

Scheid, F. [1968], *Theory and Problems of Numerical Analysis,* McGraw-Hill, New York.

Schrage, L. [1979], "A more portable Fortran random number generator," *ACM Tran. On Math. Software,* **5,** 132–138.

Schultz, M. H. [1973], *Spline Analysis,* Prentice-Hall, Englewood Cliffs, N.J.

Shampine, L. F., and M. K. Gordon [1975], *Computer Solution of Ordinary Differential Equations,* Freeman & Co., San Francisco, CA.

Smith, G. D. [1965], *Solution of Partial Differential Equations,* Oxford University Press, London.

Stetter, H. J. [1973], *Analysis of Discretization Methods for Ordinary Differential Equations,* Springer-Verlag, Berlin.

Stewart, G. W. [1973], *Introduction to Matrix Computations,* Academic Press, New York.

Street, R. L. [1973], *The Analysis and Solution of Partial Differential Equations,* Brooks/Cole, Monterey, CA.

Traub, J. F. [1964], *Iterative Methods for the Solution of Equations,* Prentice-Hall, Englewood Cliffs, N.J.

Whittaker, E., and G. Robinson [1924], *The Calculus of Observation,* Blackie, London; [1944] 4th ed.; reprint [1967], Dover, New York.

Wilkinson, J. H. [1963], *Rounding Errors in Algebraic Processes,* Prentice-Hall, Englewood Cliffs, N.J.

Young, D. M., and R. T. Gregory [1972], *A Survey of Numerical Mathematics,* vols I and II, Addison-Wesley, Reading, MA.

Answers to Selected Problems

PROBLEMS 1.1 **6.** $n(n + 1)/2$ **9.** n multiplications and n additions or subtractions

PROBLEMS 1.2 **1.** $\cosh 0.7 \approx 1.25517$ **2.** $e^{\cos x} = e - ex^2/2 + \cdots$ **3.** 18 terms
6. $\cos(\pi/3 + h) \approx (1/2)(1 - h^2/2) - (\sqrt{3}/2)h$ **7.** $m = 2$
8. $\sqrt{1.0001} = 1.00004\ 99987\ 500625$

SUPPLEMENTARY PROBLEMS (CHAPTER 1) **1.** 38 terms **3.** 10^{10} terms
5. 0.70711295 **7.** 1.00099949983, 0.000599999857 **11.** 10 terms **12.** 7 terms
13. 100 terms

PROBLEMS 2.1 **1.** $(441.681640625)_{10}$ **2.** $(3940)_{10} = (111101100100)_2$
3. $(27.45075341\ldots)_8$, $(113.1666213\ldots)_8$, $(71.24426416\ldots)_8$
4. $(1145.32)_8 = (613.40625)_{10}$ **5.** (c): $(101111)_2$, (e): $(110011)_2$, (g): $(33.72664)_8$

PROBLEMS 2.2 **5.** 1.00005, 1.0 **7.** β^{1-n} **8.** 3×2^{-48}, approximately
9. 4×2^{-48}, approximately

COMPUTER PROBLEMS 2.2 **2.** On MARK 60, $S \leqq 33.2710$

COMPUTER PROBLEMS 2.3 **4.** $|x| < 10^{-15}$

SUPPLEMENTARY PROBLEMS (CHAPTER 2) **3.** 9% 10. 0.99202 0915
11. $-\frac{1}{2}x^3 - \frac{1}{2}x^4$, -9.888×10^{-7} **13.** -2.5 **17.** 0.0402%
19. 6032/9990, 6032/10010

PROBLEMS 3.1 **1.** $\pm n\pi/2$, $n = 0, 1, 2, \ldots$ **4.** 0 **6.** 0.6191

PROBLEMS 3.2 **7.** $(1 - \pi/4 + x)/\sqrt{2}$

COMPUTER PROBLEMS 3.2 **2.** 0.32796 77853 31818 36223 77546
3. 2.09455 14815 42326 59148 23865 40579 **7.** 0.47033 169

COMPUTER PROBLEMS 3.3 **2.** 1.36880 81078 21373 **4.** 20.804854

SUPPLEMENTARY PROBLEMS (CHAPTER 3)
13. $x_{n+1} = [(m - 1)x_n^m + R]/(mx_n^{m-1})$, $x_{n+1} = x_n[(m + 1)R - x_n^m]/(mR)$
14. $0 < \omega < 2$ 19. 5.04396002

SUPPLEMENTARY COMPUTER PROBLEMS (CHAPTER 3) **2.** 3.13108

PROBLEMS 4.1 **4.** 0.00010 00025 0006 **7.** 56739

COMPUTER PROBLEMS 4.1 **1.** 0.94598395, 0.94723395

PROBLEMS 4.2 **1.**

	2	1	1/2	1/4
L	0	0	1/2	3/4
U	2	2	3/2	5/4
T	2	1	1	1

7. $n \geqq 3057$ **9.** $-(b-a)hf'(\ )/2$ **10.** 1.61×10^6

PROBLEMS 4.3 **8.** $L \approx \{[\phi(h/2)]^2 - \phi(h)\phi(h/3)\}/\{2\phi(h/2) - \phi(h) - \phi(h/3)\}$
11. $(2h/45)[7f(a) + 32f(a+h) + 12f(a+2h) + 32f(a+3h) + 7f(b)]$

COMPUTER PROBLEMS 4.3 **3.** $R(8, 8) = 0.499969819$ **5.** $R(8, 8) = 0.765197687$
7. $R(6, 1) = 1.813799364$ **8.** $2/9 = 0.22222\ldots$

SUPPLEMENTARY PROBLEMS (CHAPTER 4)
4. $x_{n+1} + n^3(x_{n+1} - x_n)/(3n^2 + 3n + 1)$ **8.** $-3h^5 f^{(iv)}(\xi)/80$ **10.** 0.7833, 0.7854
16. (a) $-\frac{1}{2}h^2 f'(\xi)$ and $-\frac{1}{2}(b-a)hf'(\xi)$ (b) $-\frac{1}{6}h^3 f''(\xi)$ and $-\frac{1}{6}(b-a)h^2 f''(\xi)$
17. $\alpha = \gamma = \frac{4}{3},\ \beta = -\frac{2}{3}$ **18.** $A = b - a,\ B = \frac{1}{2}(b-a)^2$
22. $w_1 = w_2 = h/2,\ w_3 = w_4 = -h^3/24$ **23.** $A = 2h,\ B = 0,\ C = h^3/3$

PROBLEMS 5.1 **7.** 0.38099 0.077848 **10.** $2 + x(-1 + (x-1)(1 - (x-3)x))$

PROBLEMS 5.2 **2.** 2.6×10^{-6} **5.** 1582

PROBLEMS 5.3 **7.** $\mathcal{O}(h^2)$

COMPUTER PROBLEMS 5.3 **3.** 0.2021158503

SUPPLEMENTARY PROBLEMS (CHAPTER 5) **1.** (a) $x^3 - 3x^2 + 2 - 1$
11. $\ell_3(x) = -(x-4)(x^2-1)/8$
13. $-(7/24)(x-2)(x-3)(x-4) + (11/4)(x)(x-3)(x-4)$
$-\ (28/3)(x)(x-2)(x-4) + (63/8)(x)(x-2)(x-3)$
16. $-1 + (x-1)(\frac{2}{3} + (x-2)(\frac{1}{8} + (x-2.5)(\frac{3}{4} + (x-3)(11/6))))$ **17.** $n = 10$
27. 1.5727 **29.** $-\frac{3}{5}x^3 - \frac{2}{5}x^2 + 1$ **30.** 3.03×10^{-13} **31.** 1.25×10^{-5}
36. (a) $-2h^2 f'''(\xi)/3$ (b) $-h^2 f^{(iv)}(\xi)/3$ **37.** (a) $-h^2 f^{(v)}(\xi)/4$ (b) $-h^2 f^{(vi)}(\xi)/6$
45. 0.85527, 0.85526 0.87006, 0.87004 **51.** $[f(x_2) - f(x_1)]/(x_2 - x_1)$

PROBLEMS 6.1 **1.** (a) $x_1 = 58,\ x_2 = -19.37,\ x_3 = -26.84$
(b) $x = 2.096,\ y = -1.288,\ z = -0.05769$

6. $\begin{bmatrix} 1 & 0 & 0 & 0 \\ 1 & 1 & 0 & 0 \\ -1 & 1 & 1 & 0 \\ 1 & -1 & 1 & 1 \end{bmatrix} \begin{bmatrix} 1 & 0 & 0 & 1 \\ 0 & 1 & 0 & -2 \\ 0 & 0 & 1 & 4 \\ 0 & 0 & 0 & -8 \end{bmatrix}$

7. (a) $\begin{bmatrix} 1 & 0 & 0 \\ 1 & 1 & 0 \\ 3 & -1 & 1 \end{bmatrix} \begin{bmatrix} 2 & 0 & 0 \\ 0 & -2 & 0 \\ 0 & 0 & 3 \end{bmatrix} \begin{bmatrix} 1 & -\frac{1}{2} & 1 \\ 0 & 1 & -\frac{1}{2} \\ 0 & 0 & 1 \end{bmatrix}$ (b) $x = (-1, 2, 1)^T$

12. $\begin{bmatrix} \alpha^{\pm m} & \pm m\beta\alpha^{\pm m - 1} \\ 0 & \alpha^{\pm m} \end{bmatrix}$

PROBLEMS 6.2 **3.** $N(N+1)$ **5.** .005¢, 5¢, \$46.30, \$46,296.30

COMPUTER PROBLEMS 6.2 **2.** $x_{ij} = -i(n+1-j)/(n+1) = x_{ji} \quad (i \leqq j)$
4. $(3.46, 1.56, -2.93, -0.43)^T$ **9.** 16.5344×10^{-7}

PROBLEMS 6.3 **1.** $8N - 7$

COMPUTER PROBLEMS 6.3 **2.** $d_1 = 1,\ d_i = 1 - (4d_{i-1})^{-1}$ for $2 \leqq i \leqq 100$

SUPPLEMENTARY PROBLEMS (CHAPTER 6)

2. $$\begin{bmatrix} 1/4 & 5/2 & 7/4 & 1/2 \\ 4 & 2 & 1 & 2 \\ 1/2 & 0 & 5/9 & 17/9 \\ 1/4 & 3/5 & 27/10 & 1/5 \end{bmatrix}$$

6. $(\frac{1}{3}, 3, \frac{1}{3})^T$

13. $$M = \begin{bmatrix} 1 & 0 & 0 & 0 & 0 \\ 0 & 1 & 0 & 0 & 0 \\ 0 & -2 & 1 & 0 & 0 \\ 0 & 0 & -2 & 1 & 0 \\ -4 & 0 & 0 & 0 & 1 \end{bmatrix}, \quad U = \begin{bmatrix} 25 & 0 & 0 & 0 & 1 \\ 0 & 27 & 4 & 3 & 2 \\ 0 & 0 & 50 & -6 & -4 \\ 0 & 0 & 0 & 0 & 0 \\ 0 & 0 & 0 & 0 & 20 \end{bmatrix}$$

16. (a) $$\begin{bmatrix} 1 & 0 & 0 & 0 \\ -1/4 & 1 & 0 & 0 \\ -1/4 & -1/15 & 1 & 0 \\ 0 & -4/15 & -2/7 & 1 \end{bmatrix} \begin{bmatrix} 4 & -1 & -1 & 0 \\ 0 & 15/4 & -1/4 & -1 \\ 0 & 0 & 56/15 & -16/15 \\ 0 & 0 & 0 & 24/7 \end{bmatrix}$$

17. 192

SUPPLEMENTARY COMPUTER PROBLEMS (CHAPTER 6)

4. *Case 4.* $$\begin{bmatrix} 536 & -668 & 458 & -186 \\ -668 & 994 & -854 & 458 \\ 458 & -854 & 994 & -668 \\ -186 & 458 & -668 & 536 \end{bmatrix}$$

6. (a) (3.75, 90°), (3.27, −65.7°), (0.775, 172.9°)
(b) (2.5, −90°), (2.08, 56.3°), (1.55, −60.2°)

PROBLEMS 7.1 **9.** $n \geqq 1{,}110{,}721$ **10.** yes

PROBLEMS 7.2

7. $$S(x) = \begin{cases} [-5(x-1)^3 + 12(x-1)]/7 & \text{on } [1, 2] \\ [6(x-2)^3 - 5(3-x)^3 - 6(x-2) + 12(3-x)]/7 & \text{on } [2, 3] \\ [-5(x-3)^3 + 6(4-x)^3 + 12(x-3) - 6(4-x)]/7 & \text{on } [3, 4] \\ [-5(5-x)^3 + 12(5-x)]/7 & \text{on } [4, 5] \end{cases}$$

12.
$$S(x) = \begin{cases} 2 - (x+1)^2 & \text{on } [-1, 0] \\ 1 - 2x & \text{on } [0, 1/2] \\ 8(x - 1/2)^2 - 2x + 1 & \text{on } [1/2, 1] \\ -5(x-1)^2 + 6x - 5 & \text{on } [1, 2] \\ 12(x-2)^2 - 4x + 10 & \text{on } [2, 5/2] \end{cases}$$

PROBLEMS 7.3 **2.** $f_n(x) = \cos(n \text{ Arccos } x)$

SUPPLEMENTARY PROBLEMS (CHAPTER 7) **2.** $p(x) = 0.1927x^3 - 0.0175x^2 + x$
12. $n - k \leqq i \leqq m - k - 1$ or $n \leqq i \leqq m - 1$ **25.** $a = 3$, $b = 3$, $c = 1$
28. $a = 3$, $b = c = 0$, $d = -1$

PROBLEMS 8.1 **8.** $x^{(iv)} = 18xx'x'' + 6(x')^3 + 3x^2x'''$
9. $a_0 = a_1 = a_2 = 0$, $a_3 = \frac{1}{3}$, $a_k + ka_{k+2} = 0$, $k \geqq 2$.

COMPUTER PROBLEMS 8.1 **2.** (b) $x(1.75) = 0.632999983$ (c) $x(5) = -0.2087351554$
4. (a) error at $t = 1$ is 1.8×10^{-10} **11.** Si $(1) = 0.9460830703$

PROBLEMS 8.2 **8.** $h = 2^{-10}$

COMPUTER PROBLEMS 8.2
3. (b) $n = 7$, $x(2) = 0.82356\ 78972$ (RK), 0.8235678970(T)
(c) $n = 7$, $x(2) = -0.4999999998$ (RK), -0.5000000012(T)

PROBLEMS 8.3

2. $\dfrac{\partial}{\partial s} x(9, s) = e^{252} \approx 10^{109}$

SUPPLEMENTARY PROBLEMS (CHAPTER 8) **10.** $a = 1$, $b = c = h/2$

COMPUTER PROBLEMS 9.1 **8.** 0.898

COMPUTER PROBLEMS 9.3 **10.** 0.6617

SUPPLEMENTARY COMPUTER PROBLEMS (CHAPTER 9) **1.** 0.14758 **2.** 2/3
3. 7/16

PROBLEMS 10.1 **1.** $\mathbf{F}(2, 1, -2) = -15$, $\mathbf{F}(0, 0, -2) = -8$, $\mathbf{F}(2, 0, -2) = -12$
2. $\mathbf{F}(9/8, 9/8) = -20.25$ **4.** *Case* $n = 2$: $\hat{x} = (3a + b)/4 + \delta$ if $a \leqq x^* \leqq b'$,
$\hat{x} = (a + 3b)/4 - \delta$ if $a' \leqq x^* \leqq b$
7. $n \geqq 1 + (k + \log \ell + \log 2)/|\log r|$ **8.** $n \geqq 51$ **12.** exact solution $F(3) = -7$

PROBLEMS 10.2 **1.** (a) minimum exists, (b) no minimum, (c) no minimum,
(d) minimum exists if $ca > b^2$. **2.** $x = \frac{1}{4}$, $y = \frac{9}{4}$ **3.** $1 + x - xy + \frac{1}{2}(x^2 + y^2)$
5. The first component of $\mathbf{G}$ is $y^2z^2 \sin 2x$. **7.** $x_1 = -19/30$, $x_2 = -1/5$

SUPPLEMENTARY PROBLEMS (CHAPTER 10)
5. (b) $\frac{3}{2} - \frac{1}{2}x_2 + 3x_1x_2 + x_2x_3 + 2x_1^2 - \frac{1}{2}x_3^2 + \cdots$

PROBLEMS 11.1 **4.** (a) $(1.5, 0)^T$
5. (a) $x = (9, 0)^T$, $c^Tx = 18$, (c) unbounded, (f) no solution,
(h) $x = (0, 21/4)^T$, $c^Tx = 21/4$
6. $x = 24$, $y = 32$, $z = -124$

COMPUTER PROBLEMS 11.1 **3.** \$13.50

COMPUTER PROBLEMS 11.2 **2.** $x = (0, 0, 5/3, 2/3, 0)^T$ **3.** $x = (0, 8/3, 5/3)^T$

COMPUTER PROBLEMS 11.3
3. $p(x) = 1.0001 + 0.99782x + 0.51307x^2 + 0.13592x^3 + 0.071344x^4$

PROBLEMS 12.1 **1.** $f(x) = 1$ **3.** $a = (1 + 2e)/(1 + 2e^2)$, $b = 1$
5. $a = 2.1$, $b = 0.9$

7. $$c = \sum_{k=1}^{m} y_k \log x_k \Big/ \sum_{k=1}^{m} (\log x_k)^2$$

PROBLEMS 12.2

2. $$\begin{cases} w_{n+2} = w_{n+1} = 0 \\ w_k = c_k + 3xw_{k+1} + 2w_{k+2}\,(k = n, n-1, \ldots, 0) \\ g(x) = w_0 - (1 + 2x)w_1 \end{cases}$$

12. n multiplications and $2n$ additions/subtractions

PROBLEMS 12.3 **3.** $c = 0$ **8.** $x = -1$, $y = 20/13$ **9.** $c = 24/\pi^3$, $c = 3$

SUPPLEMENTARY PROBLEMS (CHAPTER 12) **5.** $29/35 + (2/7)x^2$
6. $a = b = 1$

PROBLEMS 13.1
2. $x(t + h) = (1 + (1/2)h^2 + (1/24)h^4)x(t) + (h + (1/6)h^3 + (1/120)h^5)y(t)$
$y(t + h) = (1 + (1/2)h^2 + (1/24)h^4)y(t) + (h + (1/6)h^3 + (1/120)h^5)x(t)$
3. $\mathbf{F} = (1, x_3, x_4, 2x_3 + \log x_4 + \cos x_2)^T$ $\mathbf{X}(0) = (0, 1, -3, 5)^T$

PROBLEMS 14.1 **6.** $\phi(z) = z$ **7.** $\phi(z) = \sqrt{9 + 6z}$
8. $\phi(z) = (e^5 + e + ze^4 - z)/(2e^2)$

PROBLEMS 15.1

2. $$\frac{1}{r}\frac{\partial}{\partial r}\left(r\frac{\partial u}{\partial r}\right) + \frac{1}{r^2}\frac{\partial^2 u}{\partial \theta^2} = 0$$

9. $$\begin{bmatrix} 0 & 1 & 0 & 0 & & & \\ -1 & 0 & 1 & 0 & & & \\ 0 & -1 & 0 & 1 & & & \\ & & \cdot & \cdot & \cdot & & \\ & & & \cdot & \cdot & \cdot & \\ & & & & -1 & 0 & 1 \\ & & & & 0 & -2 & 2 \end{bmatrix}$$

PROBLEMS 15.2 **1.** −0.23

COMPUTER PROBLEMS 15.3 **5.** 18.41° 13.75°
41.47° 36.60° 24.41°
69.41° 66.77° 61.05° 53.01° 51.00°

Index

Note: A page number followed by a number in parentheses prefixed by "#" refers to a problem on the given page. For example, 131(#7) refers to problem 7 on page 131.

Formulas and Definitions from Integral Calculus

$$\int x^{\alpha} dx = x^{\alpha+1}/(\alpha + 1) + C$$

$$\int e^{x} dx = e^{x} + C$$

$$\int x^{-1} dx = \ln |x| + C$$

$$\int \sin x \, dx = -\cos x + C$$

$$\int \cos x \, dx = \sin x + C$$

$$\int \sec^2 x \, dx = \tan x + C$$

$$\int \sec x \tan x \, dx = \sec x + C$$

$$\int \frac{dx}{\sqrt{a^2 - x^2}} = \text{Arcsin} \frac{x}{a} + C$$

$$\int \frac{dx}{a^2 + x^2} = \frac{1}{a} \text{Arctan} \frac{x}{a} + C$$

$$\int \sinh x \, dx = \cosh x + C$$

$$\int \cosh x \, dx = \sinh x + C$$

$$\int \tan x \, dx = \ln |\sec x| + C$$

$$\int \sec x \, dx = \ln |\sec x + \tan x| + C$$

$$\int (x^2 + a^2)^{-1/2} \, dx = \ln |\sqrt{x^2 + a^2} + x| + C$$

$$\int (x^2 \pm a^2)^{1/2} \, dx = \frac{x}{2}\sqrt{x^2 \pm a^2} \pm \frac{a^2}{2} \ln |x + \sqrt{x^2 \pm a^2}| + C$$

$$\int \ln x \, dx = x \ln |x| - x + C$$

$$\int \sin^2 x \, dx = \frac{x}{2} - \frac{\sin 2x}{4} + C$$

$$\int \cos^2 x \, dx = \frac{x}{2} + \frac{\sin 2x}{4} + C$$

$$\int \frac{dx}{x(ax + b)} = \frac{1}{b} \ln \left| \frac{x}{ax + b} \right| + C$$

$$\frac{d}{dx} \int_a^x f(t) dt = f(x)$$

$$\int u \, dv = uv - \int v \, du$$

$$\int F'(g(x))g'(x) \, dx = F(g(x)) + C$$